More Praise for

Beyond Human Nature

"In this invigorating look at what shapes human judgment, philosopher Jesse Prinz comes down solidly on the side of nature.... However you lean on the nature-nurture debate—or even if you think it has gone away—you will enjoy the challenges here."   —*Nature*

"An excellent guide to the current state of play.... Early in the book [Prinz] tells us that only 'a tiny fraction of articles in psychology journals take culture into consideration'. So it is time to redress the balance, and Prinz does it with insight, learning and above all a wonderful eye for the weaknesses in biological reductionist arguments.... From start to finish this book is a fine, balanced, enormously learned and informative blast on the trumpet of common sense and humane understanding."
          —*New Statesman*

"A compelling new book making the case against nature.... [A] bold position, and one with real implications for such big-ticket issues as the influence of genetics on intelligence and the effect of biology on gender inequality.... *Beyond Human Nature* succeeds by delivering serious arguments that challenge the reader without seeming dense or highfalutin. It is, in other words, a nutritious meal at a moment when so much science writing has made us sick with empty calories."
          —*Daily Beast*

"As abstract as the notion of human nature is, its stakes are extremely high. [Prinz argues] biological influence is overstated and cultural factors are more important." —*New York Times Book Review*

"Compelling arguments that cover a vast range of human behaviours."
—*Guardian*

"Prinz is clearly a nurture guy in the nature-nurture standoff. He draws on a range of research, from neuroscience to anthropology to psychology, to argue that DNA is not destiny and that experience helps along and can even supplant whatever is inborn. Empiricists, you can cheer. I'm betting that this is sophisticated but accessible reading for the Pinker/Damasio/Dennett set." —*Library Journal*

"Explores the origins of knowledge, language, thought and emotion, and argues that there is not one human nature, but many."
—*Financial Times*

"Ten years ago, the human genome crumbled into its component parts. Here, we were told, was the blueprint for humanity. A child is a recipe written at conception, pre-programmed with the genes for cancer, the genes for depression, obesity, politics and sexuality. . . . But New York philosopher Jesse Prinz wants to call a halt to the 'century of the gene'. In a new book, *Beyond Human Nature*, he gathers the arguments of a growing number of scientists who take the side of nurture against nature, in a backlash against the tyranny of DNA."
—*Sydney Morning Herald*

"A stimulating contribution to a perennial debate."
—*Booklist*, starred review

"*Beyond Human Nature* is a powerful probe of innate abilities and cultural influences, and is a powerful pick for any health or science holding from general-interest to college-level audiences."
—*Midwest Book Review*

"A big book arguing that there is 'little reason to think that biology has a major impact in accounting for human differences.' [Prinz] patiently examines the arguments given for attributing this or that trait to genetic inheritance, and tries to show either that the research is methodologically flawed, or that the conclusion is not supported by it."
                                                                —*Prospect* magazine

"Prinz's passion for his subject is evident, and his positions well-researched. . . . He presents some compelling arguments, and he is unafraid to take on popular beliefs to make his points."
                                                                —*Kirkus Reviews*

Beyond Human Nature

How Culture and Experience
Shape the Human Mind

JESSE J. PRINZ

W. W. Norton & Company
New York • London

First published in Great Britain in 2012 by Allen Lane, an imprint of
Penguin Books, under the title *Beyond Human Nature:
How Culture and Experience Shape Our Lives*

Manufacturing by Courier Westford
Production manager: Devon Zahn

Library of Congress Cataloging-in-Publication Data

Prinz, Jesse J.
Beyond human nature : how culture and experience shape the human mind /
Jesse J. Prinz. — 1st American ed.
p. cm.
Includes bibliographical references and index.
ISBN 978-0-393-06175-8 (hardcover)
1. Mind and body. 2. Thought and thinking. I. Title.
BF161P745 2012
155.2'34—DC23
2012033406

ISBN 978-0-393-34789-0 pbk.

W. W. Norton & Company, Inc.
500 Fifth Avenue, New York, N.Y. 10110
www.wwnorton.com

W. W. Norton & Company Ltd.
Castle House, 75/76 Wells Street, London W13QT

1 2 3 4 5 6 7 8 9 0

For my brother, Tommy Dog

Contents

Where Do Values Come From?

Preface

Writing in the seventeenth century, the English philosopher Thomas Hobbes argued that human beings are by nature selfish and belligerent.[1] We would gladly kill each other for personal gain, says Hobbes, and only a strong government can curb this basic instinct. In the state of nature, before such governments were established, there was inevitably war of all against all. A century later, Swiss-born Jean-Jacques Rousseau defended the opposite conclusion: human beings are by nature solitary and peaceful.[2] Competition, conflict and war result from the emergence of society, not from our natural dispositions. A third perspective was suggested by Rousseau's sometime friend the Scotsman David Hume. He concurred that human beings are peaceful, but denied that they are solitary; we are naturally social, benevolent and sympathetic to each other's needs.[3]

These are competing theories of human nature, theories of how we as human beings are biologically disposed to behave. The debate continues into the present day. Some authors use modern social science to defend the Hobbesian view of natural viciousness,[4] and others defend the Humean view of natural virtue.[5] But there is something dubious about the search for human nature. Why assume that human beings are any one way? There is considerable variation in human behaviour. We all know people who are kindhearted and others who are, well, not. In light of this, it might make sense to say that there are human natures, rather than one human nature.

This move to the plural, human natures, may be problematic as well. The phrase assumes that most variation is biologically based. But this may not be true. Are some people nice or nasty by nature? Perhaps. But human traits may also reflect the influence of nurture.

Hobbes, Rousseau and Hume acknowledged this. Hobbes believed that a totalitarian state could tame the beast within. Rousseau argued that society corrupts. Hume believed each nation has a natural character; he says the Swiss are honest, the French are funny, and the English are smart. These differences derive from 'the nature of the government, the revolutions of public affairs, the plenty or penury in which the people live, the situation of the nation with regard to its neighbours, and such like circumstances', and not physical causes.[6] It must be noted, though, that Hume's cultural explanation of human diversity is limited to whites; despite recognizing the dramatic effects of circumstance, he can't bring himself to accept that the races are equal.

These philosophers were themselves products of their times. Hobbes's unflattering view of human nature may stem from the fact that he lived through the Thirty Years War, one of the most violent conflicts in European history. Rousseau and Hume, by contrast, lived during the Enlightenment. Rousseau was raised a Calvinist, though he later abandoned that denomination, and his ideal of the noble savage may have been a reaction against the strong Calvinist emphasis on original sin. As Enlightenment thinkers, Rousseau and Hume were also impressed with the advances of science, and this undoubtedly contributed to their preoccupation with the idea that human beings have a nature that can be scientifically discovered. Our own era is also marked by an abiding faith in science, and the contemporary fascination with human nature may, in that respect, be a consequence of culture.

Once cultural plasticity is acknowledged, we are confronted with the question of whether it makes sense to talk about human nature, or even human natures. We human beings certainly have biological traits that distinguish us from other species, and there is also biological variation within the species. Biological traits surely contribute to human behaviour, and no behaviour would be possible were it not for our biological constitution. But our capacity to change with circumstance demonstrates that biology is not the complete story. This much should be obvious. Equally obvious is the fact that every human being is situated somewhere. None of us exists without circumstances, and therefore none of us acts by nature alone. From the start of life, we are moving beyond nature, and our transcendence of natural determination is our most striking trait. Human beings are genetically more homogeneous than chimps, but behaviourally more diverse than any other species.

This book concerns the cultural impact on human variation. It is, in part, a critique of approaches that oversell the role of biology, but my more central goal is to explore the rapidly flowering field of cultural psychology. We are gaining new insights into the ways that people differ, and the social forces that affect our lives. Each of us is a cultural product. Our values, our lifestyles and even the ways we think and feel have been strongly influenced by our locations in history and geography. The study of the human mind is fundamentally the study of place. If we want to know why some people wage war and others aim for amity, it is not enough to know that both capacities exist within our species. We must understand the circumstances that make us peaceful or pugnacious.

I began plotting this book when I was a fellow at the Center for Advanced Study in Behavioral Sciences at Stanford. My colleagues there included sociologists, criminologists, linguists, geographers, economists, psychologists and neuroscientists. Some of these fields, like sociology and geography, tend to emphasize social factors in explaining behaviour, while others, such as linguistics and neuroscience, tend to emphasize human nature. Fields like criminology and psychology exhibit competing emphases within them. Thus, I was thrust into the front lines of the nature–nurture debate at Stanford, and I benefited tremendously from that experience. I am grateful to the Center and my wonderful colleagues there.

I am also grateful to many others who helped shape this project. That includes audiences in three continents and hundreds of researchers whose work I draw on throughout. Many of these researchers find their way into the citations, but some of my intellectual heroes hover silently in the background. I mention here Franz Boas, whose pioneering work in anthropology has been an inspiration to many who try to establish universal human dignity though the study of diversity. At a more technical level, the manuscript benefited from careful copyediting by David Watson, Richard Duguid, Rachel Bernstein and Cressida Gaukroger. Cressida gave me invaluable feedback on content as well, and I will always be in her debt. My deepest gratitude goes to those who have exercised patience and support through the mood swings and delays that were too frequent during the writing of this book. That includes my editors, Will Goodlad and Angela Vonderlippe, my friends and my family. As always, my partner, Rachel Bernstein, endured the brunt of it with unwavering kindness and understanding. Every word deserves her thanks for that. My parents, Phyllis and Jonathan, also put up with me unflinchingly, and they taught me the importance of nurture. My brother, to whom the book is dedicated, has also been an inspiration; he embodies the human capacity to be transformed by the complex latticework of culture while also contributing to it in new and ever-surprising ways.

I

The Nature–Nurture Debate

Try this simple experiment. Go into a crowded cafeteria and look at what people are wearing. See if you can find two people wearing exactly the same thing. Count the number of styles you see: corporate types in suits, preppies in polos, academics in tweed, metrosexuals in black, hipsters in tight jeans, old men in cardigans, jocks in jerseys. No two people will have the same shirts or shoes, and if they did, that would be embarrassing. If you are feeling gregarious, start talking to people. Determine whether any two people share all the same opinions, interests and pastimes. Some are liberal, some are conservative; some are religious, some are atheists; some like water-sports and read crime novels; some like swing dancing and read Jane Austen; just about everyone has a different job description. Now ask people about their food preferences, their favourite movies and, if you dare, their sexual fantasies.

When you carry out this exercise, you will find dazzling diversity. There will be plenty of overlap, of course, but the range of human variation at a single cafeteria would make a troupe of chimpanzees look as undifferentiated as a school of minnows. If we look beyond the local cafeteria and explore gathering places in different cultures, the variation is absolutely staggering. There are profound differences in dress, beliefs, thinking styles and values. Some people think without numbers, some feel things we never feel, and some have even tasted human flesh.

Variation is not random. Each of us is the product of multiple influences. Our biographies, religious affiliations, material resources and geography all contribute in systematic ways. If you meet a stranger on a train and want to guess her political orientation, you might ask her where she lives, how much money her household earns and how often she goes to church. With just a few demographic facts, you'd probably make a pretty accurate guess. And if you want to know why a particular demographic subculture has the values it does, you can often gain insight by looking into history. There are reasons why some people are conservative and others are liberal. There are also reasons why some people are cannibals and some are kosher.

SHIFTING FOCUS

Until recently, cognitive scientists have neglected cultural differences. The term 'cognitive science' refers to the collection of academic disciplines that try to collectively investigate how the mind works. When the term was officially adopted in the mid 1970s, six disciplines were included: anthropology, computer science, linguistics, neuroscience, philosophy and psychology. Since the inception, however, anthropology has hardly played any role. Anthropologists rarely incorporate discoveries about the way the mind works in their research, and those who study mental processes rarely apply their techniques to members of different cultural groups. Linguists ignore the languages of other cultures because they erroneously assume that language has no influence on thought. Psychologists do all their experiments on university students and assume that they can extrapolate from this select group to all other minds around the globe. Neuroscientists rarely attempt to study brain activity in members of different cultures; they assume that all brains function in the same way, despite overwhelming evidence that much of the brain is not pre-wired.

By ignoring cultural variation, researchers end up giving us a misleading picture of the mind. We end up with the idea that psychology is profoundly inflexible. This outlook grossly underestimates human potential. It leads to the view that our behaviour is mostly driven by biology. Mainstream cognitive scientists give the impression that human traits are 'innate', 'genetic' or 'hardwired'.

I will refer to researchers who emphasize biological causes of behaviour as 'naturists', because they place their bets on the nature side of the nature–nurture debate. Numerous books and articles have been published in recent years defending the naturist cause.[1] Human court-ship has been explained in evolutionary terms, political preferences have been explained by genetic inheritance, and depression is chalked up to a chemical imbalance in the brain. These explanations miss the obvious fact that courtship, politics, and psychiatric maladies vary across cultural boundaries. Myopic focus on our own cultural group has promoted an undue faith in biological determinants of behaviour.

The chapters ahead are intended as a corrective. To counterbalance the current trend of evolutionary and genetic explanation, I will explore the exciting new frontier of cultural psychology. Breaking away from the orthodoxy, some researches have been investigating the psychological impact of cultural differences. The results of these investigations are fascinating. Consider:

- People raised in Western countries tend to see the trees before the forest, while people from the Far East see the forest before the trees.
- In south-east Asia, there is a common form of mental illness, unheard of here, in which people go into a trancelike state after being startled.
- Compared with Northerners, people in the American South are more than twice as likely to kill someone over an argument.

Research on cultural variation is leading to a new understanding of how the mind works. Until recently, psychologists frequently claimed that the mind is heavily constrained by innate biological mechanisms. Now, we are beginning to realize that the mind is far more malleable than we had appreciated. Again, consider:

- Your colour vocabulary can affect how similar two hues appear.
- Large IQ gaps between two people can be cut in half after four years of college.
- If you have an identical twin, there is a 50 per cent chance that you will have significantly different personalities.

The widespread differences in perception, intelligence and tempera-ment can often be traced to cultural, rather than biological, causes. Such discoveries illustrate a new appreciation of global variation within the social sciences, and a new understanding of disparities here at home. The way we think – not just what we think about – depends on experience and socialization.

This perspective can be called nurturism. Nurturists agree with naturists that biology matters. If we want to understand human behaviour, it is important to remember that human capacities and motivations are biologically constrained. But, when we focus on those biological constraints, we miss out on the headline news. What makes our species most interesting is that we exhibit astonishing variation. We are the only creatures on the planet that can radically alter their biological programmes. Nurturists emphasize flexibility, while naturists emphasize fixity.

Naturism is not just misleading; it is potentially dangerous. It has been used to keep various groups down, and it vastly underestimates human potential. When we assume that human nature is biologically fixed, we tend to regard people with different attitudes and capacities as inalterably different. We also tend to treat differences as pathologies. We regard people who think differently than we do as defective. We marginalize groups within our borders and we regard the behaviour of foreigners as unnatural or even subhuman.

Of course, most of the scientists who defend naturism do not have malevolent intentions, and they resent the fact that biological explanations of behaviour have been labelled politically incorrect. Naturists have to endure venomous attacks from left-wing groups that regard their programme as intrinsically racist or sexist. In response to these charges, naturists argue that political values should not forestall scientific inquiry. If behaviour is strongly influenced by biology, those influences must be scientifically investigated even if they make us uncomfortable and even if some people will use science to promote vicious ends. Moreover, nurturism can be every bit as dangerous as naturism. If we assume that all undesirable behaviour is the result of bad rearing, for example, we may blame the wrong people and miss out on opportunities for intervention. If we assume that human beings are infinitely flexible, we might initiate social programmes that are doomed to fail.

With all this I concur. We should never deny facts because we don't like them. The problem with naturism is not that it is politically incorrect.

The problem is that naturist conclusions are too often based on inadequate science. They tend to overlook evidence for variation and base strong conclusions about biological sources of behaviour on limited evidence. By ignoring cultural factors, naturists reify aspects of behaviour that are shaped by experience.

I will not argue that biology is irrelevant to human behaviour. That would be ridiculous. We need very sophisticated biological resources to be as flexible as we are. Nature and nurture conspire together. One must keep both in view. But, if we are interested in a full understanding of human behaviour, then nurture is especially important. The nurturist perspective has been underrepresented in scientific publishing. Books that focus on biological contributions to behaviour greatly outnumber books that look at human psychology from a cross-cultural perspective, and only a tiny fraction of articles published in psychology journals take culture into consideration.[2] If publishing patterns are any indication, both professional and lay readers are captivated by the idea that what we do can be explained in biological terms. But in focusing on these biological reductions, readers are missing out on some of the most fascinating, surprising and illuminating facts about human behaviour. I will compensate for this imbalance by telling the nurture side of the story.

IS THE NATURE–NURTURE DEBATE OVER?

I just suggested that both nature and nurture contribute to behaviour. This has become a platitude. Everyone seems to think that nature and nurture constitute a false dichotomy. Even microscopic organisms, like the lowly nematode worm, are influenced by both nature and nurture. These tiny creatures are capable of associative learning. In the lab, they will show a preference for chemical environments like those in which they have found food. Worms that have found food in sodium solutions will gravitate towards sodium, and those who have found food in chloride will gravitate towards chloride. Their taste in food is determined by nature, but their knowledge of where to find it is driven by nurture.

When presented with examples like this, it should be obvious that nature and nurture are not mutually exclusive. It doesn't make sense to ask whether human beings are a product of nature *or* nurture. Obviously, the answer is both. A failure to appreciate that has led to the construction of two straw men in the nature–nurture debate. The first straw man is the biological determinist. Very few people actually believe that DNA is destiny. Even dyed-in-the-wool naturists prefer a more nuanced formulation. DNA, they say, significantly increases the chance of certain behavioural outcomes. Naturists think that biology exerts a very strong influence on how the mind works, but they take the influence to be probabilistic rather than deterministic. Some human traits (such as whether a person has Down's syndrome) may be genetically determined, but the majority of traits (such as whether a person has schizophrenia) are merely *promoted* by genes.

The second straw man is the blank slate. In a book called *The Blank Slate*, Steven Pinker defends naturism against nurturist critiques. The mind is not a blank slate, he claims. Nurturists would emphatically agree. If a blank slate is entirely devoid of content, then no one takes this metaphor seriously. Blank slates don't learn anything. You can write a thousand words on a chalkboard, and it still won't understand what they mean. Nurturists who use the blank slate metaphor approvingly certainly don't take it too literally. They agree that nurture can have an impact, but only in virtue of our nature.

If naturists and nurturists agree that both biology and experience matter, then where is the dispute? Shouldn't they stop quarrelling and sign a truce? That would be nice, but there are still many important battles to be fought. Between the poles of nature and nurture, there is a vast spectrum of possible positions. Everyone agrees that the truth lies between the poles, but major theoretical and ideological differences remain.

To see what's at issue, let's contrast two familiar human capacities. First, consider our capacity to distinguish colours. Red and green objects reflect light waves that differ in length, and the human visual system contains cells that respond differently to those wavelengths. In this respect, our capacity to distinguish colours is heavily based on nature; we have biological mechanisms that are designed for that purpose. A trained artist might learn to notice colour distinctions that others of us fail to see, but that does not prove that distinguishing colours is a learned capacity; it is rather a natural capacity that we can learn to improve. Second, consider our capacity to play baseball. Baseball depends on having certain biological capacities; we must be able to run, track moving objects, clutch oblong objects and swing. But we are not born with mechanisms specifically designed for playing baseball. Our capacity to run, for example, serves many different purposes (fleeing, hunting, jogging, catching trains). To play baseball, we need to learn how to put a collection of general capacities to a new use. Playing baseball has natural prerequisites, but it is learned through nurture.

Disputes between naturists and nurturists typically take two forms. Sometimes they disagree about whether a particular psychological capacity is like colour or like baseball. Consider language. Naturists think that language is a natural capacity: we have mental mechanisms that are designed for acquiring language. Nurturists who think about language recognize that there are biological prerequisites for language learning, but they do not think these prerequisites were *designed* for language. Some nurturists argue that general-purpose pattern recognition abilities are sufficient for language acquisition. For naturists, language is like colour, and for nurturists, language is like baseball. That is a debate about the nature of the biological contribution to language: is the biological contribution *specific* to the domain of language or more *general*?

This is about 'innateness'. Capacities that our minds acquire through psychological mechanisms that are specialized for that purpose are said to be innate. Distinguishing colours is an innate capacity; playing baseball is not. Naturists and nurturists disagree about what's innate and about the extent to which innate capacities can be affected by experience. Nurturists think experience can radically alter our innate machinery.

Notice that this way of characterizing the debate between naturists and nurturists is consistent with the claim that all capacities are influenced by both nature and nurture. No one believes in blank slates or biological determinism, but there is still much room for debate. In the remainder of this chapter, I will discuss some of the major fault lines. We will get a vivid picture of what the nature–nurture debate is all about. Naturists and nurturists agree that both factors are essential to human psychology, but they have fundamental disagreements that cannot be resolved by any simple compromise.

THREE NATURE–NURTURE DEBATES

The alliterative juxtaposition of 'nature' and 'nurture' traces back to the works of Richard Mulcaster, a sixteenth-century educational reformer. Mulcaster thought that educators should work to enhance our natural capacities: 'whereto nature makes him toward, but that nurture sets him forward'.[3] Mulcaster was the first to use these terms, but not the first to speculate about how nature and nurture contribute to behaviour. That question was the subject of a long-standing debate that flared up a hundred years after Mulcaster's lifetime, but began centuries earlier.

The nature–nurture debate has a long history. During some periods, people have favoured the view that nature exerts more influence on the mind than nurture; then opinion shifted the other way. The character of the debate has changed somewhat with each shift, and contemporary disputes are the product of this complicated intellectual history. Three of those disputes have been especially influential in shaping the current debate.

The first dispute concerns knowledge. The key question was posed by the ancient Greeks: are there *innate ideas*? An innate idea is an idea or belief that we have without learning. Innate ideas arise in us without need for observation or instruction. Plato was one of the first defenders of innate ideas. He believed that we come to the world equipped with an understanding of ideas of love, religion, morals, mathematics and many complex domains of inquiry. These ideas are not necessarily present at birth; toddlers do not dissertate on justice. But they can be awoken in us through reflection. His student Aristotle challenged this view, and argued that experience is the source of all knowledge. The entire history of Western philosophy can be viewed as a set of elaborate footnotes on this seminal debate.

The nature–nurture battle is equally enduring. Where Rationalists and Empiricists disagreed about the origin of ideas, parties to the second battle in the debate disagreed about the origins of *individual and group differences*. Some people are smarter than others, some are nicer, some are more artistic, and some are more gregarious. We occasionally even find that such difference are correlated with groups. Men are more physically aggressive than women, for example. The origin of such differences is often obscure, but it is tempting to think they are innate. Some folks are naturally clever and talented, and others are naturally dumb and inept. Shakespeare seems to have liked the idea that some people are born evil. In *The Tempest* he describes Caliban as, 'a born devil, on whose nature nurture can never stick'. In the nineteenth century, this idea became a serious scientific conjecture. Its most influential proponent was Francis Galton, one of the great pioneers in psychology. Galton developed many research techniques that are still in use today: the statistical notion of correlation, the use of questionnaires and twin studies. He was also the person who discovered that people can be identified by their fingerprints. More relevantly, Galton popularized the phrase 'nature versus nurture'. He studied all aspects of the mind, but he had a special interest in heritable character traits. Galton believed that human differences have a biological origin. His critics, like the behaviourist psychologists of the early twentieth century, adamantly disagreed. Recently, this debate has been rekindled by modern genomics, and it remains one of the most controversial issues in the human sciences.

The third nature–nurture debate originates with the work of Galton's cousin, Charles Darwin. Darwin proposed that many of the traits observed in nature evolved through a process of natural selection. Traits would appear by random mutation, and those that increased the prospects for survival were most likely to be passed on. Darwin believed that some human psychological traits might be explained along the same lines. This suggestion became the focus of a heated debate in the 1970s, when E. O. Wilson began waving the banner of 'sociobiology'. Wilson was an accomplished Harvard entomologist, who had discovered that ants navigate by pheromones. He was interested in the parallels between the social organization of simple creatures like insects and that of more complex creatures like us. This led him to publish *Sociobiology: A New Synthesis* in 1975. Most of the book discusses animal behaviour, and it has been heralded as the most important book on that topic ever written. But Wilson stirred up a maelstrom of controversy by including a thirty-page chapter on human beings at the end of his 600-page tome. There he proposed that altruism, aggression, caste systems and the sexual division of labour are all biologically based. In more recent years, Wilson's approach has been refined by a group of researchers who call themselves evolutionary psychologists. Prominent practitioners include Steven Pinker, David Buss, Geoffrey Miller, Gerd Gigerenzer, Leda Cosmides and John Tooby. They combine Wilson's approach with theories that have come out of the contemporary cognitive sciences. They propose that the mind is like a computer with a large number of highly specialized programmes that have been shaped by natural selection for specific purposes. This may sound like a plausible and benign suggestion, but it has sparked a maelstrom of criticism. Wilson was verbally and physically assaulted by progressive students in the 1970s, and his contemporary heirs are lambasted by the guardians of political correctness. There have also been more credible objections, which we will have occasion to explore.[4]

Strictly speaking, these three nature–nurture debates are independent. The first battle, about innate ideas, flared up long before people had formulated the concepts of genetic inheritance and evolution. The second battle, about innate individual and group differences, is silent on the question of innate ideas and usually silent on natural selection. The third debate, about evolutionary psychology, bears a resemblance to the debate about innate ideas, but evolutionary psychologists move beyond ideas to include innate emotions, rules of inference and social dispositions, and they posit Darwinian explanations of how all this innate machinery promoted fitness in our ancestors.

Despite the differences, these three debates concern the same fundamental question: how is human psychology influenced by nature and nurture? In each case, there is a pro-nature view and a pro-nurture opposition. Thus, the debates revolve around opposing conceptions of what human beings are like. This deep philosophical divide tends to promote a theoretical division into basic positions:

THOROUGHGOING NATURISM
The mind comes furnished with an extensive inventory of innate ideas and rules for reasoning about different kinds of ideas in

different ways. This constitutes our universal human nature, shaped by the forces of natural selection. In addition to these universals, there are individual differences in character traits, vocational dispositions and aptitude. These differences are not necessarily adaptive, and they can be affected by experience, but they are largely the result of heritable differences in our genes.

THOROUGHGOING NURTURISM

The mind comes furnished with few, if any, innate ideas, and the innate rules of thought can be used for a wide range of different cognitive capacities. Most of our specific capacities are learned, and the cognitive differences between humans and our close animal relatives stem largely from small improvements in the general-purpose mechanisms that we share with them. Character traits, vocational dispositions and aptitude may be influenced by our genes, but they also are heavily influenced by experience.

Notice that both positions are compatible with the view that the mind is shaped by nature and nurture. Thoroughgoing naturists are not biological determinists, and thoroughgoing nurturists do not presume that the mind is a blank slate. Nevertheless, the views are dramatically opposed. One could try to split the difference by adopting a naturist position on some issues and a nurturist position on others, but the thoroughgoing positions constitute stable and coherent packages. Someone who is inclined to think that biology places strong constraints on psychology is likely to appeal to biological factors to explain both human universals and human differences. Someone who is inclined to focus on learning and human flexibility is likely to think that experience is responsible for many of both our shared and our divergent psychological traits. In *The Blank Slate*, Steven Pinker defends a fairly thoroughgoing naturism and, in this book, I will be defending a fairly thoroughgoing nurturism. These duelling perspectives invite very different explanations of behaviour, and they encourage a very different picture of human potential.

Perhaps the truth lies somewhere in between. Perhaps the divisions are just a matter of emphasis. Perhaps we should all learn to get along. Perhaps. But the attitude of reconciliation isn't always helpful. Too often, weary combatants in the nature–nurture debate concede that nature and nurture are equal partners, and then go on to emphasize one side at the expense of the other. One cannot advocate a truce while giving lip service to one side. Some attempts at reconciliation are a bit more successful, such as those presented in Matt Ridley's *Nature Via Nurture* Peter Richerson and Robert Boyd's *Not by Genes Alone*. These authors take both sides seriously and emphasize places where biology and environment interact. These are valuable contributions to the nature–nurture debate, but reconciliation can be taken too far. Scientific debates are rarely settled by compromise, and when we look for 'biocultural' explanations of behaviour we dilute the interest of both sides. If some traits are strongly influenced by the genes, then it is potentially harmful to waste energy identifying cultural factors that might explain a tiny per centage of the variance. And, if some traits are strongly influenced by culture, the involvement of genes is no more informative than the involvement of quarks – sure, we need genes to behave, but genes add little explanation of why we behave one way rather than another. I happen to think there are numerous cases where the nurturist perspective sheds more light than the naturist perspective, and that this is singularly the most interesting and important fact about the human species. If I am right, then describing human behaviour as a conspiracy of nature and nurture is true but terribly misleading. It is as futile as doing a genetic analysis of Picasso's remains to help determine why he was inspired by African masks when painting *Les Demoiselles d'Avignon*. Biological explanations can contribute to psychology, but, in many cases, we have more to gain from looking at other influences.

Arguing for the primacy of nurture over nature will be the burden of this book. In this climate, nurturism may seem like a radical position. With the Human Genome Project and modern neuroscience, we are learning new facts every day about biological factors that contribute to behaviour. Nurturists are often depicted as opponents of science, and, all too often, the nurturist perspective is most vocally championed by postmodernists and cultural theorists who are sceptical of all efforts to scientifically investigate the human mind. My view is that science offers resounding support for the nurturist perspective. Nothing about our current knowledge of the brain, genes or psychology should lead us to think naturists will win any of the battles I have been describing.

The fact that naturism is so popular should really be surprising. Causal observation suggests that biology must have a fairly limited role in explaining human behaviour. Any trip to a cafeteria would confirm that. Unlike all other creatures, human beings are radically varied. If we were to compare two chimpanzees from two different locations and two different centuries, the differences between them would be minuscule. If we were to compare two people from different times and places, we could fill volumes with descriptions of how they differ. There would be many similarities too, of course, but the differences would be staggering. It is an obvious and fundamental fact about human beings that nurture exerts a profound influence on how we think and behave.

HUMAN NATURE, THE VERY IDEA

The nature–nurture debate is a debate about human nature. Are we, by nature, primarily driven by innate, evolved and genetically controlled traits, or are we primarily driven by experience? But the debate is also about the very idea of human nature. The human sciences are sometimes compared to natural sciences, where the goal is not to recount particular events, but to uncover universal laws. On this model, the human sciences are in the business of determining the laws that govern human behaviour. Thoroughgoing nurturists resist this picture.

The concept of human nature refers to things that human beings do naturally, in virtue of our biological constitution. Consequently, debates between naturists and nurturists can usually be reframed as disagreements about human nature. For example, naturists say we have a natural faculty for language,[5] and some nurturists deny this.[6] Even debates about human differences, as opposed to universals, can be seen as debates about what we do naturally. Naturists say that cognitive differences between the sexes are built into the nature of our species.[7] Some nurturists deny this, saying men and women are pretty much the same from the neck up, prior to the influence of socialization.[8]

At a deeper level, nurturists question whether human nature should be the primary focus of the human sciences. Suppose that we are capable of behaving in ways that are not heavily constrained by our biology. It follows that human behaviour will vary dramatically from place to place. For example, we might find that some societies are monogamous while others are polygamous. These kinship patterns are human, but they are not natural. That is, there is nothing about our biology that impels us to become monogamous or polygamous. Of course, the fact that marriage conventions are not part of our nature does not make them less worthy of scientific investigation. In fact, variations of this kind are among the most interesting aspects of our species. Nurturists think that a tremendous amount of human behaviour is like this, and, consequently, the study of human nature leaves much out. A full science of humanity should actively explore the ways in which we go beyond human nature.

Many academic fields regularly look beyond human nature. History, sociology, anthropology and literature would be unthinkable if they restricted themselves to what we do naturally. It is odd, then, that the cognitive sciences have had such a biological bias. Fortunately, that is beginning to change. In the last decade, there has been a widespread effort to develop a cultural approach to psychology, which disbands with the traditional agenda of finding universal laws and focuses, instead, on human differences. That is the story that I want to tell here. It is a story that begins where biology leaves off. In other words, it is the human story.

Where Do Traits Come From?

2

Putting the Genome Back in the Bottle

Hidden in every cell of the human body, there is a genetic instruction book that determines how many limbs we have, the arrangement of our internal organs and the colour of our eyes. The instruction book guides the process that transforms each us from humble zygotes to vastly complex multi-cellular organisms. Minor differences in the instructions lead to major differences in form. It shouldn't surprise anyone to learn that genetic differences can also influence psychology and behaviour. But this truism is sometimes presented in a deeply misleading way, promoting the idea that human behaviour is genetically controlled, and that environment plays a negligible role. Don't believe the hype.

Naturists sometimes suggest that there are genes that code for specific psychological traits. Headlines spread the word by reporting that scientists have discovered a gene for this and a gene for that. 'Gene for X' talk is very seductive, but almost always misleading. This hypothesis gives the impression that human psychology is under direct genetic control, and it is a small step from that idea to the idea that DNA is destiny. In reality, there is virtually no solid evidence linking specific genes to specific behavioural traits. The relationship between genes and behaviour is indirect and complex. Variations in certain genes correlate to some degree – usually negligible – with behavioural traits, but many other factors make contributions as well. Crucially, genes do not exert any influence on their own. The activity of a gene depends on many other genes, on biological materials outside the genome and ultimately on factors outside the organism.

The search for genes that affect behaviour is generally pursued by molecular geneticists using 'linkage studies' to correlate bits of DNA with observed traits. That hard task has been possible only since the birth of modern genomics. There is also an older method for correlating genes and behaviour that boasts a much more impressive success record: heritability studies. Population geneticists estimate genetic contribution to behaviour by measuring correlations between traits and familial relatedness. If a trait runs in families, it's more likely to be genetic, they say. And they can draw this inference without worrying about identifying the genes that do the work. This research has resulted in startlingly high heritability scores for a wide variety of traits – everything from temperament to views on tax policy. It turns out, however, that the inference from heritable to genetic is highly problematic.

It is important to correct these misconceptions about the relationship between genes and behaviour. Gene-talk is so common these days that public opinion is increasingly comfortable with the idea of genetic determinism. Scientists who work in genetics do not make this mistake, but, in conveying their ideas to the general public, something gets lost in translation. It's time to set the record straight.

PROTEAN PROTEINS

To understand the connection between genes and behaviour, it is useful to start with the basics.[1] We need to begin by getting clear on what genes actually are. The biological programme for any living creature is encoded in its DNA (deoxyribonucleic acid). DNA is constituted by a sequence of chemical bases (adenine, thymine, cytosine and guanine). Each base is paired up with a complementary base; adenine pairs with thymine and cytosine pairs with guanine, for example. These pairs are like rungs in a spiralling ladder – the famous double helix – linked together by sugar and phosphate molecules. A collection of three bases (called a codon) can encode instructions for making an amino acid. There are twenty different kinds of amino acids, and some stretches of DNA contain instructions for how to string together amino acid sequences, also known as polypeptide chains. A gene is simply a stretch of DNA that carries instructions for making a polypeptide chain. Such stretches are also called structural sequences. Most stretches of DNA are not structural. Some stretches of DNA, called regulatory sequences, regulate the behaviour of genes, by, for example, indicating where a gene starts and stops or by serving as landing sites for molecules that turn genes on and off. More about that in a moment. In the human genome, 98 per cent of the DNA is neither structural nor regulatory. It is called junk DNA.

Amino acids are important because they are the primary building blocks of proteins, and proteins are the primary building blocks of cells. Genes produce organisms by guiding protein production, but they do not do it alone. Here's how it works. DNA resides within the nucleus of a cell. That cell might be a single-celled organism, a single cell in a multi-celled organism or the single-celled zygote that will become a multi-celled organism. But DNA is not the only stuff inside the nucleus. In addition there is a chemical called RNA, or ribonucleic acid. RNA comes in several varieties, and the RNA inside the nucleus is called mRNA, because it serves as a messenger for the genes. It does this by a process called transcription: mRNA is chemically similar to DNA, so it is able to latch on to DNA bases and make copies. These copies travel then out of the nucleus into the cytoplasm of the cell. When mRNA enters the cytoplasm, it is met by molecules called ribosomes, which can read mRNA like a code. As ribosomes move down a string of mRNA, another form of RNA, called tRNA, transfers amino acids to the ribosomes, and these are strung together. This process is called translation. Certain codons instruct ribosomes to stop reading, and, when such a codon is reached, the string of amino acids is released. Thus, polypeptide chains are born. Once formed, molecules surrounding polypetides cause them to fold into three-dimensional structures, called proteins. Some proteins are built up from multiple polypeptides. Consequently, a single gene can be used in the production of more than one protein. Indeed, there are about 30,000 different proteins in the human body, which may exceed the number of genes.

Proteins are the main ingredients making up our cells. A typical body cell will comprise 10,000 different proteins. Most cells have a complete copy of the genome, yet they are very different. The reason for that is some of the genes are like switches. Within any given cell, some genes will be turned on and some will be turned off. This process is called gene regulation, and it can occur at various different stages of protein manufacture including transcription and translation. The DNA inside a cell comes into contact with proteins from that cell, from neighbouring cells, or from the environment, and these can determine which genes get put to use. Sometimes this is achieved by transcription factors, which are proteins that attach to regulatory sequences of DNA and either increase or decrease the transcription of specific genes.

Regulation is essential for the differentiation of cells in embryonic development. In the earliest stages, all the cells in an embryo are alike, but eventually they are differentiated into skin cells, muscle cells, hair cells, blood cells, brain cells and all the other lovely cells that make up an organism. Human beings have about 200 different kinds of cells in their bodies. Embryonic cells are able to diversify because they are exposed to a chemical environment that is not uniform throughout. Asymmetrically concentrated chemicals assign different fates to different cells. Initially all cells have the potential to become blood, brawn or brain, but the chemical environment quickly fixes their fate by toggling their genes on and off. Nurture, in a chemical sense, is already influencing nature at the earliest stages of development.

The DNA in a human being (the human genome) consists of about 3 billion bases. Genes vary in size. Some comprise a few hundred bases, and some comprise a couple of million. In the early 1990s, an international group of scientists set out determine the entire sequence of bases in human DNA, and, ultimately, to identify all the genes contained therein. This was the Human Genome Project. By 2001, the sequence was completed. Determining the exact number of genes is very difficult because we don't always know where one gene ends or where another begins. Early estimates predicted that we would find as many as 140,000 genes in the human genome. When the sequence was finished, estimates dropped dramatically to only 30,000. In October 2004, the number fell again, and the current estimate is that human beings have between 20,000 and 25,000 genes. This is a humbling discovery. Fruit flies have 13,600 genes, and tiny worms have 19,000. If that weren't bad enough, mustard grass has 25,300 known genes. We may have fewer genes than a weed!

The numbers are even more startling when you consider how many genes we share in common with simpler creatures. We share 31 per cent of the genes found in yeast, 40 per cent of the genes in worms, 50 per cent of the genes in fruit flies, and an estimated 98.5 per cent of the genes found in chimpanzees.[2]

Apparently, it doesn't matter how big your genome is; it's what you do with it. Genes create organisms by guiding the production of proteins, and proteins make cells and drive cellular behaviour. By regulating the quantity, order and arrangements of proteins produced, one can generate a boundless variety. Similar genes can produce totally different creatures, just as a single box of Lego can be used to produce a castle or a rocket, and everything in between. With genes, things are a little bit more constrained, but not much. We share with most plants and animals a collection of genes called homeobox genes – hox for short. Hox genes code for amino acids that become transcription factors, the proteins that turn other genes on and off. Hox genes regulate the genes that make body parts. Starfish, flies, chickens and people have different body plans because the hox genes shared across these species differ subtly in where and when they do their work. The hox genes in animals with short rib cages stop rib production sooner than they would in animals with long rib cages. In snakes, hox genes allow ribs to be formed down the entire length of the body. Other hox genes determine how many limbs you have and where they grow. A hox mutation in fruit flies causes legs to grow from flies' heads in place of antennae. In human beings, hox mutations result in polydactyly, the growth of extra fingers.

This illustrates how a similar stock of genes can lead to startling differences. Changes in hox timing can affect how many ribs, limbs and vertebrae a creature will end up with. The differences between us and other creatures originate in subtle differences in our genes. Genetic differences can also explain variation within a species. Hox mutations lead to extra limbs. Other genetic mutations can result in other birth defects and diseases. Genetic differences also underlie healthy variations. Hair colour, eye colour, skin colour, skull shape, body type and most other aspects of appearance have genetic causes.

BIOLOGY AND BEHAVIOUR

Since genes can have such dramatic impact on our bodies, it is natural to think they can have an impact on psychology and behaviour. Genes can impact all of our organs including the brain, and genetic conditions that affect the brain can certainly influence our psychological abilities. Our capacity to learn more than earthworms, mice and monkeys derives from the fact that our genes are different from theirs. Genes also provide us with a variety of instincts (such as suckling our mothers and walking upright), basic drives (such as hunger and thirst) and reflexes (such as flinching and gagging). There is no debate about whether genes influence behaviour; controversy concerns the exact nature of that influence.

At one extreme, there are people who think many human traits are strongly determined by our genes. People in this camp say that our attitudes, preferences, personalities and behavioural tendencies are genetically influenced, and that those influences far outweigh any other factor. On this view, someone can be a born criminal or a born believer. On the other extreme, there are people who think that genes furnish us only with very general learning capacities. The range of possible attitudes, personalities and behavioural dispositions may be fixed by our species, but which ones we have will be entirely a function of experience. Genes contribute a range of options, but they do not bias us towards one option over any other. Between these two extreme views, there is a spectrum of intermediate positions. Most researchers think that genes do bias us towards certain outcomes, but that experience can alter and even override those biases.

Both naturists and nurturists usually claim to occupy an intermediate position. The difference between the views is a matter of emphasis and a matter of degree. The difference is especially pronounced when it comes to explanations of human differences. Nurturists like to identify experiential causes of human variation. Naturists prefer to offer genetic explanations. Both explanations can be taken too far. Here I will focus on the excesses of naturism, because, in these days of the genome, they are getting more airtime. There has been a tremendous amount of excitement about genes that control various aspects of our mental lives, but, on close analysis, there is actually very little evidence for a link between genes and psychological traits. Some of our traits may owe a big debt to our genes, but, arguably, the majority do not.

Do Genes Cause Traits?

Let's begin with a clear case of genetic influence on psychology: colour vision. We see colours because there are photosensitive cells in the retinae called cones. Most human beings have three kinds of cone cells, which are sensitive to three different ranges of lightwaves: short, medium and long. The colour you see depends on the proportion of lightwaves in these different ranges. Blue objects reflect most short lightwaves, green objects reflect medium waves, and red objects reflect long waves. The cone cells are generated in accordance with a genetic recipe that produces colour-absorbing pigments. The most common form of colour blindness, which affects one in twenty men in the United States, is caused by genetic abnormality in a single pigment gene. As a result, the medium-wavelength cones in these men respond like long-wavelength cones, and objects that are red and green become indistinguishable.

Genetic abnormalities can cause dramatic colour deficits as well. The most extreme case is achromatopsia. Achromatopes don't see any colours at all, and as a result their vision is very poor, because colour-sensitive cells are crucial for seeing the world in sharp detail – think of how things blur at night, when colour cells are less responsive. Achromatopsia can be caused by a focal injury to the part of the brain that processes colour information, but it can also be caused by an abnormality in the genes that produce cone cells. Oliver Sacks describes a Micronesian island called Pingelap in which as many as 10 per cent of the inhabitants can't see colours.[3] Years ago a typhoon reportedly wiped out most of the island's population, leaving twenty survivors, some of whom carried a recessive abnormality in a gene that contributes to colour perception. The abnormality is in a gene called CNGB3, which codes for a component of the receptors in cone cells. If a person inherits two copies of this gene, they will not be able to perceive colours. The present-day population of Pingelap is descended from the small group of typhoon survivors, so achromatopsia is far more common there than in populations that are more genetically diverse.

This simple example shows that genes can affect psychology, and it is easy to get excited about such findings. But we shouldn't get carried away. In colour blindness, we find genes that code for components of our visual system. That is very different from finding a gene that codes directly for a psychological trait. When we move beyond the sense organs into the mind, traits become much more complex, and the link to genes is less direct. It is important to realize that there is almost never a one-to-one mapping between genes and psychological traits. Instead, psychological traits can be influenced by many different genes, and by interactions between genes and the environment. In most cases, the environment contributes as much or more than genes.

We often read in the headlines that scientists have discovered the gene for a particular psychological trait. This way of talking is usually based on three fundamental fallacies. I will call these the Fallacy of Genetic Causation, the Fallacy of Genetic Necessity and the Fallacy of Genetic Sufficiency. Each of these fallacies promotes the idea of genetic determinism. Let me explain.

The Fallacy of Genetic Causation is the mistake of thinking that a gene somehow codes for, and is thus directly responsible for, a particular psychological trait. It is extremely important to exercise caution when talking about what genes code for. First of all, what genes really code for are amino acids, so it should raise eyebrows when someone says there is a gene for a psychological trait. What people usually mean is that those amino acids are implicated in the production of cells which play a specific and predictable role in the production of behaviour. So, at best, talk of genes for behaviour is an exaggeration. It would be better to say that there are genes that have a predictable impact on behaviour. Second of all, the impact that a gene has on behaviour is often an accidental byproduct of the fact that it has an impact on something other than behaviour. For example, genes that affect metabolism can sometimes influence behaviour as a result. When such genes are found it's seriously misleading to describe them as genes for behaviour, because that implies that the gene directly causes us to behave in a certain way.

Consider alcoholism. We often read about scientists investigating the genetic basis of alcoholism. We are told that alcoholism is a genetic disease, and that there is a gene for it. When we read headlines like this, it's hard to resist thinking that there is a little stretch of DNA that causes a person to be born with an alcohol addiction. This is far from the truth. Genes that have been implicated in alcoholism are not genes for alcoholism; they are genes that make a small contribution to the probability that someone will become addicted to alcohol under certain environmental conditions. Scientists have been especially interested in genes that influence alcohol tolerance. High tolerance to alcohol can make an indirect contribution to addiction. People with a high tolerance can drink more before getting drunk. Suppose that such a person likes getting drunk. There are dozens of reasons why people like drinking: some drink because they are down on their luck, some drink out of peer pressure, some copy behaviour they see at home, some like to overcome their inhibitions, and some just enjoy the pleasure that intoxication naturally produces. A person with a high alcohol tolerance who also likes to get drunk will have to drink a lot more than a person with low alcohol tolerance. Consequently, he or she will expose him- or herself to much more alcohol, and that will increase the likelihood of developing a physical dependency. People who are genetically better at metabolizing alcohol are statistically more likely to expose themselves to doses that result in addiction. Alcoholism is not caused by a gene. It's caused by drinking a lot, and genes are among the many factors that can make a small contribution to how much a person drinks. Availability of alcohol and an enthusiasm for getting drunk are much more important factors, and neither of these is genetically determined.

The case of alcoholism can also be used to illustrate the Fallacy of Genetic Necessity. This is the idea that you need to have a particular gene in order to have a particular psychological trait. All the talk about genes for alcoholism might lead one to think that you can't become an alcoholic unless you have a particular gene. This is a mistake. First of all, different genes can have the same effects. Tolerance to alcohol can be influenced by how alcohol is broken down by the digestive system or by how alcohol influences information processing in the brain, and each of these two factors might be influenced by a large number of genes. Consequently, no single gene can serve as the red flag for alcoholism. Moreover, alcoholism can occur in individuals who have none of the genes that promote high tolerance. If a person with low tolerance drinks heavily, the risk of alcoholism is very high.

Even psychological disorders that are much more closely tied to genes can have non-genetic causes. Consider schizophrenia. Patterns of inheritance lead researchers to believe that schizophrenia usually requires a genetic abnormality. There are probably many genes that can contribute to schizophrenia, and in most people with schizophrenia one or more of those genes is present. But this is not always the case. Schizophrenia occasionally occurs in individuals with no family history of the disease. Non-genetic causes may include drug use, epileptic seizures, other brain injuries and infections *in utero*.

Are Genes Sufficient for Traits?

Schizophrenia can also illustrate another mistake that people frequently make in talking about the connection between genes and behaviour: the Fallacy of Genetic Sufficiency. When we read that there is a gene for a psychological trait, we tend to think that the gene is sufficient for that trait – we think the trait is inevitable if someone has the gene. But an individual gene is never sufficient for a trait. Every gene depends on other genes and on contributions from the environment. By ignoring these other factors, we exaggerate the role of nature, and we underestimate the role of nurture. Schizophrenia is a case in point. If you have a genetically identical twin with schizophrenia, your chance of getting the disorder is only 50 per cent (more on twins below). That's a high probability, but it is startlingly low given the widespread assumption that schizophrenia is genetic. We should say instead that it is a disorder derived from gene–environment interactions.

Scientists don't fully understand the environmental factors that contribute to schizophrenia. In some cases, the variation may be organic in nature. Complications during birth may have an influence on whether one of two twins with a genetic predisposition for schizophrenia becomes symptomatic later in life. But social factors may also play a role. There is evidence that individuals with a genetic predisposition for schizophrenia are more likely to develop the disorder if they are raised by parents who exhibit communication deviance.[4] Parents are said to be communicatively deviant if they tend to be excessively vague in conversation, or if they shift topics unpredictably, or if they tend to prevent some topics from being discussed. Other studies have linked schizophrenia to social isolation and urbanization.[5] Other factors will certainly emerge in future research. The key point is that there is no evidence that genes are sufficient for schizophrenia. The environment plays a role as well.

Gene–environment interactions are commonplace in other species. Siamese cats have dark faces and paws because the genes that determine their fur colour are affected by temperature. If you shave a patch on the back of a Siamese cat and keep that patch under ice, the hair will grow back dark. In some species, appearance can be altered by a change in social environment. Consider cichlids, a family of African fish. In some cichlid species, alpha males are brightly coloured and have distinctive eye stripes. If the alpha male dies, the other males fight it out, and the victor develops bright colouration and eye stripes in a matter of seconds. Social organization has an even more dramatic impact on the blue-headed wrasse. These fish swim in small groups with one male and several females. If the male dies, the largest female in the harem changes her sex and becomes leader of the group. In the face of such examples, we shouldn't be surprised to find that many human genes are sensitive to environmental conditions. By learning about these, we can exert considerable influence on traits that are heavily influenced by the genes. For example, there are a number of ways in which a person can be genetically predisposed towards obesity, but we can determine whether those predispositions are expressed by carefully controlled diets. Likewise, we may ultimately be able to reduce the incidences of schizophrenia by identifying and intervening with environmental triggers.

I have tried to stack the deck in favour of naturism by focusing on psychological traits that have been correlated with genes. Even when such correlations exist, the genes in question usually don't directly code for psychological traits, usually aren't necessary for those traits and usually aren't sufficient for those traits. The phrase 'gene for X' encourages us to think that some psychological traits are genetically determined. It's easy to read headlines and draw the conclusion that certain genes are necessary and sufficient causes of human psychological traits. That conclusion is almost always mistaken. Consequently, it is misleading to indulge in 'gene for X' talk. Even schizophrenia, which correlates very well with genetic factors, depends on the environment. Alcoholism is even less directly tied to the genes, and we will see other examples of allegedly genetic traits that owe much more to the environment.

These days, committed naturists concede that DNA is not destiny, but they still exaggerate the importance of genes. They have given up on genetic determinism, but they are fond of saying that genes *predispose* us to behave in certain ways. In some cases, such talk is warranted. Most people who develop schizophrenia have a genetic predisposition. But this may be the exception and not the rule. Consider alcoholism. Some studies report that male children of alcoholics are four times more likely to become alcoholics than children of non-alcoholics.[6] These numbers hold up even when the children of alcoholics are adopted by non-alcoholic parents. That sounds like a genetic predisposition, but the story turns out to be much more complicated. First of all, children of alcoholics who are adopted may be exposed to alcohol in the womb and they may be given up for adoption in mid-childhood, after being raised during early life in alcoholic homes. Second of all, the claim that children of alcoholics are four times more likely to become alcoholics is based on studies that are methodologically flawed. The authors used an anachronistic definition of alcoholism, and they actually found that their control group, adopted children who were not born to alcoholic parents, were more likely to be problem drinkers than children born to alcoholics. Other adoption studies come up with lower numbers, and, for women, alcoholism often shows no evidence of biological inheritance at all. Finally, even if being born to an alcoholic quadruples your chance of developing an addiction, your chances may still be very small. By one plausible estimate, 18 per cent of the sons of alcoholics become alcoholics, as compared to 5 per cent of the sons of non-alcoholics. So the overwhelming majority of sons of alcoholics do not become addicted to alcohol, and a significant number of sons of non-alcoholics do develop an addition. It would be a great exaggeration to say that children of alcoholics are disposed to become alcoholics. When we read in the newspapers that there is a genetic predisposition, we infer that children of alcoholics are likely to follow in their parents' footsteps. The opposite is the case. When naturists trade in talk of genetic determinism for talk of genetic predispositions, they still exaggerate the link between genes and behaviour.

Consider one more example. In the early 1990s, a Dutch research group did a genetic analysis of a family whose male members were chronically violent and aggressive.[7] They discovered that these individuals have a gene called MAOA-L. The same gene was later found to have higher frequencies in American gang members than non-gang members, and it has also been found in over half the Maori, the indigenous people of New Zealand, who are notorious for violent conduct and were once a warrior culture. Unsurprisingly, MAOA-L became known as the warrior gene.

This label is extremely misleading.[8] First, it commits the Fallacy of Genetic Causation. MAOA-L does cause violent behaviour. Like all genes, it codes for proteins, and, in particular, it codes for a protein that breaks down neurotransmitters such as dopamine, noradreneline, and seratonin. The MAOA-L variant does so less efficiently than other variants, however, and that leads these neurotransmitters to accumulate, which can have widespread inpact on behaviour. The effects can include violence, but also attention deficit disorder, depression and anxiety. The so-called warrior gene is also neither necessary nor sufficient for aggressive behaviour. Many violent gang members lack the gene, and many who have the gene do not join gangs or engage in violence. Indeed, it is highly common – found in one-third of white men. Within the New Zealand sample, the link between MAOA-L and violence depends on social factors; it correlates with higher levels of violence only among those who endured childhood abuse. In other words, there is no gene that makes people into warriors. The warrior gene affects a large family of common neurotransmitters, and this can have a wide range of unpleasant consequences in suboptimal social environments. It is not sufficient for any behaviour, nor does it cause a behavioural predisposition in ordinary, healthy environments.

Do All Traits Have Genetic Correlates?

When naturists advertise links between genes and behaviour, they also promote another kind of misconception. They distract away from the fact that the vast majority of psychological traits could never be coded in our genes because they are learned. For example, there is no particular gene that predisposes me to believe that Canberra is the capital of Australia. Beliefs about geography are acquired through maps and schoolbooks. Likewise, I prefer jazz to disco, but that probably isn't the result of any genetic predisposition, and I don't know how to waltz, but I am not suffering from a detrimental mutation in a waltzing gene. There are an unbounded number of beliefs, preferences and abilities that have no genetic causes. Of course, if I had gerbil genes instead of human genes, I couldn't memorize national capitals, but that doesn't mean I have a gene for geographical knowledge. Genes are a precondition for human learning, but they may not be directly responsible for anthing we do.

We have considered a continuum of cases. First, there are cases like colour vision that are under the control of a small number of genes that determine what parts of the rainbow we can see. In that case, it makes sense to say there are genes for seeing colours. This way of putting it is a little loose, since the genes really code for amino acids that make proteins that are used in the construction of cells that carry information from eyes to brain. But it's okay to abbreviate this by saying there are genes for seeing colours, because, in a normal, healthy environment, those genes will ensure that colour-sensitive cells are created. Second, there are cases like schizophrenia. In this case, it's a little misleading to say there are genes for schizophrenia, because non-genetic factors may be equally important, but we can say that genes make a major contribution to schizophrenia. Unless you sustain brain damage, having these genes is a prerequisite for developing the disorder. Third, there are cases like alcoholism. Talking about genes for alcoholism is very misleading. The genes that people describe this way can statistically increase the likelihood of developing an addiction to alcohol, but they do so indirectly, by affecting alcohol tolerance. Having high alcohol tolerance is neither necessary nor sufficient for becoming an alcoholic, and environmental factors are probably much more important. And, finally, there are cases like music preferences or beliefs about geography. If you are congenitally tone deaf or profoundly retarded, you may never develop a taste for jazz or learn to list state capitals, but that doesn't mean there are genes dedicated to these specific outcomes. No genetic mutation would prevent a person from acquiring a taste for jazz while leaving all other aspects of music perception in task.

For any trait, we can ask: is it more like colour vision or more like music preferences? Naturists and nurturists disagree about the scope of genetic explanation. Naturists think that a lot of psychological traits are like colour vision. Nurturists think that colour vision is the exception. When it comes to psychology, most traits will owe at least as much to the environment, and often genes won't be any part of the story. Appealing to genes to explain how we learn about geography is no more useful than appealing to oxygen. We need to breathe to learn about geography, but breathing is a precondition for most things that we do. It doesn't directly explain how we memorize state capitals. Like oxygen, genes don't control geographical knowledge.

It is easy to come up with examples of psychological traits that are not genetically controlled, and extraordinarily difficult to come up with examples that are. If you are not convinced, try to list all the beliefs, preferences and skills that you can think of in thirty minutes and then see how many of them look like the kinds of things that might be promoted by specific genes. Chances are the items on your list that have a direct link to genes will be totally swamped by those that don't. If you were to list the psychological traits of a worm, or a mouse, or a pigeon, the balance would shift the other way. Like other creatures, we certainly have some genetic predispositions, but human psychology owes more to experience. Genes do not dictate what you believe, what political values you have, what occupation you pursue, what clothing you wear or what you eat for breakfast. Every minute of your waking life you are probably doing something that you were not genetically disposed to do: getting dressed, reading the morning paper, driving to the office, writing emails, sending bills, shopping, eating processed food, watching television and changing light bulbs. Ask yourself how much of this daily routine could be illuminated by a complete analysis of your genome.

To settle the nature–nurture debate, it's not enough to point out that we do millions of things that are not genetically controlled. We ultimately want to be able to figure out what aspects of behaviour owe a significant debt to our genes. It's pretty obvious that ticklishness is largely under genetic control and equally obvious that knowledge of state capitals is not. But there are lots of traits between these extremes that are less clear, and the battle between naturists and nurturists often concerns these cases. Examples include psychological disorders, personality traits, sexual preference, gender differences and intelligence differences. Naturists and nurturists tend to agree that these things are influenced by biology, but they disagree about the extent of those affects. How do we decide where to draw the line? How do we decide which aspects of psychology are closely linked to our genes?

Ultimately, the best thing we can do is look for genes. There's no substitute for a smoking gun. If you want to show that a trait is genetic, find the gene. Scientists try to do this, and every week they fill journals with studies that purport to link genes to behaviour, but these studies are often problematic. In many cases, they are difficult to replicate, and, even in the best cases, the genes that are identified only account for a small amount of the variance. One reason for this is that most complex traits are multigenic: they are influenced by more than one gene. Identifying a single gene for a single trait is often impossible. Another problem is that the links between genes and behaviour are indirect, so we can find genes that are correlated with psychological traits without having any idea how or whether they are causally related to those traits.

It will be a very long time before we have a good understanding of how most genes work. In the meantime, we need a way to make educated guesses about which traits are most likely to be genetically controlled. There's no sense in searching for the genes that promote a trait unless we have good reason to think that trait has a genetic basis. For example, it would be a waste of money to search for genes that correlate with the belief that Canberra is the capital of Australia.

WHERE TRAITS COME FROM

We all have hunches about which traits are most likely to have a strong genetic influence. Height is likely to be under heavy genetic influence, and hairstyles are not. But we need to move beyond hunches and find a scientific way to figure out how likely it is that genes are contributing and how much. Population geneticists have developed an ingenious method for doing this – a technique that does not require directly studying the genes. The great hope in biology is that population genetics and molecular genetics can work together. Population genetics can turn hunches into data by quantifying the degree to which each trait is likely to be genetically influenced, and then the molecule people can come in and find the actual genes that are doing the work. At this stage, population genetics is far ahead of molecular genetics in terms of establishing a relationship between biology and psychology. We still don't have clear cases of individual genes causing psychological traits, but we have mountains of data suggesting the traits are biologically influenced. Some of the results are truly astonishing. Population geneticists have claimed that everything from personality to political preference depends on biology. Should we believe them?

Amazing Twins

To quantify the genetic contributions to human behaviour, scientists must look for correlations between genetic traits and psychological traits. To do that, researchers must measure variation in genetic similarity. If genetically similar individuals have more in common than less similar individuals, despite few relevant differences in their environment, then researchers conclude that genes are making a contribution. Twins play an important role in this research.[9] Monozygotic twins (popularly called 'identical twins') emerge from the same sperm and egg. They share 100 per cent of their genes. When we find similarities in monozygotic twins that cannot be explained by appeal to shared environmental variables, we have evidence for a genetic contribution. For example, if monozygotic twins are reared in separate homes, then they will have different life experiences. They will attend different schools, have different friends and be raised by parents with different personalities and values. If separated monozygotic twins are psychologically similar, then that similarity may have a genetic basis.

Everyone has heard amazing stories about monozygotic twins who were separated at birth.[10] When these twins are reunited as adults, they discover surprising similarities. For example, one pair of separated twins who met as adults discovered that they had both named their dogs 'Toy'. A second pair of separated twins discovered that they both cross their eyes when excited. A third pair each had a miscarriage in the same year. It turned out that another pair of twins both owned body-building gyms. And still other pairs have discovered that they use the same shampoo.

Stories like this tend to make our spines shiver. We are astonished by the eerie similarities between separated twins. But the astonishment is usually unwarranted. Anecdotal examples of behavioural convergence tell us nothing about the biological bases of behaviour. First of all, many of the anecdotes focus on behaviours that are totally unlikely to be genetically influenced. It would be ludicrous to suppose that there is a gene for naming dogs 'Toy'. The fact that a pair of separated twins gave their pets the same name has nothing to do with their biology. Second of all, many similarities between separated twins are not statistically anomalous; by sheer coincidence, any two unrelated individuals will have many things in common. Suppose you and a randomly chosen individual each listed your pets, interests, habits, ticks, pastimes, favourite products, employment histories and other idiosyncratic biographical details. If we were to compare these lists there would probably be many points of overlap. By coincidence, you might have majored in the same subject in college, you might like some of the same movies, and you might even have had pets with the same name. None of this would be very surprising. Likewise, it should not be surprising that separated twins find points of biographical overlap. These coincidences strike us as amazing because we assume that the similarities are caused by biology. Once we realize that any two people have many similarities, that illusion goes away.

To establish a biological influence on behaviour, one needs to show that monozygotic twins have *more* similarities than people who are less closely related. Anecdotes don't tell us that. That's where the science called behavioural genetics comes in. Behavioural geneticists have devised ways to mathematically quantify how much monozygotic twins have in common. The basic strategy is to compare twins to pairs of individuals who are less closely related. They sometimes use twins who have been reared apart for this purpose, but those cases are rare. More often, behavioural geneticists look at monozygotic twins who are reared in the same household, and they compare them to dizygotic twins reared in the same household. Dizygotic twins (sometimes called 'fraternal twins') emerge from two eggs and two sperm. Like ordinary full siblings, they share 50 per cent of their genes. But like monozygotic twins, they have very similar life experiences. Comparing monozygotic twins to dizygotic twins is illuminating, because they grow up in environments that are equally similar. If monozygotic twins are different from each other in any way, then those differences must be due to the environment, and, in particular, they must be due to environmental factors that are not shared by the twins (such as unique life experiences). If dizygotic twins are more different from each other than monozygotic twins, then those differences must be due to their genetic differences, and, correlatively, the comparative similarity between the monozygotic twins must be due to genetic similarity. Using this logic, behavioural geneticists can compute the amount of variation in a trait that correlates with genes. On the standard formula, you subtract the correlation between dizygotic twins for a given trait from the correlation between monozygotic twins, and then multiply by two. The resulting number is called the 'heritability' of that trait. Heritability can be defined as the amount of variance that is correlated with the genes (or other biological materials).

Caveat 1: Heritable Versus Genetic

Before looking at heritability studies, two preliminary warnings are in order. In the press, the term 'heritable' is sometimes mistakenly treated as a synonym for 'genetic'. This is confused in many ways. Heritable traits need not be genetic and genetic traits need not be heritable. Here are three differences between these two constructs: [11]

First, heritability is a measure of variance. It refers to the amount of variation in a group that can be explained genetically. If there is no variation, there is no heritability. Thus, heritability measures can dramatically underestimate the genetic contribution to a trait. Suppose that all human beings have hearts. Having a heart is clearly the result of having certain genes, but it is not heritable, because there is no variation.

Second, heritability is a measure of correlations, not causes. To say that a trait is heritable is to say that it varies along with a genetic trait. That does not mean it is caused by the generic trait. Consider lipstick. Overwhelmingly, lipstick is worn by women, not by men. Being male or female is a genetic trait; it is determined by genes. Therefore, whether or not you wear lipstick is extremely well correlated with a genetic trait. Wearing lipstick is, therefore, highly heritable. But wearing lipstick clearly isn't genetic. Thus, heritability can dramatically overestimate the genetic contribution to a trait.

Third, heritability is a population statistic, not a feature of the traits in a particular individual. Heritability is a measure of how much variation in a group of people is correlated with genetic variation. If some trait T is 50 per cent heritable, that emphatically does not mean that, for any individual with that trait, 50 per cent was caused by the genes. To see why, notice that heritability can vary depending on the group. Suppose we are measuring how many fingers people have. Within the general population, most variation in finger number is attributable to genetic defects, so finger number is a highly heritable trait. But suppose we are looking at a population of miners, who often lose fingers on the job. In that population, environment will explain much more of the variation in finger number, so the heritability score will be very low. In both populations genes are contributing in the same way to the number of fingers a person has, but heritability differs. Similarly, heritability can change over the lifespan. Reading skills in young children may be highly heritable, because most of the variation will be correlated with biological differences in reading capacity. By adolescence, however, children who are not as naturally adept at reading will have caught up with their precocious peers through schooling, and now the majority of the variance will derive from differences in access to education – an environmental variable.

These three differences between heritable traits and genetic traits must be borne in mind when considering research on twins. When population geneticists quantify the heritability of a trait using twin research, there is still an open question about what role genetic factors play in driving that trait. Heritability can overestimate the genetic contribution, underestimate the genetic contribution or conceal the extent to which the impact of the genetic contribution can be overridden by environmental variables.

Caveat 2: Troubles with Twin Studies

Twin studies are sometimes challenged on methodological grounds. The logic of twin studies is simple: if monozygotic twins are more alike than dizygotic twins, that difference must be due to the fact that they share more genes in common. But there is another possibility. It could be that monozygotic twins are simply treated more similarly. Identical twins are often dressed alike and confused for each other. People may expect them to be more alike and treat them accordingly. They are also equally attractive or unattractive. There is a large body of research suggesting that outward appearances affect how people treat you. For example, people who are attractive benefit from a 'halo effect' – they are presumed to have other positive traits such as trustworthiness, intelligence and willing personalities. Couldn't it be that the increased similarities in monozygotic twins all stem from how they are treated? In that case the similarities would be environmental, not genetic.

Behavioural geneticists reject this interpretation. They think research has ruled out the possibility that monozygotic twins are treated more alike or that such treatment matters. However, the two studies that they cite most frequently are inconclusive. In one study, John Loehlin and Robert Nichols purport to show that there is no difference in how similarly monozygotic twins and dizygotic twins are treated.[12] The conclusion is based on a questionnaire given to twins' parents. The results showed that parents overwhelmingly recalled treating twins alike regardless of whether the twins were identical or not. But parental recall in a questionnaire is unreliable. Parents may not want to give the impression that they gave preferential treatment to one of their non-identical twins, so they may exaggerate the similarities. Loehlin and Nichols also claim that monozygotic twins who are treated more similarly are not more alike. But, once again, this failure to find a correlation may stem from the fact that parents' assessments of similarity in treatment are unreliable. Most parents want to give the impression that they treated their monozygotic twins equally, but the few who say otherwise may want to convey that they were promoting individuality in the twins, a normative ideal that would be less frequently expressed by parents of dizygotic twins, who may already assume their children are distinct. In any case, the fact that the parental judgement of treatment similarity was so high for both groups of twins in this study is inconsistent with other measures of treatment similarity, and it does not adequately assess whether monozygotic twins were treated more similarly outside the home.

In another frequently cited study, Sandra Scarr and Louise Carter-Saltzman investigated twins who had been mistaken about their zygoticity – that is, monozygotic twins who were misidentified as fraternal, and dizygotic twins who were misidentified as identical.[13] They reasoned that, if monozygotic twins are treated more similarly, then twins believed to be monozygotic should be treated more alike; and, if similar treatment causes similar behaviour, then twins mistaken for being monozygotic should be more alike than dizygotic twins who are not misclassified. They say that twins misclassified as monozygotic are not more alike than dizygotics and conclude that the similarities between monozygotic twins do not result from similar treatment. But a close look at the data suggests that this conclusion is overstated. They actually found that mistaken beliefs about zygoticity *did* have an impact on personality scores. They also found that twins who looked more alike performed more similarly in an intelligence test, regardless of zygoticity. They also didn't rule out the possibility that twins who mistakenly believe they are identical may nevertheless look less alike than twins who really are identical, and they may have other physical differences, including differences in energy level and health, which can indirectly impact psychological measures.

These difficulties do not prove that monozygotic twins are treated more alike or that such treatment affects behaviour, but they do suggest that we cannot rule out these possibilities conclusively.[14] Twin studies may inflate the genetic contribution by pinning differences between monozygotic and dizygotic twins on biology, when environment could play a role.

To get around this worry, some researchers have investigated monozygotic twins who were raised apart from each other. Sometimes parents give their twins up for adoption, and, given that adoptive parents often don't want the burden of two children, the twins can end up in separate homes. There are two extensively studied registries of twins reared apart: one in Minnesota and the other in Sweden. Researchers working with both sets claim that monozygotic twins reared apart show higher levels of similarity than dizygotic twins reared apart, and, importantly, their degree of similarity is comparable to monozygotic twins reared together. This latter result would rule out the possibility that monozygotic twins are rendered more alike by the fact that parents treat them similarly.

These results are striking but must be regarded with caution. The main problem is that the twins in these studies have considerable contact with each other before their similarities are measured, and this may inflate their degree of similarity. The twins in the Minnesota studies volunteered for the research or were identified to researchers by friends and family. Twins who feel strong personal connections may be more interested and willing to volunteer for these studies. Moreover, some of these twins were raised in the same neighbourhoods, or by relatives in the same family, and some spent up to four years together prior to adoption, or had contact during childhood. By the time these twins came to be studied, they had spent an average of 112.5 weeks in contact with each other. That means that they could have spent entire summers (three months) together for almost ten years. In that time they could have influenced each other, and they could have been influenced by others who, struck by their physical similarity, treated them as if they were alike in many other respects. The Swedish study has similar limitations. These twins had lived together for an average of 2.8 years prior to separation, and had been apart for an average of 10.9 years. But 75 per cent had some contact during that separation, and most were in their sixties by the time of the research. Thus, there had been half a century in which contact could have been extensive.

For these reasons, studies of twins reared apart should not be interpreted that differently from studies of twins reared together. The word 'apart' implies that the twins had no contact, but that is misleading to say the least. Even so, monozygotic twins who were separated for some portion of their childhood do show corresponding differences. In certain dimensions of intelligence and personality, they are significantly less similar than twins reared together, suggesting some impact of the environment.

Suppose we found a perfect sample: twins who had zero contact prior to testing. Would their behavioural similarities be entirely attributable to genes? Presumably not. For one thing, similarities in appearance and health could account for some of their behavioural similarities. As noted, twin research has not eliminated the possibility that people who look alike are treated alike, and there is much reason to think that is the case. In addition, adoption agencies may work hard to put monozygotic twins in similar homes, which could inflate the contribution of their genes.

This last point raises a more serious problem with twin research, which we will have occasion to revisit again below and in the next chapter. Whether reared apart or together, twins are raised in very similar socio-economic and cultural settings. In that sense, their environments are more or less alike. It's trivially true that, if environments are alike, then residual differences are biological, not environmental. Thus, by looking at twins who are raised in similar circumstances, we may be systematically and dramatically overestimating the contribution of biology by controlling for the environmental variables that are likely to exert the most influence. These studies mask such basic contrasts as rich/poor, urban/rural and conservative/liberal, given that these dimensions tend to be the same for twins, even when they are given up for adoption. And, of course, the studies also mask effects of national culture, because twins are raised in the same country.

To be absolutely clear, I am not suggesting, as some critics have, that twin studies are useless or that biology has no impact on traits. Quite the contrary. I think twin studies do show that biology can have a sizeable impact on some traits, when certain environmental factors are fixed. That is interesting and important. But it certainly doesn't follow that DNA is destiny, for there could be environmental factors that affect behaviour profoundly. If we had cross-national twin adoption studies, for instance, we might find that correlations between monozygotic twins become negligible or drop to rates lower than for dizygotic twins raised in the same country.

The Family Inheritance

Behavioural geneticists have measured the heritability of many human traits, and the results are often striking. Many traits turn out to get high heritability scores. The numbers come out around .50 for a wide range of traits, which means that half the variance measured in these studies correlates with biological relatedness. If you are outgoing, there is a 50 per cent chance your identical twin is outgoing too. Such findings have promoted the conclusion that human behaviour is heavily influenced by biology. Before we get too excited, however, let's look at some representative results in more detail.

Consider personality. Psychologists have identified five personality dimensions that are especially important in human behaviour: openness (which basically means being interested in new experiences), conscientiousness (being disciplined and dutiful), extroversion (being outgoing), agreeableness (getting along nicely with others) and neuroticism (being emotionally unstable). In introductory psychology courses, students remember this list by the mnemonic acronym OCEAN. Everyone scores somewhere on each of these five dimensions, and geneticists have shown that close relatives tend to have similar scores. Reviewing numerous studies, Thomas Bouchard, who heads up the Minnesota twin research, finds that heritability scores for the Big Five traits range from .58 for neuroticism to .40 for conscientiousness.[15] It looks as if half of our personality is programmed in our genes.

These results may be less impressive than they initially appear. First of all, it's really not all that surprising that biology contributes to traits of this kind. If biology can influence any aspect of psychology, personality looks like a good bet. Indeed, it's quite a discovery that 50 per cent of the variance in personality owes to experience. It's fascinating that a person who is completely neurotic can have an identical twin who is as cool as a cucumber. If there is any headline news in these findings, it's that experience can decide whether you are a saint or a jerk, a social butterfly or a wallflower, an adventurous thrill seeker or an uptight prude. It's comforting to know that some experiences can radically change our approach to life.

Second of all, personality traits are not perfectly stable across the lifespan. Looking at a variety of different traits over a 50 year period, psychologist Norma Haan and her collaborators found that correlations ranged from .37 (for the outgoing/aloof dimension) to .14 (for the dimension warm/hostile).[16] Research on the Big Five traits suggests that correlations vary across the lifespan. During childhood, correlations across the years are as low as .35, and high levels of stability don't seem to emerge until people are in their 50s, at which point correlations across the years get up to .75.[17] The average correlation across several-year periods during the lifespan is .50. They are sometimes predictably keyed to changes in life events. For example, conscientiousness seems to rise in young adulthood, when people begin to take on greater professional and personal responsibilities, and traits can be affected by factors such as marriage and employment status.[18] This suggests that the environment can bring about significant changes in personality.

The impact of the environment can also be established by taking a cross-cultural perspective. Heritability scores systematically overestimate the biological contributions to personality by filtering out the impact of culture. Participants in the twin studies live in similar conditions and similar settings. As a result, the impact of environment is reduced. But what if environments are varied? A naturist approach might lead us to think that personality traits will be distributed equally in all cultures. That is not the case. Cross-cultural studies show that differences abound. North Americans are more extroverted and less neurotic than people in east Asia. Africans are more agreeable than western Europeans. People in the Middle East are more conscientious than people in eastern Europe. South Americans are more open to new experiences than south Asians. And so on. For every dimension of personality, there are regional differences, and these are sometimes enormous. Some of these differences have been correlated with economic and political variables, such as gross domestic product and the prevalence of status hierarchies. If you live in a poor, highly stratified country, you are more likely to be introverted.[19]

Findings like these don't always conform to stereotypes of national character. Those tend to be pretty inaccurate, but sometimes the clichés have a kernel of truth. This comes out most profoundly in moral dimensions of personality, which vary considerably across cultures. Robert Levine and collaborators wanted to see if some people are more helpful, so they went out into the streets of twenty-three major cities around the world and put this to the test.[20] For example, they watched to see who would help a blind person cross the street or pick up a pen for a stranger who dropped it. The most helpful people in the world turned out to be Brazilians in Rio de Janeiro. Least helpful were folks in Kuala Lumpur, with New Yorkers close behind. This doesn't mean you'll never find a crook in Rio (just watch the film *City of God*) or an altruist in New York (remember 9/11?), but it does mean that the disposition to aid strangers can be influenced by geography. Likewise for the disposition to harm. In a famous study of obedience, Stanley Milgram found that 65 per cent of Americans were willing to administer potentially lethal doses of electricity to a stranger if a psychologist in a lab coat instructed them to do so.[21] In Germany, the same experiment resulted in obedience rates of 85 per cent.[22] The number dropped down to 44 per cent in Australia, where people pride themselves on being anti-authoritarian, and the least obedient of all were Australian women, who complied a mere 18 per cent of the time.[23]

These aspects of moral conduct bring us to the topic of criminality, another domain where behavioural geneticists have claimed biology can shape behaviour. Crime was an early preoccupation among genetic determinists. In the nineteenth century, the founders of the field began to speculate that evil might be inherited. The criminologist Cesare Lombroso said you could tell a crook by his face: shifty eyes, jutting jaws, flat noses and fleshy lips were all telltale signs. We've moved a bit beyond this method, but contemporary geneticists think we shouldn't dismiss Lombroso completely. The heritability of criminal behaviour has been confirmed in numerous twin studies. There have also been impressive adoption studies, where adopted children are shown to be uninfluenced by the crimes of their adoptive parents, but somewhat prone to criminal behaviour if their biological parents are. Some people are born to be bad.

This chilling conclusion can be tempered a bit if we look more closely at the findings.[24] Studies have not been able to confirm a genetic contribution to violent crime.[25] Only property crimes seem to be affected, and the effects only seem to show up in adulthood. Juvenile offenders show little correlation with their biological parents – an important fact since much of the worst crime is committed when people are young. There may also be some methodological problems with this research. We've already seen some concerns about twin studies, but the best research on criminology has been done using adopted children and comparing them to their adoptive and natural families. The difficulty here is that children who are born to criminal parents may be harder to place in adoptive homes; they may wait longer in foster care and they may be placed in less stable environments, perhaps with adoptive parents who know their family history. With less adequate care, any minute genetic predispositions may amplify.

Notice I am not denying that there are predispositions. It may be that there are. I am simply saying the size of the genetic influence may be small. It may also be that the contribution is indirect. It turns out that criminal behaviour correlates with neuroticism and a diminished ability to exercise self-control. Thus, there are no crime genes. There are just genes that can make a person angry and capricious and, in the wrong setting, this can increase the probability of anti-social behaviour. Knowing this is vitally important, because it prevents us from making Lombroso's mistake of treating evil like eye-colour. No one is born bad. Rather than stigmatizing the offspring of petty criminals, we should use our knowledge of the underlying personality variable to find effective environmental interventions.

If you think this talk of environmental interventions looks like hopeless liberal idealizing, then have a look at crime statistics. There is extraordinary variation in crime, and we know some of the societal variables that make a difference. In the United States, crime rates are twice as high in cities as they are in non-urban areas. Interestingly, this is true for both violent crimes and property crimes, even though only the latter are highly heritable. If genes were making a big difference, we might expect property crimes to remain fairly constant across communities. They don't. If you compare by state, property crimes are three times more common in Arizona than North Dakota. Cross-nationally, the contrasts are even greater.[26] In India, there are about 1.6 crimes per 1,000 people, and in the United Kingdom, there are 85.5, a 53-fold increase. Between Yemen and Dominica, there is a 103-fold increase.

There is also well-known variation in criminality within communities. For example, much ink has been spilled about crime rates among African Americans in comparison to other groups, even living in the same metropolitan areas. There are more African American men in prison than in college. Naturists are sometimes tempted to conclude that the ethnic difference is correlated with a genetic difference that increases the disposition to crime. This racial interpretation is a non-starter. We've seen that there is little evidence for the heritability of violent crime, and biological influences hardly show up in young offenders; given the prevalence of violent crime and youth crime in black communities, it's just not plausible that genes are playing a meaningful role. Moreover, as we'll see in the next chapter, race is not a biologically meaningful category.

African American crime is clearly linked to environmental variables. Poverty is known to have a significant impact, and interventions, such as providing subsidized middle-class housing, have a measurable impact on crime rates. African American communities are also plagued by other problems, such as isolation from upwardly mobile social networks, reduced education quality, elevated rates of substance abuse, frequency of broken homes and high population density. Together, such factors can be a toxic mix, even when some of them have only modest effects on their own. The upshot is simple: even if crime can be influenced by biological factors, sociological variables may have much more impact. Heritability studies filter these out, giving the impression that genes are doing 50 per cent of the work in causing people to steal cars, when socio-economic status has a far greater impact.

The lessons that we've been drawing from work on personality and criminality apply equally well to other areas where high levels of heritability have been recorded. Consider the astonishing discovery that divorce is highly heritable.[27] There is presumably no divorce gene, but personality traits may have some impact on people's ability to maintain healthy relationships over time. Still, to pin divorce on the genes is to neglect the huge impact of culture. Divorce rates skyrocketed in the twentieth century, and rates vary by region and many other factors. Conservatives divorce less often than liberals, the rich divorce less often than the poor, and, yes, Catholics divorce less often than Protestants.

Speaking of religion, behavioural geneticists have shown that biology contributes to theology. How religious you are seems to depend on your genes. This has encouraged faith in the idea that there is an innate religion module in the brain that makes some of us into sceptics and others into true believers. Calvin and Nietzsche just had their god-spots set on a different switch. Like other traits measured by geneticists, heritability for religiousness tends to come in around just shy of .50. An estimate of .44 is given in one reliable study.[28] Now suppose you meet someone and know nothing other than the fact that her identical twin is an atheist; there is no simple way to compute individual probabilities from heritability, but you might take a leap and guess that she has a 44 per cent chance of being an atheist too, even if she and her twin were separated at birth. That's remarkable, but it's important to bear in mind that national origin is a better predictor. According to a recent BBC poll, 85 per cent of Americans believe in God. If these numbers are accurate it means that, if you know someone is a Yank, you can guess that there's an 85 per cent chance that he or she is a theist – almost twice the predictive power of genes. Plus there is considerable cross-cultural variation. Only 52 per cent in the United Kingdom claim to believe in God, and the number of theists hovers near the ceiling for residents of Indonesia. If we consider other measures of religiosity, the spread is even larger. 91 per cent of Nigerians say they attend religious services, and only 7 per cent of Russians do. Given this, the .44 heritability score is very misleading. Suppose you took a set of identical twins, separated them and raised them all across the globe. You might have one twin raised in London and the other in Jakarta. The correlation between relatedness and religiosity would probably drop dramatically, perhaps down to 0. Is religion influenced by the genes? That depends on what other factors you hold constant. Within a community, yes, but cross-culturally, perhaps not.

One of the most intriguing recent twin studies concerns political values. The authors captured headlines by showing that one's views about a wide range of political issues, from taxation to immigration, are highly heritable.[29] This is mind-boggling on the face of it, since no one would postulate a gene for opposing high taxes. Personality variables, such as authoritarianism, may be driving these effects. In any case, the numbers are clearly inflated by twin methodology, which underestimates environmental influence by examining twins who are raised in similar settings. If you take two twins born in the rural American South and find that their politics differ, then it's not unlikely that something about their biological temperament is to blame, since the major environmental factor (region of rearing) has been kept constant. Raise these twins in different settings, and heritability scores would drop off. We need only look at an election map to see that geography has a huge impact on political values. Other variables include income, ethnicity, age and religion. For example, 75 per cent of black women in California voted to ban gay marriage. If you want to guess who someone will vote for, a few demographic variables can be more predictive than a genetic analysis.

In summary, the numbers handed to us by behavioural geneticists tell us how much variance correlates with relatedness if you keep demography and nationality fixed. People from the same background differ a bit, and those differences may be partially attributable to biology. But, if we look across groups, these biological differences may pale in comparison. The idea that half of our behaviour owes to our genes is a fallacious inference. Genes may explain some variation within groups, which means, if we already know that you are a poor, black, urban woman, we may get some predictive mileage out of learning your biological history. But, if we already know you are a poor, black, urban woman, we can probably already guess how you'll vote in the next election.

Rethinking Environmentalism

One of the most startling claims made by behavioural geneticists is that parenting has very little impact on behaviour. Every twin study measures both biological and environmental influence. Biological influence is ascertained by comparing correlations between monozygotic and dizygotic twins. If the heritability score is .50, that means half the variation in the distribution of a trait in the tested population is correlated with biological relatedness. A .50 heritability score also entails that half the variation correlates with environmental factors. One of the most startling conclusions from twin studies concerns the nature of these environmental factors. For many traits, geneticists have concluded that parenting makes no significant difference.

Here's how they come to that conclusion. The environment that a pair of twins has experienced can be divided into two parts: the parts they both experienced together, and the parts each one experienced independently. These are called the shared and non-shared environments. For example, if both twins were raised by kind and loving parents, that's something they share, and if one twin was mugged while walking home at night, that's something they don't share. To calculate the contribution of the shared environment, researchers subtract the heritability score from the correlation between monozygotic twins. Monozygotic twins have the same genes and the same parental upbringing, so any similarity that exceeds the biological contribution is attributed to that upbringing. To calculate the contribution of the non-shared environment, researchers take the correlation between monozygotic twins and subtract it from 1. The number 1 corresponds to the total amount of variance in the trait (if calculating with percentages, 1 could be replaced by 100 per cent). Since monozygotic correlations are the sum of biology and upbringing, any variance in a trait beyond these two factors must be due to factors that are not shared. Using these formulas, behavioural geneticists made an astonishing discovery: shared environment usually makes a negligible contribution to variance in our traits. How neurotic you are has virtually nothing to do with how you were raised.

This discovery sent shock waves through psychology. It implied that parents don't matter very much. Judith Harris, a textbook writer, gained celebrity by writing a book bringing these results to light and showing that the orthodox assumptions in developmental psychology are based on a fallacy, which she calls the Nurture Assumption.[30] Psychologists assume that parents can affect behaviour, and they have hundreds of studies to back it up. But the studies simply show correlations between how parents behave and how their kids behave, and those correlations could be entirely genetic. Without controlling for biological relatedness, developmental psychologists assume that parents shape their kids' behaviour by nurture, but, given the failure to find any impact of shared environment, it must actually be nature that's doing the trick. Harris sums this up with what she calls the 50-0-50 rule: 50 per cent of the variance in a trait owes to genes, zero to home environment, and the remaining 50 per cent to non-shared environment.

This conclusion left Harris with a puzzle. What is this all important non-shared environment that co-conspires with genes to shape our behaviour? Harris's answer is both ingenious and provocative. Half of what we do depends on our genes, and the other half depends on our peers. Kids don't pay much attention to their parents, but they would do just about anything to win approval from their classmates. Harris proposed that peer pressure plays a major role in shaping behaviour, and parents should stop worrying so much about how to raise their kids and think a bit more about whom their kids are hanging out with.

There is something importantly right about Harris's proposal. Social conformity is one of the major factors driving human behaviour, and our desperate efforts to mirror our peers can have an enduring impact on behaviour. It is plausible that peer pressure is at least as important as parental guidance. By shifting attention to peers, Harris also helps us see that behaviour can be impacted by the *social* environment. The groups we affiliate with make a difference. In making this point, Harris helps us understand the means by which culture can influence a person's behaviour. In this chapter, and in the chapters that follow, I say that culture causes us to act in certain ways. This may sound obscure. Culture is not an entity; it's an abstraction. It refers to a set of socially transmitted beliefs and values. How can these things ever affect what you or I do? The answer is that beliefs and values are in the minds of the people we interact with. When we meet people, we learn about those beliefs and values. If we are interacting with a group of people whose friendship and respect we desire, we may end up adopting their beliefs and values. Peer groups fit this description to a tee. But Harris may also underestimate the role of parents and overestimate the contribution of genes. Let me voice three concerns.

First, it is a mistake to equate shared environment with what goes on at home and unshared environment with what goes on outside the home. Shared just means something that two siblings experience in the same way. But many things that happen at home are different for siblings. For example, if there is a divorce or a parent death, it may affect two siblings differently. This is part of non-shared environment. Also, if parents treat two siblings differently, then parental treatment is non-shared, rather than shared. Sometimes children who have small genetically based personality differences end up being treated in dramatically different ways at home. A colicky baby or a mildly hyperactive five-year-old can be hard for parents to handle, and that can affect parenting style in a way that has long-term implications. Likewise, factors outside the home, like drug use or good report cards, can affect behaviour in the home, and the way parents react to these things will count as non-shared environment if there is a sibling who said no to drugs or skipped a few more classes. The method of quantifying non-shared environment cannot distinguish peer influence and differential parenting, so the conclusion that non-shared environment matters more than shared environment does not entail that differences in parenting techniques don't matter. Conversely, kids may come under very similar social pressures at school, so peer influence may end up counting as shared environment. The truth is, we need to do a lot more research to figure out what the relevant aspects of non-shared environment are.

Second, the claim that shared environment has little impact does not entail that parenting has little impact. It merely entails that, within the populations used in twin research, parenting styles are pretty similar. If most parents within a population raise their kids in similar ways, then, trivially, variance in that population will not be due to parenting styles. To really test for parental impact, we would have to do a study that directly compared two different parenting styles on pairs of related individuals. Harris appreciates this fact. She does not think that parenting has no impact on behaviour. Research plainly shows that kids who grow up in abusive homes can have bad outcomes regardless of genetic factors. Harris is mainly trying to establish that there is a wide range of non-abusive parenting in which variation of style makes no difference to outcomes. But that claim needs further support. Most twin studies make no effort to measure such variation, and the parenting styles may be pretty uniform in the populations tested in twin research. Most parents are pretty caring, non-abusive and conscientious. Having parents that fit that description is probably vitally important for how we turn out, so it would be dangerous and absurd to infer from twin research that parents don't matter. We can infer only that the impact of different styles of conscientious and caring parenting has not been adequately measured within behavioural genetics and may contribute less to variance than differences in peer groups.

Finally, Harris is too quick to endorse twin research. She welcomes the idea that genes are responsible for half of the variance in our traits. That, we have seen, is a mistake. Twin research shows that genes correlate with variance if we keep major environmental factors such as culture and demography fixed. As someone who has done so much to emphasize the impact of peers, she should also appreciate that culture matters. Within a community, peers may be fairly similar. The peer pressure on one identical twin is probably pretty similar to the peer pressure on another, because values within a community are relatively uniform. If Harris admits that peer pressure can shape 50 per cent of our behaviour within a community, she should also recognize that the impact could be vastly greater across communities. If one person is raised in a wealthy, homogeneous Indiana suburb, and another, her identical twin, goes through the government-provided school system in a less affluent part of New York City, they will experience very different social pressures. The differential impact on their interests, values and traits might be exponentially greater than the impact of two peer groups in the Indiana suburb. And if these twins were raised in different countries, the differences could be greater still.

BEYOND BIOLOGY

The take-home lesson is that the search for genetic causes of behaviour has left us with little reason to think that biology has a major impact in accounting for human differences. Efforts to identify the genes that drive behaviour have met with limited success, because everything we do involves many different genes and interactions with a compliant environment. Efforts to correlate behavioural similarities with genetic relatedness have been more successful, but such correlations systematically overestimate genetic influence and underestimate environmental variables. Biology does affect behaviour, but its contribution to human variation may be modest in comparison to the impact of our social environments. In the next chapter we will see how these lessons play out in more detail by looking at an important human ability that has been attributed to our genes.

3

Get Smart

Of all the psychological traits, the one that stirs up the most controversy is intelligence. Everyone agrees that intelligence can be affected by the genes. The fact that humans are smarter than dogs is clearly a consequence of our biology. Everyone also agrees that differences in human intelligence can be genetic. Some people are congenitally retarded, and extreme forms of genius are likely to be genetically based as well. But what about the vast majority of us who lie somewhere between Einstein and Tweedledumb? Genius and retardation are rare conditions, which may result from genetic mutations. Are the differences between people who fall in the normal range distinguished by our genes? Is a run-of-the-mill dullard biologically different from a garden-variety whiz kid? And if so, are those biological differences fixed, or might they be altered by experience? These questions become even more heated when we turn from individual differences to differences between groups. Do biological differences in brainpower come pre-packaged with biological differences in pigmentation? These are touchy topics, and naturists have felt considerable heat for defending positions that are politically incorrect. I don't think we should let politics arbitrate in this case, however. I think naturists simply get the science wrong. While some differences in intelligence may be linked to biology, most people have pretty comparable biological endowments. If we want to find an explanation for group-wide social inequity, then we would be better off studying the negative effects of poverty and the positive effects of cultural practices that encourage learning.

INTELLIGENCE TESTING,
AN UGLY LEGACY

Behavioural genetics was founded in the nineteenth century by Darwin's cousin, Francis Galton. Some people are born geniuses, Galton claimed, some are born to be officers in the military, and some are born criminals. Galton believed that these traits could literally be read off a person's face. He developed a photographic technique whereby photos of one genius, officer or criminal could be superimposed on another. The composite photographs would reveal the prototype for each type. Bad guys have a characteristic appearance, he claimed, and so do good guys. It's a bit mystifying that Galton could have come to this conclusion, because a glance at his composite photographs points to the opposite conclusion. People from all walks of life look similar, and when different faces from the same social class are blended together, the result always looks like a bland generic face indistinguishable from the composite for every other class. If you examine a composite photo of several officers and compare it to the composite of several low-ranking soldiers, the faces are eerily alike. The features

Figure 1. A Galton composite comparing the features of officers and privates, compiled using photographs obtained by Darwin's son.[1]

are so similar that one might even think each composite was a snapshot of the same man. Galton was strangely oblivious to his own data, though he later abandoned the technique.

Galton's composite photography never caught on, but many of his other innovations did, including the use of fingerprints in criminal investigations, various psychometric techniques and the study of how traits run in families. It is this latter research programme that launched behavioural genetics, though in Galton's hands it took on a rather sinister cast. Galton launched the eugenics movement. Eugenics means good breeding, and the movement was an effort to improve the human race by encouraging people with valued characteristics to procreate and discouraging those with undesirable traits. Galton suggested that we identify especially promising candidates for good breeding by tracking scholastic achievement, physical health and familiar histories. Those of good stock should be encouraged to intermarry and have many children. He suggested that good houses be offered to desirable couples at very low expense. As for the less desirable individuals who lived lives of petty crime, Galton noted that it would be, 'a great benefit to the country if all habitual criminals were resolutely segregated under merciful surveillance and were peremptorily denied the opportunity to produce offspring'.[2] Of course, this was little more than a grotesque racialization of class. Those who entered universities at the time were generally wealthy, and those who committed crimes were generally destitute. Galton believed that success in life was the result of biological inheritance, ignoring the obvious fact that financial inheritance was the major determinant. In Galton's utopia, the rich get richer and the poor are mercifully extinguished.

One ironic feature of Galton's programme is that his own family had been upwardly mobile. His grandfather was born into modest means but then amassed a fortune as an arms dealer (despite his Quaker faith). If poverty were the result of bad genes, such social escalation should be impossible. Galton explained his family's success by appeal to hereditary intelligence. This had a plausible ring to it, given his numerous accomplished relatives, but the logic is flawed, since a family that is genetically superior should be able to trace back many generations of greatness. Galton's view that his own family was genetically gifted was part of a more general tendency to interpret people with his background as superior to others. He said that whites are more able than blacks, Christians are more innovative than Jews, and the English are more intellectually flexible than the Spanish, Portuguese and French (because freethinkers were killed off during the Inquisition and Counter-Reformation).[3] Galton even claimed that women in London are more attractive than elsewhere, based on the careful 'beauty maps' he created, by recording the location and date of the attractive and repellent women he encountered in his travels. It is no coincidence that Galton praised the genes of those he considered like himself and cast aspersions on the biological constitution of others. Eugenics was, from its inception, an instrument for promoting the in-group at the expense of all others.

The eugenics movement had one positive effect, however. In the early twentieth century, birth control was difficult to obtain, and condoms were still predominantly only used by prostitutes. In the United States, it was illegal to send information about birth control across state lines. As a result, women had little knowledge of, or access to, birth control devices. Advocates of birth control used eugenics to change public opinion. The morbid fear of bad genes proved greater than the morbid fear of recreational sex. Margaret Sanger helped to decriminalize the dissemination of information about birth control by arguing that rampant breeding among 'unfit classes' was 'the greatest present menace to civilization'.[4] Birth control came into wide circulation because it was regarded as a powerful weapon against the spread of bad genes. With this change, women liberated themselves from the fetters of premature motherhood and were able to compete with men in the workplace for the first time in modern history. A major advance, to say the least.

But, during the same time period, eugenics also began to have catastrophic effects. In the first decades of the twentieth century, countries in North America and Europe introduced laws that allowed governments throughout the Western world to sterilize women who were deemed genetically inferior. In most cases, these women were poor. Uneducated women were labelled dumb, psychologically troubled women were labelled insane, and women who broke laws were labelled criminally deviant. Poverty itself was regarded by many as sufficient grounds for sterilization, because financial hardship was a sign of bad

genes. In the United States, there were 64,000 involuntary sterilizations, in Sweden there were 62,000, and Germany took the prize with
400,000 sterilizations. Germans also gave out 'mother's cross' medals
when women of good stock had many children: a mother of four was
awarded a bronze cross, a mother of six got silver, and a mother of
eight could proudly display a mother's cross in gold. Women of less
desirable stock were sterilized, and it was forbidden for an Aryan to
marry anyone who was more than a quarter Jewish. Realizing that
such laws would not lead to the rapid elimination of Jews and other
people deemed undesirable, the Germans inaugurated the most brutal
genocide campaign in human history (though by no means the first or
the last).

This frenzy over eugenics was driven by the intuitive plausibility of
the idea that good traits can be biologically inherited. Those who
know anything about breeding farm animals or show dogs know that
one can exercise some control over an offspring's characteristics by
carefully selecting the parents. The problem is that this intuitive idea
collapses into dangerous pseudo-science in the case of human beings.
There are no human breeds or races. The categories by which we divide people into 'racial groups' have little or no meaning biologically.
For one thing, there is vastly more biological variation within racial
groups than between them, whereas biologically defined categories
tend to have greater internal uniformity.[5] The features we use to classify people racially are often superficial, and do not correlate with
deeper biological similarities.[6] For example, dark skin pigmentation is
shared by sub-Saharan Africans and by some indigenous peoples of
New Guinea, who are genetically closer to east Asians.[7] Even within
Africa, there is vast variation, more so than in Europe or in any other
continent.[8] The majority of African Americans can trace their origins
to western Africa, but close to 30 per cent come from elsewhere, and
many also are of European descent. Likewise, 30 per cent of Americans who self-identify as white have some non-European ancestry.[9]
One can identify biologically meaningful ethnic groups by tracing
genetic markers geographically, especially for regions that have been
isolated,[10] but these classifications cross cut ordinary conceptions of
race.[11] Thus, if racial groups have any biological meaning, they are
not the ones we use in everyday life. Those categories, such as black,

white and Asian, are social constructs, and do not entail biological coherence.

More to the point, the human traits of interest (such as being a criminal or being an officer) have little to do with biology. Biological influences on behaviour are usually small, indirect and dwarfed by environmental factors such as culture. This was the message of J. B. Watson, the founder of a movement in psychology known as behaviourism. Behaviourists believed that human abilities are determined by histories of reinforcement. Like Pavlov's dogs, which would salivate at the ringing of a bell, human beings could be encouraged to act in desirable ways by a careful training regimen of punishments and rewards. In 1930, Watson boasted:

Give me a dozen healthy infants, well-formed, and my own specified world to bring them up in and I'll guarantee to take any one at random and train him to become any type of specialist I might select – doctor, lawyer, artist, merchant, chief, and yes, even beggarman and thief, regardless of his talents, penchants, tendencies, abilities, vocations, and race of his ancestors.[12]

This claim contrasts sharply with the views of eugenicists, who were exerting considerable influence on public policy at the time. Fortunately, Watson was able to exert some influence. Years earlier, he had lost his post as chair of the Johns Hopkins psychology department due to a scandalous affair with his research assistant, who later became his wife. Banished from the academy, Watson pursued a career in advertising and, in his free time, wrote popular books on childrearing, which encouraged parents to shape their children's behaviour through behavioural conditioning. Along with his earlier academic writings, these works inspired one of the most influential psychologists of the twentieth century, B. F. Skinner.

Skinner took Watson's ideas about conditioning even further, and developed new techniques that could be used to shape behaviour by rewarding success. He found that punishment was far less effective and helped promote humane approaches to childrearing. Skinner believed that good conditioning could lead to a utopian society, which he described in his acclaimed novel *Walden Two*. He wanted to replace good breeding – the approach of eugenics – with good rearing. By the 1960s, Skinner was the most famous living scientist in America, and eugenics, which was associated with the Third Reich, had fallen out of favour.

Around the same time, however, behaviourism came under attack. Behaviourists had overestimated the role of reinforcement in shaping behaviour. It turns out that there are considerable differences across species in susceptibility to conditioning. Sometimes conditioning an animal to perform a very simple task is impossible, as Keller and Marion Breland discovered in 1961, when they tried to condition a raccoon to put coins in a piggybank.[13] Some animals exhibit behaviours that cannot be reshaped by conditioning. Conditioning animals to perform complex tasks presents an even greater problem. You can't condition a pigeon to write a sonnet. Human behaviour can be conditioned, but only within strict limits. Many of us would refuse to do certain things (say, take another life), even if offered exorbitant rewards. Moreover, we often behave in ways that are indifferent to real rewards while pursuing rewards that are entirely imaginary, as in the case of the suffering artist who skips meals to pursue a pipedream of artistic glory. If human behaviour were simply a function of conditioning, we should expect much less variety. Watson and Skinner had few resources for explaining why individuals raised in similar environments, such as fraternal twins, do not end up exactly alike. Clearly, our behaviour is not entirely driven by experience. Behaviour depends on personal interests, talents and aspirations, along with biologically influenced aspects of temperament and ability. For these reasons, Watson's and Skinner's conclusions were ultimately rejected, and the stage was set for a return to biology.

In the 1970s, researchers began again to actively explore biological effects on behaviour. They revived and updated a technique that had been developed by German psychologists early in the twentieth century: heritability estimation through the comparison of monozygotic and dizygotic twins. In the decades that followed, behavioural genetics began to thrive. Most researchers in this field are primarily interested in *individual* differences, why one person's traits differ from another's, but some are interested in group differences, the study of how traits may vary as a function of ethnicity, gender and other categories that are deemed to be natural. Those who pursue this line of work are the true heirs to Galton. They rarely promote selective breeding, but they do think that biology is to blame for variation in human groups.

THE IQ CONTROVERSY

Measuring Intelligence

Galton's first major contribution to behavioural genetics was his book *Hereditary Genius*, which tried to show that intellectual talent runs in families. The eugenic extension of this thesis was the conjecture that certain groups may be smarter than others, and, in the climate of the nineteenth century, with scientific racism and a growing underclass of factory labourers, this idea had strong appeal. One problem, however, is that there was no reliable method of measuring intelligence, so the hypothesis could not be put to the test. That changed in 1904, when Alfred Binet and his assistant, Theophile Simon, developed an effective intelligence test for the French government, which became known as the Simon–Binet scale. The notion of an 'intelligence quotient', or IQ, was born.

Alfred Binet did not think intelligence was biologically fixed, but proponents of eugenics seized upon his test and began using it for their agenda. In his view, IQ can be changed by education. But members of the eugenics movement saw things differently. They believed intelligence was inherited and unchangeable. By 1907, some parts of the USA were beginning to practise the compulsory sterilization of 'imbeciles', and versions of Binet's test were used for screening. Henry Goddard, who had first translated Binet's test into English, advocated restrictive immigration policies to filter out people of inferior intellectual stock. This eventuated in immigration policies that were biased against Italians, Asians and Jews, among others. Goddard's work was ultimately discredited, but IQ testing became deeply entrenched in American culture.

In the 1970s, with the rise of modern heritability testing, Goddard's ideas got a big boost. Researchers established that performance on IQ tests is highly heritable. Results vary from study to study, but a common estimate is that intelligence has a heritability of .50.[14] That means that 50 per cent of the variance in a given population correlates with genetic variation. The contribution of shared environment, such as schooling, is considerably lower. For example, in a study of pairs of adopted children (who are genetically different but reared in the same environment), there was a correlation of .29, suggesting that shared environment accounts for 29 per cent of the variation.[15] Interestingly, this number reduces over the lifespan. When pairs of adopted siblings are past school age, shared background accounts for only a tiny percentage of the correlation between them. Unshared environmental factors, such as individual biographical experiences, account for more, but no one knows exactly which experiential factors play a role. Parents have been encouraged to pump up their infants' ultimate IQ performance by playing Mozart in the crib, but there has been virtually no empirical support for the validity of this advice.

What should we infer from the fact that intelligence has a heritability of .50? First, we should *not* infer that genes explain most of the variance. They explain half the variance. There is another 50 per cent that has not been adequately explained. If one is trying to decide where to stand on the nature–nurture debate, a .50 heritability score should not sway you in one direction over the other. Such numbers bring naturists and nurturists to a stalemate.

There are also more serious problems with drawing strong naturist conclusions from the research on IQ. Most heritability studies are done within relatively homogeneous populations, with similar IQ scores. Heritability estimates can be interpreted as showing that, within such homogeneous samples, about half of the variance correlates with genetic factors. But this doesn't show much about the role of genes in determining inheritance if there isn't much variance. If two individuals score 103 and 115 on an IQ test, the spread between them is 12 points on the test, but there is also an overlap of 103 points. Suppose we concede that half the spread results from biological factors. So one of these individuals had a 6 point biological advantage. But heritability scores do not tell us how to account for the massive overlap. How did these individuals end up getting those 103 points that they share? One possibility is that people have similar genes, so they end up with relatively similar levels of IQ. But an equally viable explanation is that people in these studies have similar experiences. Heritability studies almost always look at people who live in the same country at the same time in the same social class and under very similar schooling, because education is largely standardized.

Schools teach skills that are relevant to IQ performance, so it is possible that the widespread similarity in IQ within a culture reflects education. If you take a teenager who has gone to school and one who hasn't, rather than two who have gone to similar schools, there would probably be a huge discrepancy in IQ scores. If so, IQ is mostly determined by education.

This suggestion is not meant to imply that biology plays a negligible role in intelligence. Apes can't pass human IQ tests. Something about our biology allows us to learn maths and vocabulary. But what explains the size of a person's vocabulary or the specific maths skills that have been mastered? The obvious answer is education. IQ tests that measure skills like this are largely measuring classroom experience. To do algebra problems, you need to take a class in algebra. If you test one person who has taken the class and another who hasn't, the disparity will be embarrassing. Passing the test depends on education. But now suppose we find a disparity between two students with similar backgrounds who take the same algebra class. It is possible that genes account for half their difference in scores. But it certainly doesn't follow that their ability to pass an algebra exam is 50 per cent genetic. Passing the test is 100 per cent a function of their having taken the class, together with the generic cognitive abilities that make us human. Likewise, most of your performance on an IQ test may result from learning, even if biology accounts for some portion of the difference between your scores and the scores of another person who was educated in the same country during the same time period under the same socio-economic conditions.

If the .50 heritability score is only capturing the differences in IQ that remain after homogeneous schooling, then we should expect heritability to drop precipitously in populations that have more erratic educational experiences. Consider people who are very poor. In poor communities schools vary in quality, and motivation to learn is heavily influenced by parental and peer values, diet, inspiring teachers and dozens of other environmental factors that vary from person to person. Consequently, in poor communities, the heritability of IQ drops from .50 to .10.[16] That means only a small amount of IQ variation among the poor can be credited to the genes.

In summary, it is misleading to infer from high heritability scores that a trait is largely determined by genes. At best, we can say that, within populations that have consistent and stable environments, the small differences in IQ that remain are partially attributable to genetic differences.

Intelligence and Race

One of the great tragedies of IQ testing is that researchers have used their results to argue fallaciously that certain groups of people differ in intelligence. Claims of this kind have been made many times over the years. The most recent example came in 1994, when Richard Herrnstein and Charles Murray published *The Bell Curve: Intelligence and Class Structure in American Life* (New York: Free Press). Herrnstein and Murray make several startling claims. They say that men are capable of greater extremes of intelligence than women, that white people are more intelligent on average than people of colour, and that east Asians and Jews are on average more intelligent than Christians. These claims are consistent with widespread pre-theoretical assumptions that are usually dismissed as hateful bigotry: women aren't as smart as men, black people are dumb, and Jews and Asians are dangerously clever. Herrnstein and Murray give these attitudes an air of scientific respectability by presenting a hundred pages of charts and tables documenting measurable differences in intelligence.

Herrnstein and Murray also reveal their politics by recommending major policy changes. Most notably, they argue that we disband affirmative action programmes. If some people are congenitally unintelligent, we shouldn't bother to give them access to better educational institutions. One might have thought the policy recommendations should go the other way: if some people are naturally less intelligent, we should take extra steps to give them a superb education so they can compete with others who have a genetic advantage, just as we should provide health benefits to those who are sick. The policy recommendations are based on two assumptions: affirmative action programmes cannot increase the intelligence of their beneficiaries, and people with lower IQ will perform less effectively in the average job.

The claims in *The Bell Curve* are highly provocative – even offensive.

But its defenders say that we should not dismiss the book simply because it makes claims that are unpalatable. If those claims were true, we had better accept them and figure out what implications they have. The best way to assess *The Bell Curve* is to look at the science. Do all those tables and charts prove that various populations have irremediable differences in IQ? Decidedly not. It turns out that all of the central claims in the book are based on faulty assumptions, bad inferences and questionable methods.

Before considering the contentious claims about race in *The Bell Curve*, it's worth considering a broader question: does IQ matter? Group differences in IQ would not be very disturbing if IQ didn't have much impact on life prospects. Herrnstein and Murray go out of their way to establish that IQ makes a big difference, but their case for this claim is very weak. They try to argue that one's financial lot in life is determined by natural cognitive abilities, not socio-economic status. This would be nice if it were true. We would like to think merit is the main determinant of success, but that, of course, is a utopian myth. Wealthy people are often born with silver spoons in their mouths, and those born into poverty often find it impossible to climb out. Herrnstein and Murray conceal these obvious consequences of inequity through the magic of statistics. Their miscalculations are carefully documented in a book called *Inequity by Design*, written by a group of researchers from the University of California at Berkeley. For example, Herrnstein and Murray assume that all economic variables make equal contributions, but this is not the case. Parental income has a much greater impact on financial well-being than parental education and parental occupation. When they did the statistics correctly, the Berkeley group found that Herrnstein and Murray had underestimated the impact of parental wealth on offspring wealth by well over 50 per cent. Herrnstein and Murray also overlooked a variety of key variables in their analyses. They did not examine how many siblings the subjects had growing up, the extent to which fathers were actively involved in their upbringing, the local unemployment rates, whether subjects were living in urban versus rural environments. All of these things can affect long-term socio-economic status. By ignoring crucial variables, Herrnstein and Murray exaggerate the role of IQ on life prospects.

Herrnstein and Murray may also mistake effect for cause. Socio-economic status can determine quality of education, and education can affect IQ. Herrnstein and Murray claim that IQ begets poverty, but this assessment is based on dubious statistics. In reanalysing the data, the Berkeley team concluded that Herrnstein and Murray over-estimate the impact of IQ on poverty by as much as 71 per cent. More often, poverty begets low IQ. A recent study found that infants trans-ferred from poor homes into affluent homes increase scores by 12 to 16 points.[17]

The Bell Curve is notorious for its claims about difference between groups. The central argument of the book goes like this. Premise 1: certain groups of people have different average IQ scores. For example, the average for American white people is 10 points higher than the average for American black people. Premise 2: IQ is heritable. Con-clusion 1: therefore, the IQ difference between white people and black people is a biological difference. Conclusion 2: therefore, that difference cannot be remedied by education reform or other corrective interventions. The problem with this argument is that neither conclu-sion follows from the premises.

Let's begin with the conclusion that differences between groups are based on biological differences. The inference is based on a simple fallacy: one cannot study traits within groups of people and then draw conclusions about differences between groups of people. Rich-ard Lewontin offers a helpful analogy.[18] Suppose we take a packet of seeds and plant half in nutrient soil and half in bad soil. Then we let the seeds grow, watering both groups and exposing both to equal light. After a few months pass, we measure how tall they have grown. Height in plants (like height in people) is highly heritable. In fact, if we focus on each group of plants separately, the variation in height will be entirely explained by differences in the seeds. Within each group, all seeds had exactly the same light, water and soil, so all within-group variation is based on the intrinsic potential of each seed. Height is 100 per cent heritable when we look within groups. But now suppose we compare the two groups. It is extremely likely that the seeds that were planted in bad soil will be much shorter than the seeds that were planted in nutrient soil. Suppose the average height of the seeds that were planted in bad soil is 15 centimetres lower than the average height of seeds grown in the good soil. Since height is heritable, and these groups are significantly different in height, we might conclude that the difference is biologically based; we might say that the seeds in the short group are biologically inferior to the seeds in the tall group. This would be an obvious fallacy. All the seeds came from the same packet. Both groups had exactly the same potential for growth. The difference between the groups is entirely attributable to an environmental difference. The plants in the short group would have been just as tall as the plants in the tall group, on average, if they had been planted in nutrient soil.

The plant case exactly parallels the IQ case. If the average IQ for white Americans is 15 points higher than the average IQ of black Americans, that difference does not show that whites are biologically smarter than blacks. The difference might be completely explained by environmental differences. Black Americans might be nurtured in the sociological equivalent of bad soil. The data are ambiguous between a genetic explanation and an environmental explanation. How, then, should we decide which explanation is right? The answer is simple. We should favour the biological explanation if blacks and whites are reared in the same environment, and we should favour the environmental explanation if there are significant environmental differences between blacks and whites. When things are presented this way, it should be absolutely obvious that the best explanation for the IQ discrepancy between whites and blacks is environmental. After all, there is overwhelming evidence for huge environmental differences between whites and blacks. In America, black people receive considerably worse health care than whites (they die five years younger on average), are much poorer (the net worth of whites is fourteen times higher than blacks), and are victims of discrimination in American workplaces (blacks are half as likely to get a callback for a job application than whites who submit identical résumés)[19] and courtrooms (blacks are three times as likely to be executed for the same crimes).

Herrnstein and Murray try to prove that IQ differences between blacks and whites are biologically based. To make that case, they argue that IQ differences between whites and blacks remain even when researchers control for socio-economic status. This finding might help the case for a biological explanation were it not the case that there are other enormous environmental differences separating whites and blacks. The fact that black people are generally much poorer and, hence, generally much more likely to be less educated and engage in criminal behaviour, has an enormous impact on how black people are perceived. We all involuntarily form stereotypes on the basis of the most salient members of a class, and we then use those stereotypes to judge all members of the class. Affluent black people are regarded as less intelligent and trustworthy than affluent white people, because the black stereotype is formed by exposure to black people who are poor. White Americans see people of colour working in low-paying service positions, or being carried off in handcuffs by police on the nightly news. These experiences have a measurable and unconscious effect on white attitudes.

Researchers have devised powerful techniques to measure unconscious racism. One method for doing that begins by presenting some subjects with words associated with a racial stereotype (e.g., jazz, Harlem, welfare, dreadlocks). After that, subjects are presented with neutral descriptions of different individuals whose race is not specified. Subjects who have been unconsciously presented with words associated with the black stereotype judge the individuals in the descriptions more negatively than subjects in a control group. They rate those individuals as more hostile and more likely to be guilty of committing a crime. In some studies, the effect is equally strong for subjects who rate very low on questionnaires that are designed to assess racism.

Prejudice inevitably exerts a negative influence on its victims. Black Americans are stopped by police on highways more frequently than white Americans; they are regarded with fear by white pedestrians; and they are regarded as less intelligent than whites by their educators. Racial prejudice influences black Americans' self-assessment and behaviour. For example, in one study a group of good black and white students were asked to take a test, and half the students were asked to write down their race.[20] White students who specified their race did just as well as white students who did not specify their race. Black students who did not specify their race performed just as well as whites, but black students who specified their race dramatically underperformed. Their scores were significantly lower than the other groups, and, in a word-completion task, they were much more likely to complete the word LA_ _ as 'lazy' and DU_ _ as 'dumb'. When black students are made aware of their race, their performance declines.

The evidence for racial bias is overwhelming. Black Americans are raised in bad soil. They are subjected to an environment that promotes inequality by chronically and pervasively conveying the message that black people have less potential than whites. The result is an erosion of confidence, a dearth of opportunities and a drop in aptitude.

In sum, we have solid evidence that black Americans grow up in an environment that significantly reduces chances of success. This can explain the differences in white and black IQ scores. There is no reason to think these differences are biologically based. Similar morals can be drawn for the reported differences between men and women, between Jews and Christians and between Westerners and Easterners. Each of these groups has very different life-experiences, on average. Those differences may account for differences in IQ. Until all cultural differences are ruled out, we should assume that IQ discrepancies are environmental, rather than genetic. This should be our default assumption, because cultural differences are known to exist, and cultural differences can have an impact on psychological traits. For example, members of different cultures have different beliefs and values.

From the gene's eye point of view, members of different 'ethnic' groups – such as blacks and whites, Asians and Westerners, Jews and Gentiles – are very similar. Indeed, within-group genetic differences are much greater than between-group genetic differences. Two randomly chosen black people may be less genetically similar than a randomly chosen black person and a randomly chosen white person. Many researchers believe that the term 'race' has no biological meaning when it comes to our species. The racial groups we talk about have insufficient genetic uniformity to be classified together. Ethnic categories are created by us on the basis of superficial features. Differences in skin colour are genetic, of course, but so are differences in eye colour, and differences in ear lobes. There is little reason to think that any of these superficial traits correlate with genetic differences in psychology.

The Jewish Question

In some cases, groups that have been isolated from others, and that are small enough to have descended from a very limited gene pool, end up with some recessive traits that distinguish them from other groups. Residents of Pingelap have high incidence of colourblindness, residents of Martha's Vineyard used to have high incidence of deafness, and Ashkenazi Jews have high incidences of certain diseases such as Tay-Sachs and breast cancer. Large racial groups, such as black and white, may have no biological meaning, but biological regularities may begin to appear in groups that descend from a very small number of 'founders'. Mightn't we find group-wide IQ differences if we forget about the black/white divide and focus on groups that have been genetically isolated by geography or creed?

This case has been made for Ashkenazi Jews. Herrnstein and Murray say that Jews who trace their ancestry to the medieval Jewish populations of central and northern Europe have higher IQ averages than other groups, and that's why they take home a disproportionate number of Nobel prizes. If Ashkenazi Jews are genetically prone to certain diseases, why not assume they are also genetically different when it comes to intelligence? Indeed, Gregory Cochran, Jason Hardy and Henry Harpending, a team of researchers from the University of Utah, have suggested that there may be a link between IQ and illness in Ashkenazi Jews.[21] Among the dozen or so diseases to which the Ashkenazim are susceptible, there is one group of diseases, including Tay-Sachs, which leads to increased storage of sphingolipids. There is another group of diseases, including breast cancer, which result from a genetic abnormality in the processes that permit damaged DNA to be repaired. The Utah researchers think that both of these groups of diseases are an accidental byproduct of mutations that were actually advantageous to medieval Jews. The genes that lead to sphingolipid accumulation might lead to an increase in the growth of axons and dendrites (the fibres that connect neurons), and the genes that disrupt DNA repair could, in theory, lead to an increase in the number of neurons. Brain size is positively correlated with intelligence, so the mutant genes underlying these disorders could actually make Ashkenazim smarter. When two of the mutated genes are inherited, the combination is fatal, but, when one is inherited, there just might be a cognitive benefit. But why were the Ashkenazim the ones to end up with these mutations? The Utah team speculates that they came under selection pressure when they began settling in northern Europe in the ninth century. In this time period, many Jews began to work as money lenders, because Christian authorities prevented them from working in other trades. Moneylending was a viable option for Jews, because Christians shunned that profession as sinful. According to the Utah team, moneylending introduced high intellectual demands. Jews who were smarter had greater chance of success in the intellectually demanding field of finance, and they were more likely to procreate. But the successful Jews were smarter as a result of genetic mutations, which also carried negative side effects, which began to proliferate.

This is a seductive story, but it is full of holes. First of all, the science is dubious. The link between faulty DNA repair and the genesis of neurons is highly speculative, and the diseases in the sphingolipid group, such as Tay-Sachs and Canavan disease, have actually been associated with neural degeneration when two copies of the mutated genes are inherited. There is no direct evidence for the claim that inheriting a mutant gene has any positive effects on the central nervous system. In any case, there are other explanations of Ashkenazi diseases. Some of them may have been selected for their capacity to protect against microbes in densely packed ghettos, and others may just be a consequence of random genetic drift. With high rates of intermarriage, it is not unusual to see the spread of recessive traits that serve no positive function.

A further problem with the Utah team's story is that there are cultural explanations for the intellectual achievements of Jews. Jewish emphasis on education dates back at least to the time of the Talmud, which was written centuries before Jews became moneylenders in Europe. To defuse this explanation, the Utah researchers would point out that Sephardic Jews, who descend from Jewish populations in Spain and Portugal, do not have elevated IQ scores. So intellectual prowess cannot be a consequence of Jewish enculturation. But this response overlooks major cultural differences between Sephardic and Ashkenazi Jews: assimilation. When Ashkenazi Jews were being forced to work in finance, Sephardic Jews were working in a broader range of fields, and they were more assimilated with the Christian and Muslim populations. Then came the Inquisition, and Sephardic Jews were forced to convert or go into exile. Many Sephardic converts continued to practise Judaism secretly, but they would have been less able to overtly promote Jewish cultural values or educate their children in Jewish schools. The Sephardic Jews who fled the Iberian Peninsula into exile were comparatively poor and, thus, less focused on education. Both groups of Sephardim may have been less successful than their Ashkenazi counterparts in promoting scholastic achievement.

Another flaw in the theory advanced by the Utah team is the assumption that medieval Jews needed unusually high intelligence to succeed. On the face of it, moneylending is no more intellectually challenging than any number of jobs available to Gentiles, including trading, farming, manufacturing, engineering, managing estates, practising law, commanding armies, running governments, doing science and teaching in schools.

To demonstrate that Jews are genetically programmed for finance, the Utah researchers cite evidence that Ashkenazim get especially high scores in mathematics and only average scores in tests of visual and spatial abilities. They say that this explains why Jews don't excel in representational painting and sculpture. This is preposterous. Jews don't have an enduring tradition of representational art, because Jewish law prohibits representations of people, and European art production was controlled by wealthy Christian patrons for centuries. Once we move into modern times, Jews are not under-represented in art. Jewish artists include Camille Pissarro, Marc Chagall, Max Beckmann, Amedeo Modigliani, Man Ray, Frida Kahlo, Mark Rothko, Roy Lichtenstein, Robert Rauschenberg and Diane Arbus. Moreover, visual and spatial skills are important in domains where Jews have proven successful. Einstein claimed to rely on mental imagery and, as the Utah team notes, half the world chess champions are Jewish. In any case, if the Utah team is right about the neuron-growing power of Ashkenazi genes, then Ashkenazim should outperform Gentiles in all psychological capacities. The fact that their scores are only above average in certain areas suggests a cultural explanation.

Indeed, there is a convincing cultural explanation for why Ashkenazi Jews excel at maths. Perhaps it's because they were forced to work as moneylenders for hundreds of years. The very fact that the Utah researchers use to argue for a genetic difference actually points to a cultural difference between Ashkenazim and other groups. Ashkenazi Jews may have encouraged their children to study maths because it was the only way to get ahead. That emphasis remains widespread today, and it may be the major source of high performance on IQ tests. In arguing that Ashkenazim are genetically different, the Utah researchers identify a major cultural difference, and that cultural difference is sufficient to explain the pattern of academic achievement. There is no solid evidence for thinking that the Ashkenazi advantage in IQ tests is genetically, as opposed to culturally, caused.

The Mutability of Intelligence

This brings us to the second dubious conclusion that Herrnstein and Murray draw in *The Bell Curve* – the claim that difference in IQ cannot be affected by education. This is based on the assumption that IQ differences are biologically fixed. That is patently false. IQ scores change throughout the lifespan, and they can be altered by experience. With each year of education, IQ scores rise between 2.7 and 4.5 points (depending on the estimate),[22] and, when the onset of schooling is delayed, IQ scores drop by 5 points a year.[23] For students in school, IQ increases during the academic year, and decreases during summer vacations.[24]

Thousands of middle-class teenagers exploit the plasticity of IQ each year when they prepare for the Scholastic Assessment Test (SAT), which is used to make college admissions decisions in the United States. In order to increase their chances of getting into a good school, students often take SAT preparatory courses, which dramatically increase scores. Students who take prep courses can expect a boost of 100 points on the SAT, which has a maximum total of 1,600 points. This is significant because SAT scores are highly correlated with IQ. Therefore, if students are increasing their SAT scores by taking a course for a few weeks, they are effectively raising their IQ score.

Further evidence for the plasticity of IQ scores is easy to find. In Japan, there is an ethnic group called Buraku, who are regarded as inferior and subjected to various forms of bigotry.[25] As with blacks and whites in America, the Buraku IQ average is about 15 points lower than the IQ average for members of the privileged ethnic group. Strikingly, however, this difference disappears when Japanese people move to the United States. Buraku living in the United States may even perform slightly better than other Japanese immigrants. This suggests that the low score in Japan is a function of environment and that the discrepancy can be eliminated by moving to a culture where there is no special bigotry against the Buraku.

The plasticity of the black/white discrepancy has also been directly tested.[26] A group of researchers looked at a group of African American college students who had been admitted to universities through affirmative action programmes. At the time of admission, these students had, on average, IQ scores that were 15 points lower than white students. This gap was cut in half by the time the students graduated. The result is especially striking given that racial prejudice is widespread in higher education. The moral is that a college education can alter IQ and shrink significant differences between groups, even in the face of enduring biases.

Another piece of evidence for the plasticity of IQ scores comes from James Flynn, a New Zealand political scientist who discovered that there have been dramatic increases in IQ over the course of the last half-century.[27] Scores have been rising by about three points every decade. In some cases, there has been a 20 point increase between one generation and the next. It is not entirely clear what is driving these changes, but the most probable factors are environmental. People are spending more time in school and receiving more training in taking standardized tests. People are also being exposed to much more information through television and, now, computers. The net result is that we are getting smarter. If IQ were genetically fixed, that would be impossible.

In fact, you can boost your IQ score in just a few minutes. Ap Dijksterhuis and his colleagues demonstrated that test performance can be raised or lowered by simply thinking about different social groups.[28] They asked one group of subjects to answer a few questions about what university professors are like, and then they asked another group of subjects to answer some questions about soccer hooligans. Afterwards, both groups were given a test on general knowledge, and the subjects who had just been thinking about university professors did significantly better. Thinking about a group of stereotypically smart people can make you smarter. Presumably, if you take an IQ test just after thinking about intelligent people you will do better than if you take the test after thinking about people who are stereotypically dumb. If IQ can be affected by momentary thoughts about professors and hooligans, think about how it might be affected by a lifetime of being told that you belong to an ethnic group that is stereotyped as smart or as slow.

DISPENSING WITH INTELLIGENCE

We have seen that group differences in IQ scores may result from differences in environmental factors, especially education. We have also seen that IQ can change with experience, which means that members of disadvantaged groups can boost performance under the right conditions. IQ can be improved. This way of putting it raises a question. What is it that we are measuring when we measure IQ? Is there some thing that can rightfully be called intelligence?

The G Factor

Proponents of IQ testing often suppose that these tests measure a single psychological capacity, rather than a variety of independently learned skills. The reason for this is simple. IQ tests quantify how well a person performs on a variety of different kinds of problems, including vocabulary, word completion, sentence memory, mathematical calculation, picture matching, auditory learning and analytic reasoning. It turns out that performance on these very different kinds of problems is highly correlated; people who are good at vocabulary tests are often also good at maths, for example. These correlations can be explained by the assumption that people have a cognitive capacity that cuts across a range of different tasks. Researchers call this common capacity the *g* factor, or sometimes just *g*, which stands for general intelligence.

The idea of *g* was first proposed by the statistician Charles Spearman in the early twentieth century. Spearman noticed that academic performance is correlated across seemingly unrelated fields and he inferred that these correlations derive from a common cause. Spearman was heavily influenced by Galton, and he proposed his theory at just the time when modern intelligence testing was being developed. In this context, it was natural to relate g with IQ testing. Spearman believed that such tests measured a general and innate capacity. To this day, the practice of IQ testing is sometimes justified by the assumption that g exists. If there is no g, then generalized intelligence tests may serve little purpose. We might replace them with independent tests for different skills, and we might dismiss the whole concept of general intelligence as meaningless.

In recent years g has come under attack. Some researchers claim it is a statistical artefact.[29] As just noted, g is postulated to explain the fact that performance on a variety of different aptitude tests is highly correlated. But there is an alternative explanation of these correlations. Perhaps standard IQ tests measure a range of different abilities that happen to all be learned in school. If so, motivated students who get training in one are likely to get training in the others. If schools made a habit of teaching fencing or macramé, performance in these domains might correlate with IQ tests as well. Motivated students would do well at all of them. Motivated students may not have any general intelligence capacity. They just have good study habits, due, perhaps, to family encouragement, abundant energy, compensation for bad social skills, cultural pressure or other factors.

The simple assumption that motivated students can do well in multiple areas may go a long way to explaining correlations in test performance without any need for g. It's also worth noting that the correlations are far from perfect. Some people are good at picture matching, for example, and bad at vocabulary. In fact, people with entirely different strengths can get exactly the same IQ score. In this respect, IQ testing is like SAT testing. Two people who have the same SAT score might have totally different strengths: one may have excelled in maths, the other in vocabulary. It would seem very implausible, in such cases, to say the two individuals have equally high 'general intelligence', even though that is precisely what IQ tests are presumed to measure.

The lack of correlations in abilities related to IQ is even more pronounced if we move away from typical high-school students and consider special populations. Autistic savants, for example, can perform at extremely high levels on some of the tasks used in IQ testing, but so low on others that they are deemed mentally retarded. In small-scale societies, we often find individuals who have navigating or hunting skills that require incredible memory, complex calculation and careful reasoning, but these same people do poorly on Western intelligence tests. The skills that correlate in Western high-school students may have little relation to each other outside of our educational system.

There is a growing body of research supporting the conjecture that intelligence is not monolithic. In place of g, Robert Sternberg has suggested that there are three distinct kinds of intelligence: analytical, creative and practical.[30] People think George W. Bush is dumb because he was a C student, but his capacity to win over the hearts of voters reflects a form of practical intelligence that his opponents underestimate. Likewise, we don't know if Picasso would have won any scholastic awards, but he was one of the most innovative artists during the most innovative century in the history of art. Howard Gardner has argued that intelligence should be fractionated even further.[31] He postulates multiple domains in which people can show variable aptitude, including language, logic and maths, spatial reasoning, understanding of the body, music skills, knowledge of the natural world, self-knowledge, social competence and insight into existential and spiritual issues. According to Gardner, each of these domains constitutes a distinct species of intelligence. IQ tests measure maths and vocabulary skills, but they leave out street smarts, leadership abilities, body mastery and artistic genius. Therefore, performance on IQ tests cannot be used to fully assess a person's abilities.

The skills that are summarized under the dubious notion of g constitute a fraction of the ways in which people can excel, and most jobs in the world require skills that are not measured by IQ tests. In response to this, defenders of IQ boast that a quarter of the variance of job performance can be correlated with IQ scores. But there is a simple explanation for this. As we have seen, IQ scores correlate with amount of education, and amount of education correlates with various socio-economic variables. The fact that IQ correlates with job performance may be attributable to the fact that people with high IQ scores have had better education and better lives at home. If this interpretation is right, then IQ proponents are mistaking correlation for cause. IQ does not improve job performance, it is merely an indicator of a person's educational and social history. We would be better off with tests for specific skills that are actually used on the job.

G is Genes and Brains

The final verdict about the reality of *g* may depend on whether we can find any biological correlates. If *g* is real, and not just an artefact of correlations between test scores, then we might expect to find some biological mechanism that accounts for it. Faith in *g* has led molecular geneticists on a search for the genes that make some people smart. This search has left a legacy of false leads, failed replications and unimpressive results. The emerging consensus is that there may be many different genes that contribute in incremental ways to intelligence, but each one is a bit player. No single gene will emerge as the key to cleverness.

When scientists began the search for smart genes in humans, they focused on disorders that result in retardation. Perhaps the genes responsible for mental disability are also responsible for mental prowess. This search has not been very fruitful. Just as oxygen deprivation can cause brain damage, the genetic causes of retardation often involve chemical changes that have nothing directly to do with intelligence. One example of this is phenylketonuria (PKU), a disorder that affects individuals who inherit two copies of a defective gene, which is ordinarily involved in manufacturing an enzyme to break down phenylalanine, a chemical found in many foods. With defective genes, phenylalanine builds up in the body and results in irreversible retardation. This outcome can be avoided with early diagnosis, by simply feeding children with PKU a diet low in phenylalanine. It's not known exactly how phenylalanine build-up causes retardation, but it is known that the genes involved are not directly responsible for making anyone smart. This illustrates an important moral. Genes that can affect intelligence are not necessarily genes for intelligence.

Another strategy for finding smart genes is to look for a genetic difference between people who score high on IQ tests and people whose scores are only average. A team of researchers at King's College London tried this method and found that there was a form of a gene called IGF2R, which was found in 32 per cent of the high-IQ subjects and only in 16 per cent of the average subjects.[32] This was headline news. Notice, however, that the numbers are not very impressive. If only 32 per cent of the geniuses in the study have the gene in question, it's not necessary for brilliance, and, if 16 per cent of the average performers have the gene, then it is not sufficient for brilliance. Still, the press announced that the smart gene had been found. This conclusion needed to be quietly retracted a short while later when the King's College team ran the same study again. They were hoping to replicate their original results, but this time around, the results were reversed: 24 per cent of the average performers had the alleged smart gene and 19 per cent of the high performers had it.

Other groups have tried to identify other smart genes. There was some excitement about a particular variation in a gene called CTSD and another called CHRM2, which is involved in the production of chemical receptors in the brain. Both of these genes only account for about 3 per cent of the variance between high performers and average performers, which is hardly sufficient for being called a smart gene. Imagine that we took a group of people who weighed 160 pounds and another group who weighed 210, and found that there is a gene that can account for 1.5 pounds of the difference between these groups. Would we call this the obesity gene? Worse still, the underwhelming results are being described as major breakthroughs by the press before they have been replicated by other labs.

The task of identifying smart genes would be easier if we could first find the correlate of IQ in the brain. Some researchers have come to believe that, when it comes to intelligence, size matters. There is evidence that brain volume is positively correlated with IQ scores. Indeed, there is .40 correlation, which is impressive. Researchers who are eager to identify smart genes are looking for genes that regulate brain size, and one team has made progress on this front by doing genetic analyses on a population of people who have normal intelligence despite having very small brains.[33] These analyses have led them to identify a gene that makes brains grow, and they think this gene may be a factor in intelligence. This idea is seriously misguided. Even if brain size correlates modestly with intelligence, having a big brain is not necessary for intelligence. Some geniuses wear small hats. Nor is it sufficient. People with autism have larger brains than typically developing individuals, but often have lower IQs. By analogy, think of leg length in athletics. Having long legs presumably correlates with athletic performance, but a lanky klutz with long legs will not succeed in professional basketball.

The idea that IQ results from big brains rests on a hopelessly simplistic theory of brain function. It used to be believed that the entire brain contributes equally to every cognitive task. If that were true, big brains might be brilliant. But everyone in neuroscience now recognizes that different brain areas do different things. So the idea that brain volume is directly responsible for higher IQ doesn't make much sense. Having a big olfactory bulb, for example, may help you smell better, but it won't make you Einstein. In this spirit, some researchers are now trying to locate specific brain areas that are implicated in intelligence. One research group thinks they have found g in a part of the brain called area 46, which is located on the side of the frontal cortex.[34] They based this conclusion on a study in which area 46 is highly activated during three different tasks that are used in IQ tests – a verbal task, a spatial task and a perceptual task. The fact that one area lights up for three very different IQ tasks suggests that intelligence depends on a common factor, g.

This study is completely unconvincing. All three tasks used in the experiments involve finding an odd-one-out, one of four items that differs from the other three. And they are all difficult odd-one-out tasks, which means, in each case, a person performing the task has to pay close attention to all four items and compare them. That requires holding a lot of information in short-term memory. Area 46 is famous in neuroscience, because it's a main player in short-term memory. It lights up when people hold long lists in their heads, for example. So it's not surprising that it's active in the odd-one-out tasks that they chose for their experiment. But there are many short-term memory tasks that aren't rocket science. If you look up a number in the phone directory and then hold it in your head for a few seconds while getting your phone, area 46 will light up. But this would hardly win you any points in an IQ competition. Conversely, there are items in IQ tests, such as analogies, text comprehension and vocabulary, that don't require much short-term memory. So it's highly unlikely that area 46 is the seat of general intelligence.

These efforts to find *g* in the genes and the brain have been unsuccessful, and that is not surprising. It's possible that *g* does not exist. The fact that some people in our culture do well on tasks that have nothing obvious in common does not entail that there is a hidden common psychological capacity at work in each of them. It's not even clear what the common factor could be, if the tasks have little in common. It's more likely that correlations in task performance result from the fact that people in our culture are exposed to a variety of different skills in school, and highly motivated learners can attain mastery of them all. Correlations between these skills do not hold up across cultures, and no biological correlate has been found. It may be time to give up on the idea of general intelligence.

LESSONS LEARNED

In this chapter, we have seen that heritability tests overestimate the genetic contribution to intelligence. This has led some to think that group differences are genetic. But there are many reasons to be sceptical of this conclusion. First, group differences in IQ correlate with social and economic differences, which offer an obvious explanation of these results. Second, IQ scores are highly mutable, suggesting that group differences can ultimately be eliminated by correcting social disparities. Third, the very notion of intelligence underlying the group differences is suspect. Intelligence is not a single capacity that can be under the control of some gene or brain area; it is rather the name we give to a range of independent skills that happen to be emphasized in our schools. The IQ controversy is an extreme example of a more general tendency to explain human abilities by appeal to biology. It is a particularly egregious case because it legitimates biases against many subjugated groups and mistakes social injustice for biological necessity.

The main lesson of this chapter is that some of the differences between people have little to do with biology and much more to do with what we learn from our environments. In the next chapter, we will switch from human differences to human universals and we will shift from genetic explanations of behaviour to approaches that postulate innate knowledge. These topics may look unrelated, but the moral is the same. It is common practice to explain human capacities by appeal to nature rather than nurture. Sometimes this is a mistake. Here we have seen that human variation can result from experience, and in the next chapter we will see that experience can also account for some human universals.

Where Does Knowledge Come From?

4

What Babies Know

After years of schooling, we come to know quite a bit about various domains of knowledge. We learn about maths, psychology, biology and physics. Education is hard work. We have to read textbooks, memorize facts and learn techniques for calculation. In the USA, kids spend about seven hours a day in school, five days a week, for nine months of the year. That's about 1,200 hours of schooling a year, and this continues for over a decade. That's a lot of instruction. But we also learn a lot of things without ever opening a book. We learn that three cookies is more than two, that insulting people makes them mad, that pet puppies grow and that a glass of milk will fall to the ground when dropped. Such commonsense wisdom constitutes what psychologists call folk knowledge – things known to most ordinary folk. Folk knowledge can be divided into domains that are a lot like the departments in a university – folk psychology, folk physics and so on – but these folk domains would arise without spending a single day in school. Where does such folk knowledge come from? Some of it we clearly get from observation. Through hard experience, we learn that too much candy causes bellyaches. It has become fashionable, however, to suppose that some folk knowledge is innate. Developmental psychologists argue that we are born with something like the structure of a university in our heads, with 'core domains' that operate under the control of 'innate principles'. In this chapter, we'll take a critical look at the evidence and explore an alternative. First, a brief philosophy lesson.

RATIONALISM AND EMPIRICISM

The history of Western philosophy has been a pendulum rocking back and forth between two opposing views. On one side, there's Rationalism, which emphasizes innate knowledge. Plato got things off the ground with his dialogue *Meno*, in which Socrates concludes that mathematical knowledge is innate after showing that an uneducated slave boy can carry out a complex geometrical proof – with a healthy dose of Socratic coaching. Plato held the bizarre view that innate knowledge is actually a form of recollection; before we are born into this world, our souls reside in a world of perfect objects (the 'Platonic forms'), recollection of which allows us to categorize and comprehend the imperfect object that we encounter in life. Few of Plato's successors accepted this explanation, but many embraced the doctrine that some ideas are innate. In the seventeenth century, René Descartes revived Rationalism and postulated a 'treasure house' of innate ideas, including innate understanding of mathematics, logic, physical matter, the laws of motion and the nature of the Christian God. Descartes also encouraged readers to mistrust perception. We can learn by observation, but the senses often deceive us, and we must use the innate power of reasoning to find order in the changing flow of experience.

On the other side of philosophy's great pendulum, we find Empiricism, which originates with Plato's student Aristotle, who insisted that knowledge is based on experience, rather than innate ideas. Aristotle said that the mind comes into the world like a tablet without writing on it, a blank slate. John Locke took up this view again in the seventeenth century, calling the mind an 'empty cabinet', which contrasted starkly with the treasure-house metaphor. Locke was reacting to Descartes' Rationalist revival, but he was also irked by the fact that Europeans were using the thesis of innate ideas as an excuse to mistreat indigenous peoples in the New World: if all people have an innate idea of the Christian God, then there is no good excuse for worshipping other gods or acting in un-Christian ways. Locke said religion should be based on evidence and reason. He also railed against innate ideas more generally and argued that knowledge begins in the senses. The Empiricist programme was pushed even farther by David Hume, who argued that we can have no conception of things that we cannot perceive. There is no thinking without prior perception.

The battle between Rationalism and Empiricism has endured because the two camps present radically different views about the mind. Rationalists say knowledge is built up on a scaffolding of innate principles, while Empiricists say all knowledge derives from experience. Occasionally philosophers try to find a middle-ground position, but these usually end up sounding like spiffed-up versions of one of the warring factions. The most famous attempt at a compromise was devised by the great eighteenth-century philosopher Immanuel Kant. Against the Empiricists, Kant argued that it is impossible to learn without some innate concepts to make sense of the chaotic input from the senses. Against the Rationalists, Kant postulates only a tiny handful of extremely general innate concepts, such as *time*, *causality* and *space*. The mind is not quite a blank slate for Kant, but it's pretty close. So close, in fact, that Kant's philosophy might be described as a sophisticated form of Empiricism. In the twentieth century, many self-described Empiricists bought into Kant's view that we need a bit more innate machinery than Locke granted – for example, we may have innate rules of logic – but they insisted that more specific knowledge is acquired through experience.

This brings out an important point about the Rationalist/Empiricist divide. It's not a debate about *whether* anything is innate, but about *what* is innate. For Empiricists, the crucial thing is generality. We have innate resources that help us acquire knowledge, but the very same resources are used to learn about very different kinds of things. We use the same mental resources to learn about biology, physics, psychology, maths and so on. These resources include our senses, some general-purpose learning rules and perhaps even a few innate concepts, like those on Kant's list. For Rationalists, the innate machinery is much more specialized. We have some innate resources dedicated to maths and others to psychology and still others to biology. In the lingo, these resources are 'domain-specific'. Rationalists don't always insist that we are *born* understanding these domains. Some knowledge is acquired over the course of development. But it is not acquired using general-purpose learning rules. Instead, we have specialized psychological tools for learning about different domains. To say knowledge is innate is to say either that it is present at birth or that it is acquired using such specialized resources.

The debate between Rationalism and Empiricism has occupied Western philosophy for two millennia. But it is not merely a philosophical debate. Over the last century, it has become a central issue in the human sciences, and, like philosophers, scientists have vacillated between the two poles. In the first half of twentieth century, psychology was in the grip of the movement called Behaviourism. Behaviourists were not traditional Empiricists because they denied the existence or scientific significance of conscious mental states, but they did buy into the central tenet of the Empiricist programme. They claimed that all knowledge is learned. Rather than postulating domains of innate principles, they claimed that we can learn everything we know by conditioning. This turned out to be untenable. If conditioning were sufficient to explain human cognitive abilities, we should be able to condition non-human animals to learn all the things that we know, because animals are highly responsive to conditioning. This raised a question. If human knowledge cannot be explained by mere conditioning, where does it come from? By the 1960s, critics of Behaviourism began to suspect that human knowledge must come from within; they proposed that the building blocks of human knowledge are innate. Thus, the pendulum began to swing back towards Rationalism. Initially, Rationalism was based largely on conjecture, because, in the mid century, there were no good techniques for measuring what knowledge, if any, is innate. The situation changed in the final quarter of the century. Developmental psychologists devised new ways of investigating the infant mind, and their discoveries led them to revise the earlier Empiricist assumptions. Since then, psychologists have been defending views that would make Plato smile, postulating a wealth of innate knowledge. The evidence is fascinating, but we may be on the verge of an Empiricist revival.

ARE BABIES LITTLE SCIENTISTS?

The world is populated by many different kinds of things. For example, there are animals, plants, human-made objects, numbers, laws of nature and minds. Features that are found in some of these categories don't crop up in the others. Human-made objects don't digest, plants don't breathe, numbers don't obey the laws of gravity, physical laws don't grow, and minds don't have square roots. Over our lifespan, we clearly acquire a lot of knowledge that is domain-specific, meaning it applies to one of the categories and not the others. In fact, each domain has given rise to its own branches of science: zoology, botany, engineering, mathematics, physics and psychology. Some of the sciences are also grouped together under even more encompassing domains (zoology and botany belong to biology) and they all also subdivide (maths includes calculus and geometry). Each of these fields focuses on a different range of things, and each class of things is governed by a distinct set of principles. The sciences try to discover what those principles are.

This division of labour is enshrined in modern colleges and universities, and it also plays a role in human thought. When we come to see something as a biological organism we expect it to seek nutrients and grow. When we hear that some oddly shaped stone found in an archaeological dig is a human-made artefact, we start to wonder about its function. We know general truths about biological kinds and human artefacts, and we use these to make inferences when we encounter specific examples.

How did we come to divide the world up in this way? How did we come to grasp the distinction between psychology and physics, for example, and how did we learn the core principles that govern these domains? Within developmental psychology, the most popular answer is that we didn't learn these principles at all; they are innate. In this view, the baby's mind is much like a university. It is subdivided into different departments each of which is governed by innate rules that can be used to categorize and comprehend the world. We supplement these rules through experience, but basic understanding of these domains is in place from the start.

This is a radical view. It implies that babies are like little scientists, who already know a lot about how the world is organized. Why would developmental psychologists believe something like that? The answer is that there have been hundreds upon hundreds of experiments that provide impressive support for this position. This may sound surprising, since infants don't spend a lot of time talking about biology and physics. Indeed, in their first year, most infants aren't talking at all. They smile and belch and stare, but they have no way of describing the contents of their thoughts. For this reason, for most of human history the going assumption has been that infants are pretty dumb. But the 1960s saw a methodological breakthrough that cast new light on the infant mind. Researchers realized they could infer what an infant is thinking by carefully recording what an infant is staring at.

The most popular technique, which became widespread in the 1980s, is called habituation. Infants, like adults, can get bored. If you show little Sally an unfamiliar object, she will stare at it for a long time, which can be interpreted as a sign of interest. But if you show it to her over and over again, she will stop looking, suggesting that she no longer finds it interesting. The breakthrough came when researchers realized they could use looking time to infer which things an infant groups together as more alike. Suppose you show Sally one potato, followed by another, and then a third. If she gets bored, that will indicate that she has recognized that these all belong to the same category. If you throw a beet or turnip into the mix, she might get excited again. But Sally can tell that potatoes are alike, and after a series of potatoes, she forms the expectation that she'll have to endure more of the same each time. A change in looking time would indicate a violation of that expectation.

This basic principle can be used to figure out what things infants classify together and also much more. Suppose you show Sally a potato and then put it behind a little opaque screen so she can't see it any more. Now remove the screen. An adult will expect to see the potato again, because adults know that objects remain in place even when they are out of view. Does Sally know this? To find out, we can set up an experiment in which she watches a potato getting covered by a screen, and then the screen is lifted. The experiment requires two conditions. In one, the screen is lifted to reveal the potato, which is just what adults would expect. In the other, when the screen lifts there is no potato; that would surprise adults and look like a clever magic trick. Now suppose we show a group of infants these two outcomes and measure looking time. If the average looking time is longer for one than for the other, we can infer that the outcome that caught their attention more is more surprising to them. Such studies have shown that infants stare longer when the potato disappears; when the potato remains, that's pretty dull, but a disappearing potato is really cool. Thus, infants are like adults: they expect objects to remain in place when hidden from view.

These techniques are truly extraordinary. They can reveal thought processes in babies whose linguistic skills are limited to 'goo goo ga ga'. Potentially, they can tell us what's innate. But one also has to be a bit careful in drawing such inferences. Take potato classification. Suppose infants can distinguish potatoes from carrots and peas. Should we infer from this that they have an innate concept of potatoes? Clearly not. If anything, it shows something about what kinds of features infants use to classify, and, in particular, it suggests that they are sensitive to colour and shape. Their innate ability to see colours and shapes is hardly a new discovery, and it's not very surprising that infants group things together using the features they can see.

The fact that infants expect objects to remain in place when un-perceived is more surprising. Indeed, it remains a controversial claim, because data of this kind suffer from a crucial ambiguity. Suppose Sally shows surprise when the potato disappears. This could mean she expected the potato to remain when the screen was placed in front of it. But it could also be that she had become bored of looking at potatoes. When the screen is lifted and there is a potato present, she may not be thinking, 'There's that potato again!' but rather, 'Oh God, not another potato!' In this interpretation, Sally's renewed interest when the potato disappears is not surprise – 'Where did that potato go?' – but relief – 'Thank God, no more potatoes.'

This point shows that looking-time experiments are open to interpretation. But that does not mean we should give up on the method. When multiple interpretations are available, the solution is to do more experiments. We now have decades of research of this kind. Each month, the leading developmental psychology journals are filled with new studies that try to establish what infants know by measuring how they look. Many of these studies are absolutely ingenious. It takes a special kind of scientific imagination to study the mind of pre-verbal infants, and we have learned a massive amount from the scientists who are clever, creative and patient enough to carry out this work.

There is no way to survey the field here, so I will focus on a few of the most influential studies. As we will see, these have been used to argue for a form of Rationalism. Infants are credited with a great deal of domain-specific knowledge. As we will also see, some of these celebrated studies provide less decisive evidence for Rationalism than is often appreciated within the field. This is not to say that these studies are useless. On the contrary, any study that gets statistically significant effects reveals something about how infants think. But such studies may not vindicate Rationalism. In fact, developmental psychology may ultimately provide confirmation of Empiricism. In the review that follows, I'll just try to show that Rationalists don't make their case. This is good news for Empiricists because Empiricism is a more economical theory. There is no reason to postulate innate machinery without powerful evidence. Thus, Empiricism should be the default position until evidence weighs in favour of Rationalism.

Infant Physics

Renee Baillargeon is one of the leading developmental psychologists, and she's a card-carrying Rationalist. In the 1980s, she set out to prove once and for all that infants expect objects to persist when out of view. Her work helped launch an active research programme on infant physics – the study of what infants know about the nature of physical objects. One of her most influential studies explores how 4.5-month-olds react to a disappearing trick.[1] In the first part of the experiment, infants watch a cardboard screen sitting flat, and then flipping over in a 180° arc, like an opening drawbridge. They watch this a few times. In the second part of the experiment, the infants see the same screen lying flat, but now a bright yellow box is placed behind it. As the screen rotates up, it covers the box and then one of two things happens. In one test condition, the screen continues rotating, but stops at 120°, which is just what should happen, because the box is in the way. In the other test condition, the screen rotates a full 180°, which is an impossible event, because the box should prevent the screen from lying flat. If infants realize that objects remain in place when they are out of view, they should be very surprised when this happens. This second test condition is *perceptually* more similar to what the infants saw in the first part of the experiment, because the screen rotates a full 180°. If infants base their expectations on perceptual similarity to prior events, they shouldn't be surprised at all. But if they understand that objects do not disappear when covered, they should be surprised when the screen continues to rotate. And this is just what Baillargeon found. Infants stare longer at the physically impossible event, suggesting that they find it surprising. Baillargeon concludes that 4.5-month-olds understand that objects persist when out of sight. She thinks this is one of several innate principles that govern infants' understanding of the physical world.

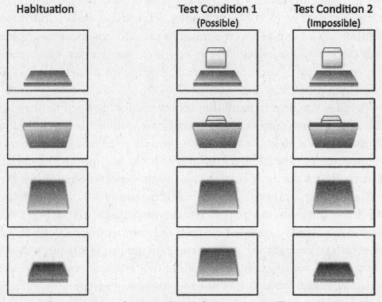

Figure 2. The sequence of events that infants see in Baillargeon's experiment (adapted with permission).

Infant physics is not limited to the principle that objects persist when unseen. Developmental psychologists have devised clever experiments to show that there are a number of innate principles. For example, infants seem to understand gravity: they are surprised when objects that lack support don't fall. They also understand that objects move as bounded wholes; if they see two parts of an object jutting out from behind a screen and moving in the same direction, they expect to see those parts connected together when the screen is removed. And infants have some comprehension of causation; they are surprised when one object seems to make another object move without physical contact between them.

These results are all very impressive, but we need to be cautious when leaping to the conclusion that knowledge of basic physical principles is innate. Consider gravity. It's true that infants expect unsupported objects to fall, but, oddly, they think even the tiniest bit of support is sufficient. If a box of cereal is hanging over the end of a table with only a centimetre on the surface and the rest perched precariously over the floor, infants expect it to stay in place. Or consider the principle that objects move as coherent wholes. This seems to be understood by infants as young as two months, but they do not show this expectation at birth, suggesting that it might be learned by observation. Comprehension of causation seems to emerge even later, showing up around six months.

We must also bear in mind that infant studies are open to interpretation. Researchers customarily interpret long looking times as indicating surprise. But we all know that infants and children take great pleasure and interest in repetition. Thus, longer looking times might initially indicate that something is *consistent* with expectations, rather than contrary to expectations ('Yay! More of the same!'). Thomas Schilling had this suspicion when he read Baillargeon's disappearing box study, so he ran the study again, but doubled the number of times infants saw the rotating screen at the start of the experiment. His conjecture was that it takes a lot to get an infant bored of a repeating event. After the infants were really bored, he showed the two test events, and, unlike in Baillargeon's experiment, his infants stared longer at the possible event than at the impossible event. Schilling's point is that we can reverse the effect of some of these studies, and it's tricky to know when long staring times reflect excitement about repetition or excitement about novelty. When you try to disambiguate these two interpretations by getting infants really bored, the evidence that they understand physical principles disappears; they seem to expect impossible events.

Even if 4.5-month-olds have mastered the principle that objects persist when out of sight, that doesn't mean this knowledge is innate.

In fact, Baillargeon provides evidence against innateness in her own study. When she tried her experiment on infants who were a month younger, half of them showed no surprise when the hidden object magically disappeared. That suggests that the principle of persistence is learned between the third and fourth month of life. In fact, it may not even be mastered until later on. Try this. Take a toy that an infant really wants to play with, and then, while the infant is watching you, place a cover over it. Amazingly, infants don't go searching for the toy, which was hidden before their very eyes. They don't figure out that they can look for the toy until they are nine months old, twice the age of top performers in Baillargeon's study! Before nine months, infants look in the right place, but don't search. The idea that objects persist seems to be learned bit by bit, not innately understood.

Infant Biology

Physics is the study of matter and the laws that govern it. Biology is the study of a special category of matter: living things. Just as some developmental psychologists have postulated an innate physics, others have postulated an innate biology, comprising a rudimentary capacity to identify living things and to understand some of the features that distinguish them from other kinds of entities.

One of the pioneers in this line of research is psychologist Frank Keil, who works with pre-school-age children rather than infants. In the 1980s, Keil ran a series of 'transformation' experiments, probing pre-schoolers' intuitions about what can be turned into what.[2] Suppose you take a coffee pot and put birdseeds in it, cut a little bird-sized hole in it and hang it on a tree outside. Is it still a coffee pot? No, kids will answer, it's a birdfeeder. When you try to change one kind of human-made artefact into another, all you need to do is change its superficial properties – in this case, the way it is constructed and used. But suppose you tried to change an artefact into an animal. Imagine taking a toy dog and covering it with fur so it looks just like a real dog, and wiring it up so it plays fetch and barks and runs around the house. Pre-schoolers know that this still wouldn't be a real dog; it would just be a good fake. You can't turn an artefact into an animal. Pre-schoolers also know that you can't turn one kind of biological organism into an entirely different kind. If you make a porcupine look like a cactus, it's still a porcupine. With artefacts, superficial appearance is what matters, but for biological organisms, it also matters where they came from and what kind of stuff they have inside. If children classified things on the basis of perceptual similarity (a view Keil dubs as 'Original Sim'), then they should believe that you can turn a porcupine into a cactus. Kids don't believe that. Kids know that biological organisms have essences that go beyond how they appear, and Keil concludes that this knowledge is innate.

Keil's research shows that young children treat biological organisms and artefacts differently, even before they have learned anything about biology in school. But, you might be thinking, pre-schoolers have already had years of experience with toys and tools and pets and houseplants. To figure out whether initial knowledge of biology is innate, it would be good to roll back the clock and see younger minds perform. What do infants know about biology?

Infants turn out to be quite good at identifying animals, which is impressive because animals differ widely in appearance. To show this, Jean Mandler and Laraine McDonough taught two different games to a group of ten-month-olds.[3] In one game, they gave a little toy dog a sip from a cup, and, in the other, they made a revving sound with a toy car. Afterwards, the psychologists gave the same infants a bunch of different toys to play with, including a variety of animals and vehicles. Without any further coaching, the infants correctly generalized: they played the sipping game with a fish, a swan and a cat, and they played the revving game with a truck, a motorcycle and a plane. This is remarkable, because the toys in each category differ widely in appearance, and some items in the two contrasting categories are very hard to distinguish. The swan toy looks a lot like the plane, but the infants played with them differently. Therefore, infants don't seem to be using perceptual similarity to classify, as Empiricists would predict, but are instead using some subtler principle of classification which is sensitive to the fundamental distinction between animate and inanimate kinds. It is tempting to conclude that infants draw this distinction innately.

Infants also seem to have beliefs about differences in how animate and inanimate entities behave. Consider the fact that animate entities can move by themselves whereas inanimate objects move only when something pushes or pulls them. Infants seem to appreciate this fundamental truth about living organisms. Amanda Woodward, Ann Phillips and Elizabeth Spelke showed seven-month-olds brief scenarios in which one object moves behind a screen and a second object comes out.[4] Then the screen is lifted and infants see one of two things: either the first object moves the second by contacting it, or the second moves without any physical contact with the first – it moves all by itself. Infants find self-propelled motion much more surprising if the objects in the study are meaningless shapes than if the objects are people. People are animate objects, and infants seem to realize that animate objects can move by themselves.

These are impressive feats. Kids in the crib may not know much about digestion or respiration, but they are already dividing the world into living things and non-living things, and they form different expectations about these categories. This has been taken to suggest that infants have an innate rudimentary understanding of the biological domain.

But this is a big leap. The fact that infants distinguish biological organisms from vehicles is impressive, but it is hardly evidence for innateness. After all, their ability to group vehicles together is as impressive as their ability to group animals together, but it would be ludicrous in the extreme to assume that evolution has furnished us with an innate concept of vehicles. It is highly plausible that both categories are learned, and learned perceptually. Notice that animals have a lot in common. They have curvy contours free from straight lines, they have rough surface textures, and they have faces. Mandler and McDonough's sipping game is easy to play with toys that have mouths. Vehicles share many features too. They have straight lines, smooth surfaces, windows and wheels. Plus, kids get a lot of practice with the category. By ten months, they've seen picture books with animals and vehicles, they've driven in cars, ambled in strollers, watched birds out the window and interacted with the family pets. So the fact that ten-month-olds can distinguish animals and vehicles – even birds and planes – can be explained by perceptual learning. Perceptual learning can also explain the fact that seven-month-olds know that people, unlike inanimate objects, can move on their own. Infants see people moving around all the time. Indeed it's one of the first things they ever see. They also see inanimate objects in motion, especially their toys. So it's easy for them to observe that people move without physical contact and inanimate objects do not.

In short, we don't need to posit an innate biology to explain how infants come to distinguish biological organisms from other things. Observational learning is the source of infant biology. Of course, there are many things about biology that are difficult to learn by observation alone, and consequently children remain charmingly naive about some fundamental biological facts. It takes children a while to learn that plants are alive, and some children mistakenly believe that cars and even buttons are alive.[5] Children also have limited understanding of illness. Four-year-olds think that bad moral behaviour is as likely to make you vulnerable to getting sick as poor diet.[6] Comprehension of biological inheritance is underdeveloped as well. Three-year-olds mistakenly believe that a black baby can grow into a white adult, and that occupation is as likely to be inherited as ethnicity.[7] Clearly some basic facts about biology do not come naturally. Facts that are difficult to observe require instruction.

What about Keil's finding that children realize that you can't make a porcupine into a cactus? This doesn't seem to be the kind of thing that could be learned by observation because porcupines and cactus can look very similar. Nor does it seem to be explicitly taught; the children in Keil's study are pre-schoolers. Might this knowledge be innate?

Probably not. First of all, infants don't seem to have any trouble imagining one kind of thing transforming into another. Fei Xu and Susan Carey showed infants a display in which a little truck drives behind a screen and a cat comes out the other side.[8] Then in the test phase of the experiment, the screen is lifted revealing both a cat and a truck or just one of those objects. Adults expect to see both because we know trucks can't transform into cats. But, before the first birthday, this fact isn't appreciated. Infants are not surprised to see one object instead of two.

Between one and pre-school age (three to five), children learn some important things. They learn that appearance and reality can come apart. A stuffed puppy isn't a real puppy, a plastic banana isn't really a banana, and Daddy is still Daddy when he puts on a silly mask. In each case, what matters is what's beneath the surface: stuffed animals don't have icky innards, plastic bananas have no fruit under their skin, and Daddy looks like Daddy when he pulls off the silly mask. With piles of toys and hours of make-believe, children learn that appearances can be a bad guide to reality. It's not at all surprising, then, that pre-schoolers resist saying a porcupine is a plant when it's made to look like a cactus; it's still a porcupine inside. The coffee pot/birdfeeder case is different for two reasons. First of all, with artefacts appearance usually is reality: a toy teacup really is a teacup. So kids don't learn to mistrust appearances in this domain. Second of all, in the coffee pot/birdfeeder case, the insides change: the coffee pot has coffee in it initially, and birdseed is put in after the transformation. So kids are happy to say it's now a birdfeeder. One interesting finding that came out of Keil's research is that pre-schoolers make systematic errors on one kind of transformation: when you make one animal look like another, they say its identity has changed.[9] A raccoon painted to look like a skunk is a skunk. This error may stem from the fact that, for all a pre-schooler knows, raccoons and skunks – unlike porcupines and cactuses – have the same kind of stuff inside. Once they enter school, they learn that this is a mistake, because each species has different stuff inside. In this way, children work their way up, using observation and instruction, to a rich understanding of biology. They don't need innate understanding of this domain.

Infant Psychology

When we see a person walking down the street with an umbrella, we usually draw two inferences. We think there is rain in the forecast, and we also think that the person believes that it's going to rain. If we've seen a forecast for sunny weather, we might drop the first inference and conclude that the person has a false belief that it will rain. We are very good at inferring beliefs from behaviour. We also infer desires (the woman in the Starbucks line must want a coffee), emotions (the giggling child must be amused), intentions (the man turning a jar lid effortfully is trying to get it open) and myriad other mental states. We are intuitive psychologists, spontaneously imagining what is going on in other people's heads. This ability is impaired in some individuals with autism and may be absent or underdeveloped, even though they are good at observational learning. This suggests that the ability is not learned, but rather innate. Further evidence for innateness comes from studies of infants.

Consider a study by György Gergely and his collaborators, in which year-old infants first spend a few minutes watching an animated ball that rolls along a surface, then leaps over a barrier and finally rolls into a second ball lying on the other side of the barrier.[10] When adults see the animation, they spontaneously attribute a goal to the first ball: it wants to make contact with the other ball. Infants seem to make the same attribution. In the test phase of the experiment, Gergely showed the infants two variations on the original animation. In one, the barrier is removed, and now the first ball rolls directly along the surface in a straight line and makes contact with the second ball. In the other variation, the barrier is also removed, but this time the first ball does not roll directly into the second. Instead, it leaps in the air in the location where the barrier had been before, and then rolls into the second. To adults, this is puzzling; if the first ball wants to make contact with the second, it should take the shortest path once the barrier is removed. Infants also seem to be surprised. They stare longer at the ball leaping over empty space than at the ball that rolls directly along the surface, even though the movement pattern of the leaping ball is very similar in appearance to the initial animation that they were watching in the first part of the study. If infants were just viewing the animation as randomly moving shapes, they should stare longer when a leaping movement is replaced by a straight one. But they stare longer at the leaping movement, suggesting that they are attributing goals to the ball, and they expect the leaping to stop once the barrier is removed.

This is an impressive study, but not decisive. By the time of their first birthday, infants have often seen balls rolling along surfaces, so the movement observed in the second test film is extremely familiar. They have never seen a ball spontaneously leap over empty space, so the movement in the first test film is entirely anomalous. This could explain why infants are more surprised by that film. Of course, the infants in the study did watch a leaping ball just before the test films, but the leaping ball in the first part of the study is leaping over a barrier. When the barrier is present, the ball looks as if it is rolling over a surface, albeit a surface with a barrier jutting out. The ball does not leap into open air, but instead hugs the surface of the barrier. If infants see this as a ball rolling over a surface, they should expect the ball to roll straight when that surface flattens out. For comparison, imagine that you are watching water drip down a window that has a wad of chewed gum stuck to it. The water will go around the gum. When the gum is removed, however, you will expect the water to go straight, because nothing is obstructing it. In making this prediction, you don't need to attribute any goals to the water. Likewise, infants may not be attributing any goals to the rolling ball.

Gergely's experiments may be flawed, but there have been other efforts to establish goal attribution in infants. Consider an influential experiment by Amanda Woodward.[11] In the first phase of her study, five-month-old infants watch as a hand reaches for one of two objects, a ball or a teddy bear. In this phase the objects are always in the same location, and the hand always reaches for the same object (e.g., the hand may reach for the ball, which is on the right). In the second phase, the two objects swap locations, and the infants see one of two test scenarios: either the hand reaches for the same object that it had been grasping before, despite the change in location, or it reaches for the other object, which now occupies the location where the earlier sought object had been. If infants attributed a goal to the hand in the initial phase (e.g., that it must want the ball), they should be surprised if it suddenly reaches for a new object. If infants merely experience the hand as a moving object with no objectives, they should be surprised if it changes its trajectory. Like adults, infants are more surprised by the former than the latter: they expect the hand to reach for the same object after that object has moved. In a clever control condition, Woodward performed the same scenarios but replaced the human hand with a shiny pole. Infants expect the pole to continue moving to the same location regardless of what object is there. Woodward concludes that infants attribute goals to animate, but not to inanimate, objects.

This study overcomes the problems in Gergely's experiment, but it suffers from another flaw. The set-up pits two visual features against each other: a relation between shapes (hand grasps ball), and a motion trajectory (hand reaches to the right). Infants may store both of these features in memory, but the most salient of the two may be stored more vividly, leading to more robust expectations. Now it may be that the relation between shapes is more salient than the trajectory, because it is more visually complex and interesting. The balance tips, however, when the moving object is vibrant and unusual. A shiny pole is a bizarre object that infants have never seen. So they may not pay much attention to other, more familiar objects when the pole is around. They may not focus on how it is related to other objects. They stare at its intrinsic properties instead, including its motion trajectory. Thus, the difference between hands and poles may have nothing to do with goal attribution and everything to do with comparative perceptual salience of relational properties versus intrinsic properties. To test this hypothesis, one could do things to make the hand look more unusual. If the hand were less familiar, it would capture attention, and its pattern of motion would become more salient than its relation to other objects. By good fortune Woodward, together with Jose Guajardo, performed this crucial control. They redid Woodward's original study, but, this time, the hand was wearing a shiny glove. Now, the infants expected the hand to retain the same movement trajectory more than they expected it to grasp the same object. Thus, they are not attributing goals; they are merely picking up on perceptual regularities and forming expectations based on which regularities are made most salient.

Let me consider one more experiment which builds on Woodward's strategy, but adds a nice flourish. Luca Surian and his collaborators showed thirteen-month-olds an animation in which a caterpillar watches as an apple and a wedge of cheese are placed behind two different barriers, and the caterpillar then crawls behind the barrier on the right, to consume the food that was placed there.[12] Then, in the test phases, the apple and the cheese are switched, and the question is, will the caterpillar pursue his food of choice in the new location or retain the same movement trajectory despite the fact that the food there has changed. So far, this is a lot like the Woodward study, but there is an interesting twist. Surian includes two versions of the test phase. In both the caterpillar is absent when the foods are switched, but in one condition, the barriers are very short. When the caterpillar arrives on the scene, it can clearly see that the foods have been relocated. In the other condition, the barriers are tall and the caterpillar cannot see that the foods have been relocated. Thus, the experiment is not only testing whether infants will attribute goals. It also tests whether infants know that a creature must *see* its goal in order to *know* where it is located. Infants must attribute seeing, knowing and wanting to perform correctly. And they do perform correctly. Infants are surprised when the caterpillar preserves its original trajectory when it can clearly see that the foods have been switched, and they are not surprised when the caterpillar preserves its trajectory when the switch cannot be seen.

Again, though, there are flaws. First of all, there is something puzzling about the results. In the condition where the caterpillar cannot see that the food has been switched, Surian's infants show *no preference* between the original movement trajectory and the new one. That is very hard to explain on the assumption that infants understand what the caterpillar is thinking. If the caterpillar is unaware of the switch, it should go to the *original* location. Infants don't form that expectation. They seem to be unsure which way the caterpillar will go. This is easy to explain if infants are going by perceptual salience rather than goal attribution. After the switch, two perceptual features from the original scenario have been pitted against each other: the rightward movement and the movement towards a particular food item. Unlike in the Woodward study, where a very familiar object (the hand) or a very bizarre object (the pole) can make relational or intrinsic features more salient, there is nothing in the Surian study to tip the balance. The caterpillar is neither bizarre nor completely familiar.

This worry does not address Surian's most important finding, however. Infants expect the caterpillar to change its earlier trajectory when the new food locations are clearly visible. Surian suggests that this can only be explained by assuming that the infants attribute a food goal to the caterpillar, and that they know it can see that its favourite food is in a new place. But there is another explanation of this result. Remember, in the perceptual saliency account, there are two salient visual features: rightward movement and movement towards a food item. A moment ago, I said these features are equally salient. But things change in the test condition under consideration. In this condition, the large barriers that were in place in the first phase of the experiment have been shrunken down dramatically. As a result, the two food items are much more visible than they were before. The inevitable result is that the food items become more salient. This, I suspect, is what tips the balance. When the foods are revealed in such an obvious way, it triggers infants' memory that the caterpillar was moving towards a particular food item a moment earlier, and that memory becomes more active than the memory of the movement trajectory. Infants are surprised when the caterpillar moves towards the other food item, not because they attribute any goal, but because food is made salient in this condition.

There are many other experiments attempting to prove that infants attribute mental states, but these are among the most influential and compelling. On scrutiny, however, they are unconvincing. Infants' looking patterns can be explained by perceptual salience, rather than mental state attribution. Actually, this is no surprise. A vast literature suggests that children are pretty bad at attributing mental states until they are three or four years old. At three, they start attributing goals, but they are still bad at attributing beliefs. In particular, they systematically fail to attribute false beliefs – beliefs that contradict what they know to be true. If a three-year-old knows that there are pencils in the cookie jar, they assume everyone else knows this too, even though they should recognize that others will erroneously believe that there are cookies in the cookie jar. False belief attribution is mastered when kids are four. Therefore, we have reason to be very suspicious of any study purporting to show sophisticated mental state attribution in infants. There is extensive evidence that these abilities emerge slowly, suggesting that they are learned.

There is one final argument for innateness that we have yet to consider. As noted earlier, individuals with autism have difficulty with mental state attribution. Autism is often characterized as a deficit in the comprehension of psychology – a kind of mind-blindness. Despite this, some individuals on the autism spectrum are very high-functioning in other cognitive domains. This suggests that general intelligence is not sufficient for learning to attribute mental states. And if general intelligence is not sufficient, then perhaps mental state attribution is not learned at all; perhaps it is innate.

The problem with this argument is that it assumes that individuals with autism have no other impairments. That is not the case. Even high-functioning people with autism have various other symptoms, which include everything from low-level perceptual abnormalities to high-level cognitive abnormalities. At the low level, people with autism are characteristically hypersensitive to sensations. Even mild sensations can seem intense, painful and distressing. At a high level, people with autism have difficulty with various executive functions, including planning, inhibition and flexibly changing goals. People with autism also have difficulty integrating information. In perception, they tend to perceive details better than whole patterns, and, when presented with information to think about, they are more likely to recall specific elements than the gist. It's still unknown whether these symptoms are bound by some common underlying cause, but it is clear that autism is not merely a deficit in mental state attribution. Indeed, many people with autism perform well on standard mental state attribution tasks, especially when their language skills are intact, and they are pretty good at understanding the minds of other people with autism – they just find the rest of us a bit puzzling. People with autism can also be very sensitive to their social surroundings, catching emotions from those around them.

The most noticeable social problem in autism is a kind of failure to connect in an immediate, non-verbal way with other people. When you encounter a person without autism, you may look each other in the eye, catch a glance, exchange a knowing smile or pay attention to what that person is looking at – what is called joint attention. These silent forms of engagement are often absent in encounters with autistic individuals. The reason for this is not fully understood, but various factors that have little to do with social cognition may be the root cause. People with autism may find it aversive to attend to another human being because of their general tendency to be over-sensitive to sensations; human beings are intense stimuli, and those moments in which you catch someone's eye may be unpleasant for people with autism. Another possibility is that the difficulty with information integration is to blame. When I see you looking at something, I need to attend to you and to what you are looking at, then I attend to how the thing you are looking at makes me feel and finally I project that feeling on to you. This is a complicated cognitive feat, which, as far as we know, no other species can achieve. Other animals follow the gaze of their conspecifics, but we do something more. We simultaneously attend to our own inner states and use that information to guess what another person is thinking or feeling. Attending to the inside and the outside at once involves integration of two different cognitive systems and this is just what people with autism have difficulty with. They may know what they are feeling, and they may be able to think about what others feel, but bringing these together here and now, while attending to what another person is looking at, may be a challenge for them.

The failure to connect with other people in this immediate way – which may result from hypersensitivity or an information integration deficit – has profound consequences. It means that people with autism are less aware of, and less interested in, the mental lives of other people, and this can impede their acquisition of mental state attribution skills. If this story is right, there is no need to postulate an innate folk psychology mechanism that is malfunctioning in autism. Highly general problems with perception and cognition result in diminished social engagement, and this, in turn, manifests itself in social awkwardness, inattention to other people and underdeveloped skills in mental state attribution.

Infant Mathematics

Let's consider one final domain that has been regarded as a department in the innate university. We all spend many hours in maths classes learning multiplication tables, solving geometrical proofs and calculating values for variables. These skills don't come naturally to everyone, and many branches of mathematics are fairly recent human inventions. But the same guy who flunked calculus will easily notice when a fellow diner surreptitiously swipes one of the four cookies on his plate. Some calculations are easy and automatic. Simple arithmetic can even be carried out by infants and a wide variety of non-human animals. This has led some authors to conclude that we have an innate number sense.

The first piece of evidence that infants have an innate mathematical ability is that they are sensitive to the number of items in a display. They can detect the difference between two and three items, for example, so that after looking at groups of three dots they are surprised to see a group of two dots. Infants are less good at discriminating large numbers, but Fei Xu and Elizabeth Spelke have shown that infants can, in fact, discriminate two large clusters of dots if one has twice as many dots as the other.[13] With large quantities, infants lose track of the exact number, but they can discriminate 2:1 ratios.

There is also evidence that infants keep track of numbers in a very abstract way. Numbers are said to be abstract because the same numerical quantity can apply to things that have little in common physically: 12 cookies in a dozen, 12 steps to fight addiction, 12 tribes of Israel, 12 apostles, 12 inches in a foot and 12 months in a year. We can apply numbers to just about anything. To show that infants have an abstract concept of numbers, Prentice Starkey, Elizabeth Spelke and Rochel Gelman devised an ingenious experiment.[14] They placed six-month-old infants in front of two photographs showing a different number of objects and simultaneously played drumbeats corresponding in number to one of the two photographs. So, an infant might hear two beats while presented with one picture of three objects and another picture of two objects. Starkey and his collaborators discovered that infants stared longer at the picture with the number that corresponded to the number of beats. They concluded that infants have a concept of number that carries across different sense modalities and different kinds of things. Infants can perceive twoness as such, not just two dots or two sounds.

Infant numerical abilities do not end there. Not only can they sense numbers, they can also add and subtract. In one experiment, Karen Wynn showed five-month-olds an object and then covered it with a screen; after that, she placed another object behind the screen in clear view of the infants.[15] When she removed the screen, infants saw either two objects (the correct outcome) or only one. They stared longer when there was only one, suggesting that they had correctly added the first object to the second and expected to see their sum. The effect also works with subtraction. If one item is removed from a small array, infants expect to see the number of items appropriately reduced.

These findings have led researchers to conclude that infants have a rudimentary understanding of numbers and arithmetic. In short, an innate mathematics. But this research – and we have only seen a small sample here – has also been challenged. For example, some critics have argued that Karen Wynn's studies may have nothing to do with arithmetic. Instead, infants may notice a change in contour size or spatial area, rather than specific numerical changes. When one object is added to another, the size of the display grows. More recent studies have tried to control for these other factors, but there is another problem with Wynn's study, which has been raised by Melissa Clearfield and Shannon Westfahl.[16] Recall from the discussion of infant physics that infants sometimes stare longer at familiar objects than at unexpected events. Preference for familiarity drops off only after repeated exposures, which eventually result in boredom. Now consider what infants see in Wynn's addition study (similar points apply to subtraction): they see one doll, followed by a screen, and then another doll appears and is put behind the screen. Wynn assumed that infants experience this as one continuous event in which one item is being added to another. But infants may experience it as a sequence of disconnected events: they see one object, then a screen, then one object again. When the screen is lifted, infants stare longer at the single object, but this may result from the fact that seeing one object is the familiar event, and they are excited to see it again, because they haven't been rendered bored by repeated exposures.

Even if Wynn's infants are keeping track of quantities, it doesn't follow that they have a maths sense. Notice that displays with different numbers of objects are perceptually different. An array of two objects looks different to an array with one. This difference has an impact on perceptual processing. The visual system identifies objects by their contours, and each object can be attended to individually, or both can be attended to at the same time. If infants are attending to two things, and one disappears, they may search for the one that has gone missing. This shows sensitivity to quantity, but it would be misleading to call this innate mathematics. It is just an innate ability to attend to multiple objects. Put differently, the infant does not care about the precise number of objects in a display, but she does notice when something she is paying attention to disappears. An explanation of this kind has been put forward by psychologists Brian Scholl and Alan Leslie. They point out that human beings can only attend to about four objects at once, and that is why infants' arithmetical ability is limited to four items; it is not an innate number sense, but rather an innate capacity to keep track of several objects at the same time.[17]

We have seen, however, that infants can perform successfully on some tasks involving large numbers. In particular they can detect when a large array has been doubled in size or cut in half. This capacity cannot be explained by attention mechanisms. But neither should we conclude that it derives from an innate mathematics. For one thing, it would be very strange if an innate system that evolved for dealing with quantities could do nothing more than discriminate when a group is doubled or halved. Keeping track of 2:1 ratios is not an especially valuable skill. For another thing, it turns out that the 2:1 ratio is pervasive in quantity estimation. Infants and adults are good at telling when images double in size or when sounds double in length or double in volume. This proportion has nothing to do with numbers in particular. It is not part of a maths sense. Rather it seems to be a fundamental feature of how the nervous system makes quantitative comparisons along any perceivable dimension.

To establish an innate mathematics, it is important to show that infants are sensitive to numbers and not just that they are capable of discriminating different quantities along a dimension that happens to be perceivable. Maths is not a matter of seeing three things or even seeing that three things are more than two things. It is a matter of seeing that the number of things in a group is three. It's about being able to assign precise numerical quantities. This difference is subtle, but important. Seeing three things and seeing that there are three things are very different abilities, because the latter, but not the former, requires a numerical concept.

For this reason, the most powerful evidence of an innate mathematics comes from the study by Starkey and his collaborators in which infants match the number of drumbeats to the number of items in a picture. This study purports to show that infants have an understanding of numbers as such. They can recognize threeness. The problem is that these studies have been incredibly difficult to replicate. In attempts to redo the Starkey experiment, infants sometimes stare longer at the picture that fails to match the number of beats, and in other attempts there is no preference.[18] Outcome is sensitive to the duration of the beats, the items in the images and presumably many other variables that haven't been identified. Furthermore, four-year-olds have tremendous difficulty matching quantities across sense modalities, casting doubt on the claim that this can be done successfully by infants. Perhaps Starkey et al. just got lucky in their study. Some as yet unknown feature of their experimental set-up led infants to match certain sounds and pictures, but it was probably not sameness of number. With slightly different sounds or pictures, the effect can reverse or disappear. Performance improves when the words and pictures are associated (such as faces and voices), but this can be explained without appeal to an abstract notion of number. Infants have learned that faces emit voices, and they have learned that, when two voices are present, two faces are as well. Searching for visual inputs that are associated with auditory inputs does not require number concepts.

This is only a small sample of the literature on infant mathematical abilities, but the points here generalize. Success with quantitative tasks in infancy does not establish an innate mathematics. Infants are sensitive to perceivable differences in quantities and they are capable of noticing when an object they are attending to disappears or when a new object is added. They can also do some matching across the senses for familiar objects, and they expect to see a matching object for each familiar sound they hear. But that does not mean that they can discern the number of objects they are perceiving. They can perceive two things, but not that there are two things.

Of course, we eventually acquire number concepts. We go from perceiving two things to perceiving that there are two things. Our ability to do this may be acquired when we learn to count. Verbal labels are an incredibly efficient way of keeping track of quantities. A language like English can label every finite number. Initially, when we learn to count, we may not realize that number words correspond to quantities. But we soon learn that each word can be mapped on to an object. We can put one potato in a pot, then two potatoes, three potatoes, four . . . Once we master number words and learn that these can be used to count, we become aware that groups of objects can be precisely quantified. At that stage, around the fourth birthday, the concept of numbers starts to take hold.

The Blank Slate?

I have been arguing against the idea that the infant mind is organized like a university, with innate knowledge domains corresponding to different departments. It is difficult to argue for a negative conclusion.

The best strategy is to review the positive evidence for the university model and critique it, but there is no way to do that inclusively, since there are literally thousands of studies designed to show innate knowledge in infants. Here, I have simply reviewed a handful of the most famous studies and indicated why they are inconclusive. Other studies have been performed to address some of the objections surveyed here, and still other critiques have been offered to address the revised studies. And so it goes in science, with epicycle upon epicycle of critique and revision. This is a healthy practice. It leads to improved experimental methods, and the cases for both Rationalism and Empiricism become more sophisticated and richly informed. Where Descartes and Locke could just speculate in vague generalities about how infants learn, contemporary psychologists are beginning to tell detailed stories of what is learned when, and how each incremental step is achieved. Even if neither extreme Rationalism nor extreme Empiricism can be sustained in the end, it is valuable to keep these positions alive, because the two sides keep each other honest, and the debate fuels good science.

I view myself as a methodological nurturist, which means that I try to assume things are learned until proven otherwise. It's scientifically useful in these Rationalist times to criticize experiments that seek to prove that knowledge is innate. One can find flaws in any study and recommend alternative interpretations of any result. This is no mere parlour game. Each objection can lead to new studies, and each study can deepen our understanding of how the mind works.

Proponents of the infant university sometimes cry foul when they listen to nurturist critiques. There is always some possible Empiricist explanation of any finding, but it seems like cheating when each study is subjected to a different objection. Without a systematic reason for doubting the Rationalist programme, these local skirmishes look like desperate attempts on the part of Empiricists to explain away a growing tide of evidence that has emerged through a massive research programme guided by a coherent theory of how initial knowledge is organized. Why should we even doubt the assumptions underlying such a productive scientific enterprise?

The answer to this question is that Rationalism focuses so much on innateness that it tends to overlook the obvious fact that the overwhelming majority of what we know is learned. It's as if developmental psychology has forgotten all about development and assumes that knowledge is already in place. There is little effort to explain how we go from the knowledge alleged to be innate to adult competence. Empiricists saw that everyone needs to work more on developing good theories of learning, and once we have such theories, the temptation to posit innate knowledge may subside. To me, the most striking fact about human beings is that human babies are so profoundly dumb. They are cute, of course, and curious, but, in terms of intellectual abilities, babies seem less like their parents and more like the family pet. Even if we grant the innate university story, infants' minds are astonishingly undeveloped given the extraordinary intellectual feats of human adults. How do we advance from drooling, babbling lumps into physicists and philosophers? This looks like the kind of enthralling scientific mystery that developmental psychologists should be labouring to solve, but all too often the focus is on how brilliant babies are, not how dumb. Empiricists and Rationalists should join forces in the effort to explain our ascension from the cradle.

The Rationalist might concede the point, but argue that Empiricist learning theories will never succeed because we need innate knowledge to make our monumental intellectual advances. We could never become physicists, they will say, if we didn't have an innate rudimentary physics. But this argument would have force only if it could be shown that rudimentary knowledge of physical principles cannot be learned, and it is here, more than anywhere, that the weakness of the Rationalist programme becomes clear. To prove that core knowledge must be innate, Rationalists would need to show that the kind of knowledge they attribute to infants is of a type that would be impossible to learn by observation. But defenders of the innate university model rarely make any effort to do this. In fact, it is easy to imagine that the knowledge they attribute to infants is learned.

Infants live in a world rich with information, and they have powerful systems of perception, association and memory, which can organize sensory inputs into separable parts, discern relations between those parts and store the results for future planning. Consider one simple example from the start of this chapter. How does an infant know that a potato continues to exist when it is briefly taken out of view? Must that knowledge be innate? Not necessarily. From the very start of life, infants who can see experience numerous occasions when an object they are watching disappears from view. In fact, this happens every single time they blink – over 10,000 times a day. When eyes close and reopen, the world remains unchanged. Hypothetically, infants could form the belief that the world disappears with each blink, but that's a pretty sophisticated inference that requires concepts of inexistence. It's very unlikely that they can entertain such an idea. In fact, they probably can't explicitly entertain the idea that objects persist. Rather, they acquire a simple expectation that, when they open their eyes, objects will be in place. Likewise, each time an object passes behind another or gets engulfed in a shadow or occluded by an infant's own hands, it reappears a moment later, so the expectation of persistence is reinforced. No one knows exactly how long it takes for this pattern of observations to sink in and generalize to new cases. The point is that infants have ample opportunity to discover that objects persist. Their expectations are informed by experience of a world in which objects rarely flicker out of existence.

Likewise for other principles. Infants see objects fall, they witness animals and artefacts in motion, they experience their own mental states and the resulting effects on behaviour and they watch as items in groups are added and subtracted. The rudiments of physics and biology seem no harder to learn than the myriad of other facts that we get by observation: that grapes are sweet, that paint is messy, and that two heaps of sand can combine to form one larger heap. Methodological nurturism says we should assume core knowledge in these domains is likely to be learned, because the world is full of information. Naturists owe us more arguments for thinking we need specialized innate knowledge to pick up on all the regularities that are manifest around us.

Am I saying the mind is a blank slate? Of course not. The human mind is equipped with powerful mechanisms for learning, and cognitive resources for putting accumulated knowledge to work in deliberation and problem solving. But the metaphor draws attention to the possibility that we may come into the world without knowledge of any object, category or domain. We are not born knowing who God is, what puppy dogs are or the basic laws of physics. Instead, we are good learners.

Consider the visual system, which is exquisitely designed to extract information from light. Vision uses light discontinuities to find edges, and it binds edges together to discern contours of objects, and it extrapolates the distance of those objects by calculating disparities between the images coming in from the two eyes. The visual system also uses temporal sequences of adjacent visual patterns to follow objects as they move. In this way, objects can be perceived, identified and tracked over time and space. The visual system can also store information in visual memory, keeping records of the frequencies with which visual features are associated. Stored records are used for classification. When we see an object, a trace is stored and matched against future objects. When a near match is found, a trace of the new instance is stored, and eventually we amass a record of many similar items. Each time we encounter a new instance, we can make predictions about it based on past cases. We are also innately disposed to generate prototypes of categories. The stored records of similar-looking stimuli are averaged together to form a representation of the most typical category instance, and this prototype can be used to facilitate future categorization. Prototypes can also be used as a kind of summary representation that can be brought before the mind to efficiently think about the category in its absence. All of our senses can do these kinds of things, and we can also store associations between our senses. Some of those associations may even be innate; infants may not need experience to learn that pointy looking objects will feel sharp when touched.

We are also born with faculties for using these inputs from the senses. We can focus attention on things, track objects over time, recall them in their absence and imagine them combined and transformed in various ways. We are extraordinary simulators. After seeing an orange cat on a green mat, we can imagine a green cat on an orange mat. When deciding whether to reach for a glass, we can recall how far our arm extends and determine whether we need to move closer before attempting to reach. Human success in manipulating the environment may derive in part from our capacity to imagine things that haven't yet happened. The blank slate metaphor misleadingly implies that all creatures have minds that are alike. The human mind may be more flexible and more capable of operating independently of the stimuli that happen to be impinging on us at any given time.

But the blank slate metaphor may have two kernels of truth. First, it might turn out that we have little or no innate knowledge, even if our capacities for using knowledge are very powerful and unique to our species. Second, it might also be the case that human learning is largely domain-general. That means we don't come equipped with one set of learning mechanisms for physics, another for maths and a third for biology. Rather, general-purpose perception, attention and memory are used across all these domains. The infant mind is not a university, pre-parcelled into specialized subfields, but an active and hungry learner that discovers different domains through observation and investigation.

No one has developed a complete account of how a general-purpose learning machine could acquire knowledge of domains as diverse as mathematics and psychology, but there is little reason, at this point, to be pessimistic about such a story. The case for innate domains has been oversold, and research psychologists should be actively exploring the possibility that babies advance from a state of total ignorance to the kind of hyper-specialized knowledge that is a trademark of our species.

PHILOSOPHY'S PENDULUM

In 1690, John Locke published his *Essay Concerning Human Understanding*. It would go on to become one of the most influential books about the mind in Western history. For 200 years it was the most widely read investigation of human psychology. It was also revolutionary, because the dominant view before Locke was Rationalism. Locke's predecessors believed in a rich stock of innate principles and mistrusted experience as a source of knowledge. Locke was not the first to challenge this orthodoxy, but he was the most effective. He launched the modern Empiricist programme in philosophy and shifted the pendulum away from Rationalism, especially in the English-speaking world.

That pendulum rotated back towards Rationalism during the last half-century. Descartes got his revenge. Scientists interested in the mind adopted the Rationalist programme with unprecedented enthusiasm. In fact, many researchers began to see the study of innate knowledge as the central task with which the science of the mind should be concerned.

We have now seen that the pendulum is on a return course to Empiricism. The assumption that we have extensive innate knowledge rests on a large body of experiments with infants that can be reinterpreted in another way. Infant knowledge emerges over time, and the things infants know can be explained by appeal to what they experience.

5

Sensible Ideas

What is a thought? That is one of the central questions addressed by the sciences that study the mind. At some level, we all know what thoughts are from first-hand experience. We report on what we are thinking all the time. We say, 'I think it's going to rain,' or 'I think I'll have the Sancerre,' or 'I think Abraham Lincoln had two Vice Presidents?' But what are these things we're reporting? Clearly they are things inside our heads. To report a thought is to report on one of our psychological states. Thoughts describe or represent things. We think about weather, wine or politics. But beyond this, things become a lot less clear. What are these things inside our heads that represent the world?

If we look outside the head, we can see that there are two main ways of representing the world. One method is to use language. Words and sentences represent things. One can write a poem about the rain, a guide to wine and a historical novel about the Lincoln administration. Words are a powerful tool for recording facts. But we also have another tool. We use images. We can film the rain, photograph a glass of wine and paint a president. The images just mentioned are visual, but we can also record the sound of rain, concoct a perfume that smells like wine and grope a statue of Lincoln. The term 'image' can be used to refer to all these sensory records.

These external representations have inspired competing theories of mental representations. Some people have claimed that thinking is a lot like writing. We describe things using the mental equivalent of words when we think. Thoughts are like sentences in the head. The alternative view says that thoughts are more like depictions than descriptions. They are sensory records of what it was like – or would have been like, or will be like – to experience something. Let's see which of these options is more plausible.

LOCKE, LEIBNIZ AND LISP

In the last chapter, the history of Western philosophy was described as a pendulum swing from Rationalism to Empiricism, and back again. The perennial feud is at the heart of the present debate as well, the debate about the nature of thoughts. As we saw, Rationalists and Empiricists disagree about where knowledge comes from. Rationalists say that some knowledge is innate, and this innate knowledge is pre-parcelled into specialized domains and provides the foundation for information we learn during our lifespan. Empiricists reject this picture. They argue that knowledge is acquired by observation, and the same methods of learning ground knowledge across diverse domains. These competing accounts of how knowledge is *attained* have also come to be associated with competing accounts of what knowledge *is*, and these result in different theories of what thoughts are.

Aristotle launched Empiricism in his book *De Anima* with the pronouncement that, 'No one can learn or understand anything in the absence of sense.' This was reformulated in medieval times by Aristotle's devoted follower Thomas Aquinas, who says in his *Summa Theologica*, 'There is nothing in the intellect that is not first in the senses.' In other words, perceptual experience is a necessary precondition for knowledge. Aristotle's famous remark continues on with a further conclusion that became equally important to Empiricist philosophers, 'When the mind is actively aware of anything it is necessarily aware of it along with an image.' What Aristotle means here is that knowledge acquired through the senses is stored in the form of mental imagery. Mental images are stored records of perceptual experiences. When we perceive things, we have sensory experiences, and these can be recorded in memory and recalled on future occasions. If all knowledge originates in the senses, it's natural to think that all knowledge takes the form of mental imagery.

John Locke resuscitated Aristotle's thesis in the seventeenth century. His *Essay Concerning Human Understanding* begins with a fierce attack on the hypothesis that people are born with innate knowledge. After that attack, he needs to say where knowledge comes from, and here he follows Aristotle quite closely: knowledge comes from the senses, and it is stored in the form of mental images. Locke calls these images 'ideas', but it might be better to call them sensible ideas, to emphasize that 'ideas' are stored copies of sensory experiences.

Locke's book had many admirers, but there were dissenters as well. Locke's most able critic was Gottfried Leibniz, the man who invented calculus (along with Isaac Newton). Leibniz was so provoked by Locke that he wrote a tome called *New Essays on Human Understanding*, which was intended as a line-by-line critique. Leibniz had fallen under the influence of Descartes, and he became one of the most influential defenders of Rationalism. Like any good Rationalist, Leibniz believed in innate knowledge. By definition, innate knowledge is knowledge that we possess prior to experience, and, if it precedes experience, it is presumably stored in some format that is not experiential in nature. Thus, Leibniz felt compelled to reject Locke's claim that we think in sensible ideas or images. Instead of images, he believed that ideas are like definitions: descriptions that can be broken down into simpler and simpler features until we get to a set of primitive features that can be broken down no further. These unstructured primitives cannot be images, according to Leibniz, because they can be possessed prior to experience. A natural suggestion is that the primitive elements that combine to form thoughts are like symbols in a language. This conclusion was attractive to Leibniz, but he did not conclude that we think using the languages we speak. Instead, there must be an innate language – what medieval Rationalists had called a *lingua mentis*.

The idea that we have an innate mental language, or 'language of thought', had been a theme in Rationalist philosophy for centuries, but it was rarely emphasized. Empiricists wrote whole treatises speculating about how we think in images, but Rationalists did comparatively little to adumbrate or defend the language of thought hypothesis, and many, including Descartes, were happy to concede that imagery plays a central role in thought. The situation changed in the middle of the twentieth century, when Rationalism had its most recent revival.

The reason for this new emphasis on a language of thought can be stated in one word: computers. When Rationalism came back into fashion in the late 1950s, computers had come on the scene. They were the newest and most important technology. Computers had played a crucial role in the Second World War and were now trickling into an increasingly large number of civilian industries. But computer scientists were not just interested in business applications. They believed that computers were potentially capable of solving any problems that we humans could solve. Pioneers devised programmes for playing chess, solving freshman calculus problems and answering the kinds of analogy questions that are found on IQ tests. The field of Artificial Intelligence, or AI, was born.

The breakaway success of AI in the 1950s resulted largely from a single innovation. Computer scientists devised a way of programming computers using simple, language-programming codes. The earliest of these were FORTRAN, IPL and its more powerful successor, LISP. Early successes in AI popularized the idea that we might some day be capable of making computers that think. But these successes also spawned another revolutionary idea. If computers can solve the kind of problems associated with human intelligence, then perhaps we solve them in just the way that computers do. Perhaps the human mind functions like a computer. Since computers of the period worked using language-like codes, researchers began to suspect that human thought might be based on an inner language as well.

This idea came into sharp focus on 11 September – not 2001, but 1956 at a conference at the Massachusetts Institute of Technology. On that day, papers were presented by George Miller, the first psychologist to measure human short-term memory capacity, Noam Chomsky, the linguist who proposed we have an innate language faculty, and Allen Newell and Herbert Simon, computer scientists who presented an IPL programme that could solve proofs in logic. 11 September is, therefore, said to be the birthday of 'cognitive science', the name that would later be given to interdisciplinary studies of the mind. Psychologists, linguistics and computer scientists were being brought together to share ideas, and that early summit spawned decades of collaboration. The specific talks given that day also did much to shape the way cognitive science would develop over the half-century that followed. In particular, it established the analogy between minds and computers.

The computer analogy fits beautifully with Miller's work on memory: the mind, like a computer, stores information and can bring several items from long-term storage into active use while solving problems (what we now call ROM, or read-only memory). The computer analogy also fits perfectly with Chomsky's views, since computers used pre-programmed rules that were linguistic in form, like Chomsky's innate grammar. Thus, in one day, cognitive science was born, the Rationalist faith in innate ideas was resuscitated, and the idea that human beings think using language-like symbols took hold.

Rationalism and the computer analogy have been deeply entrenched in cognitive science since the get go. It's not surprising, then, that there has been resistance to the idea that people think using mental images. The first generation of cognitive scientists worked hard to debunk Locke's theory of sensible ideas. They offered new arguments in favour of the hypothesis that we, like our home computers, think using an inner language.

THE LANGUAGE OF THOUGHT

The language of thought is said to have four features that distinguish it from sensory representations.

First, it is presumed to be *amodal*, which means different from the way we represent things in any of our senses. The senses are called sense *modalities* in psychology, so the word 'amodal' means not-sensory.

Second, the language of thought is said to be *abstract*, which means that it can represent features of the world that have no uniform appearance. Consider colours. Red, blue and yellow look different, but they are all colours. In the language of thought, there is hypothesized to be a symbol that represents the property of being coloured without specifying any specific colour. A sensory code would presumably lack such a symbol. If we visualize something as being coloured we need to visualize a specific colour, or a range of different colours. The language of thought can also include symbols that represent things that are too lofty to easily visualize, such as truth or justice.

Third, the language of thought is also presumed to be *unconscious*. It is not like English, French or Swahili. Those are languages we can see or hear. When a thought runs through your head in English, you hear the words. It is not amodal, but rather auditory. But the language of thought is amodal, so it can't sound like anything or look like anything. When there is a language of thought sentence in your head, there is no corresponding conscious sensation associated with it.

Finally, the language of thought is said to be *concatenative*. That is a technical term. It means that words in the language of thought combine together in such a way that each word takes the same form in every mental sentence in which it occurs. This is true in spoken languages too. The word 'birds' has the same form in the sentence 'Birds fly' and in the sentence 'Birds eat worms'. Sensory images are not like this. An image of a bird flying looks different from an image of a bird eating.

There are three main reasons why philosophers and psychologists have postulated a language of thought. The first is that the language of thought goes hand in hand with the hypothesis that we have innate knowledge. If infants had knowledge prior to experience, then it would be natural to suppose that that knowledge is not sensory in nature. If the knowledge in question could be captured in sensory imagery, then it would be simpler to suppose that it is learned rather than innate.

This argument loses force when it is discovered, as we've just seen, that the case for innate knowledge is weaker than often assumed. The kind of knowledge that has been credited to infants is exactly the kind of knowledge that infants could learn by observation, and it is also, therefore, exactly the kind of knowledge that can be grasped using mental imagery. Consider gravity. If there is innate knowledge of the principle that things fall without support, then it might be encoded in a language of thought. But this principle can be easily discovered through experience. Much to their parents' chagrin, infants have ample opportunity to watch objects fall. They topple cups over, throw food all around and drop toys on the ground incessantly just for amusement. These episodes teach infants that objects have a tendency to move downwards, and that knowledge can be stored in the form of visual images. After observing multiple objects fall, infants may come to predict downward motion for any object that isn't hanging on a wall, held in the hands or propped up on a table. Infants also experience falling in other sense modalities. They feel things slipping out of their hand and hear the impact when things hit the ground. All these sensory images can add up to an early understanding of gravity. Likewise for other principles alleged to be innate. In each case it is easy to imagine how the principles can be learned observationally and grasped by means of stored sensory records.

The second argument used to defend a language of thought begins with an analogy to the languages we speak. One of the most important features of spoken languages is that there is no upper bound to the number of sentences we can produce. With a finite vocabulary and a finite set of grammatical rules we can generate an infinite number of novel sentences. In fact, we are generating new sentences all the time. Try to take any sentence in this book and run a Google search on it. Chances are you won't find an exact match. Every once in a while, we produce a sentence that has been uttered before, but most are completely novel. Languages allow boundless productivity because they have a concatenative method of combination. Once you've mastered a set of words, you can use them to form entirely new combinations, because the words remain the same in each sentence in which they appear. If words changed form in each sentence, this would be impossible. We would need to learn a new rule for each sentence before we could form it correctly. That's what makes concatenative combination systems so powerful. This leads to a very powerful argument for a language of thought, which has been forcefully advanced by Jerry Fodor, one of the most influential living philosophers. Fodor points out that thinking, just like language, is productive.[1] Almost every thought we have is novel to some degree, and there seems to be no upper bound to the number of thoughts we can think. The fact that we can keep generating new thoughts suggests that we think using a concatenative symbol system. Mental images do not allow such productivity. If you have seen a bird flying and a dog eating, you cannot necessarily imagine a bird eating and a dog flying. To explain the productivity of thought, we must suppose that people think in a symbol system with elements that remain the same in every novel combination. We must recombine familiar mental symbols in new ways. That suggests we think in a language of thought.

This argument can be challenged. The key thing to notice is that there is no upper bound to the range of novel mental images we can form. We can imagine pink gorillas, rubber flowers and the sound of *Eine kleine Nachtmusik* played on a kazoo. Strictly speaking, imagery is not concatenative. When you combine an image of a gorilla with your image of pink things, both images change. The gorilla loses its original black colour, and the pink colour is imagined covering a surface that you've never seen it cover before – perhaps you've never seen pink hair. But that does not mean we need to see a pink gorilla in order to imagine one. We can effortlessly and automatically figure out which alterations are necessary when two images combine. A pink gorilla cannot be black, because something cannot be both pink and black at once. Of course, we will sometimes imagine things inaccurately if we haven't seen the corresponding objects. If you have never seen carnivorous plants, you might erroneously imagine that they have teeth to chew their food. Imagination is an unlimited resource, but not a perfectly reliable one. But this is not a reason to deny that we think using mental images. In fact, the hypothesis that we think using mental images correctly predicts that we will make errors based on prior experience. I'll never forget how disappointed I was when I first saw a flying fish. Moreover, imaginability can place constraints on intelligibility. If someone tells you that there are such things as carnivorous rocks, flying numbers or pink democracies, you will have no idea what that person is talking about, because you can't imagine what these things would be. The language of thought hypothesis falsely predicts that these should be perfectly intelligible ideas. Forming ideas should be as easy as forming English phrases if the hypothesis were true. But clearly it's easier to say 'pink democracy' than to conceive of what that might be.

There is still one argument for the language of thought to consider. One of the biggest challenges for the view that we think using mental images is that human thought is often very abstract. We can think about maths, logic, morality and the meaning of life. Mental images are concrete. They depict physical things in time and space. It's not clear how you can form an image corresponding to abstract ideas. What does justice look like? How can you paint the idea that life is pointless? Is there any way to visualize the basic logical principle that everything is identical to itself? The language of thought does not seem to face this problem. All these ideas are expressible in language, thus they could be expressed in a language of thought. Postulating a language of thought offers a promising explanation of our capacity to entertain extremely abstract ideas.

Or at least it seems promising at first. On close inspection, however, the language of thought hypothesis does not provide an adequate explanation of our capacity to think about abstract ideas. The problem arises because the language of thought is supposed to be made of symbols, and symbols have no intrinsic meaning. For example, the word 'hut' refers to a kind of dwelling in English but to a hat in German, and in Hungarian it's the verb to cool. It could really have had any meaning. We know the meaning by associating it with ideas. English speakers may imagine a small dwelling when they hear 'hut' and Hungarian speakers may imagine putting something in a refrigerator. How, then, do we understand a word in the language of thought? One answer is that the words are understood intrinsically without relating them to anything else. But this can't be right. Words are symbols, and symbols have no intrinsic meaning. So words in a language of thought must be understood by being related to something else, something like mental imagery. Consider the sentence 'Life is pointless'. How do we know what that sentence means? Presumably, we relate it to other things, like the sentence 'There is no reason to be alive'. But this is just another sentence, and we need to understand what it means if we are to make any progress on understanding 'Life is pointless'. We must break out of the linguistic circle somehow. Likewise, to comprehend sentences in a language of thought, we need to break out of that language and ground it in something that we can comprehend more directly.

No one has a complete theory of how we understand abstract ideas. The point I am trying to make is that the language of thought does not help us answer this question. If there is a language of thought, it cannot explain our capacity for abstract ideas, because sentences in a language of thought have no intrinsic meaning. If so, defenders of the language of thought are in no better position to explain abstract ideas than defenders of the view that we think in mental images. Both Rationalists and Empiricists owe us a theory. I will offer an Empiricist theory at the end of this chapter. For now, the main moral is that the existence of abstract ideas gives us no good argument for a language of thought.

If the foregoing suggestions are right, then all three arguments for postulating a language of thought are unsuccessful. There are also good reasons *against* the postulation. One concern is directly related to the fact that words have no intrinsic meaning. Like any words, words in a language of thought could not be comprehended without being related to something else – something that isn't linguistic in nature, such as mental images. That's not just true for abstract ideas, but for any ideas. Consider the word 'red'. For English speakers it refers to a colour, and, when we hear the word, we bring the colour to mind. If there is a synonym for 'red' in a language of thought, it too must be associated with a colour experience in order to be understood. But, if that's so, then why bother with the mental word in the first place? We should not explain thinking by postulating a language of thought if every sentence in that language is understood by relating it to something like a mental image. It would be simpler to just assume we think using mental imagery.

Another problem with the language of thought hypothesis is that there is no scientific evidence for it. There is no place in the brain where a language of thought is believed to reside. There are no brain injuries that lead to language of thought deficits. There are no thought disorders that have been accounted for by appeal to abnormality in the language of thought. Instead, there is massive evidence that mental imagery is used when we think. Let's look at some of that evidence now.

THE MULTIMEDIA MIND

The Empiricist philosophers of the seventeenth and eighteenth centuries have a very simple and elegant theory of the mind. We perceive things with our senses, and then we store copies of what we perceive to use in thought. Suppose you perceive a durian fruit for the first time. You will see its bumpy yellow skin, smell its pungent fragrance, feel its mushy flesh in your mouth and taste its sweet, garlicky flavour. This constellation of perceptions will be stored in memory, and they will allow you to recognize durians on future encounters and think about them in their absence. You might plan a trip to an Asian grocer in order to buy a durian. According to Empiricists, this plan will involve a sensory simulation. You will simulate, through mental imagery, what it's like to walk into a produce section and select a durian from among the fruit on display. You simulate the event in your head. This differs from the language of thought model, according to which you might make this plan with no accompanying imagery, by just uttering, in your mental language, I am going buy a durian.

The Empiricist theory went out of fashion in the second half of the twentieth century, but it is beginning to come back into vogue. In the past, it was defended by philosophers who relied on their own introspective reports to confirm that we think using mental imagery. Now, an emerging body of psychological evidence is offering confirmation that the Empiricists were right.

Many of the experiments in this new research programme work on a similar principle. Consider the following questions: 'Do birds have wings?' 'Are peaches sweet?' 'Do dogs bark?' These questions are trivially easy to answer because they involve familiar features of familiar objects. According to the language of thought hypothesis, we should be able to answer them without using mental imagery, because we just store this information in a giant symbolic list. We look up the mental word for 'dog' and see whether it is linked to the mental word for 'barks'. According to the Empiricist, we don't consult a list. Lawrence Barsalou, one of the major psychologists behind the Empiricist revival, says that understanding a category is a matter of attaining simulation competence: the ability to simulate what it would be like to perceive a category instance. In this view, we confirm that dogs bark by creating a polysensory simulation of a dog and listening to what sounds it makes in that simulation. Barking is not a mental word, but rather a stored acoustic record of a barking sound that we replay in imagination. Empiricists have devised numerous experiments to support this account of how people think.

In one experiment, Diane Pecher, Barsalou and René Zeelenberg asked people a series of questions about how familiar objects look, sound, taste, smell and so on and then timed how quickly these questions were answered.[2] They discovered that people are faster at answering a question about one sensory dimension, say sound, if they have just heard another question in that dimension. If they are asked whether leaves rustle right after being asked whether blenders are loud, they are pretty fast. But if they are asked whether leaves rustle after being asked whether cranberries are tart, they are slower. It is well established that there is a temporal cost when people shift attention from one sense to another, and this same switching cost shows up when people are asked these simple questions. That suggests that people are generating mental images to answer these questions, and they are slowed down when they need to imagine a feature in one sense and then switch to another sense. The language of thought hypothesis makes no such prediction.

In another experiment, Anna Borghi and her colleagues asked people questions about familiar features of cars: 'Do cars have steering wheels?' or 'Do cars have trunks?'[3] These questions should be equally easy, but, just before asking, the subjects in the experiment heard one of two sentences: 'You are driving a car' or 'You are fuelling a car'. Empiricists believe that we comprehend sentences by simulating them using mental imagery. If you have just imagined driving a car, the steering wheel will be much more vivid in imagination than the trunk, and conversely, if you have just imagined fuelling a car. Thus, Borghi predicted that people would be faster at confirming interior features of cars after the sentence about driving, and faster at confirming exterior features after the sentence about fuelling. This is just what they found.

Borghi showed that speed improved when answering questions about familiar objects. Nicolas Vermeulen and collaborators have shown that imagery can also slow performance.[4] This happens when what you are imagining differs from the feature you are being asked about. Suppose I ask you to keep a little melody in your head and then ask you whether blenders are loud. That's harder, Vermeulen showed, than if I ask you about whether lemons are yellow. The melody interferes with your ability to simulate blenders. Conversely, if I show you some shapes and ask you to remember them, it will slow down your ability to answer questions about visible features but won't slow down questions about sounds. Performance can be improved again when you are presented with images that are compatible with what you are seeing. Michael Kaschak and collaborators had people stare at movement patterns on a computer screen and then asked people to listen to sentences about objects moving in different directions.[5] Comprehension of the sentences was faster if the described movement matched the direction of the pattern on the screen.

Sentence comprehension can also influence the speed at which we recognize pictures. Robert Stanfield and Rolf Zwaan gave people sentences about hammering nails followed by pictures of nails, and then they were simply asked, 'Does the picture depict something mentioned in the sentence?'[6] People were faster if the orientation of the nail in the picture corresponded to the way a nail would have to be imagined if the sentence were visualized. Thus, if you hear about a nail being pounded into the floor, you will be fast at recognizing pictures of vertical nails, and if you hear about a nail pounded into the wall, you will be faster at recognizing horizontal nails. Similarly, Richard Yaxley and Zwaan found that people were faster at recognizing a clear picture of a moose after hearing a sentence about seeing a moose through clear goggles, and faster at recognizing a blurry picture after reading a sentence about seeing a moose through foggy goggles.[7] That's quite remarkable, because blurry pictures are usually harder to recognize than clear ones.

The psychological evidence has also been confirmed by evidence from neuroscience. It used to be thought that the back part of the brain is used for perceiving and the front is used for thinking. But we now know that the back part of the brain, where most of the senses are located, is very active when people think. Moreover, we know that the front part of the brain does not work on its own, but rather coordinates and reactivates sensory patterns in the back. Recent evidence from Linda Chao and Alex Martin has shown that reading activates the same areas as looking at pictures, suggesting that we visualize what we read.[8] Kyle Simmons and Barsalou have shown that reading object names generates activity in sensory areas corresponding to the features most associated with those words.[9] For example, reading the word 'blender' causes a lot of activation in the visual cortex, but also in the auditory cortex. It is also known that damage to these sensory areas of the brain can result in profound deficits in the ability to comprehend familiar categories of objects.

All these experiments suggest that people use imagery to think. In order to understand a sentence or answer a question, we generate corresponding mental images. When our ability to generate images is interrupted, performance on these cognitive tasks declines, and when our ability to generate images is facilitated, performance improves. This is just what the seventeenth-century Empiricists would have predicted.

Abstract Images

None of this should be very surprising. It should seem obvious from introspection that we use mental imagery in thought, and it makes perfectly good sense that we learn about familiar categories by storing polysensory images of them. Empiricism is a commonsense view. It's the view that everyone should have before they read any philosophy or psychology. The language of thought hypothesis is, in contrast, quite a departure from common sense. When you introspect, you might hear yourself speaking in English, but that experience is just a form of mental imagery; you are hearing the sounds that words make. And English words are just a shorthand for other kinds of images. As we have seen, when you hear words you also generate images of what those words represent. But there is nothing in introspection that corresponds to a *mental* language other than the ones you speak. The language of thought is said to be unconscious. Its words have no sound or shape. So there is no introspective evidence for the language of thought. Empiricism seems obvious, and the language of thought seems far from obvious.

As we have seen, the language of thought is postulated to explain things that Empiricists have difficulty explaining. Empiricism seems like it must be true when we think about cars and blenders – things we can easily imagine – but the intuition that Empiricism is right begins to waver when we consider more abstract ideas. How do we form an image of justice, truth or democracy? These things cannot be seen or tasted or smelled. So even if Empiricism offers a plausible theory of how we understand very concrete categories, it seems to do badly when we go more abstract. But recall that the language of thought also has difficulty explaining abstract ideas. I argued earlier that words – whether in English or in a mental language – can only be understood if they are related to something non-verbal, because words are arbitrary symbols with no intrinsic meaning. So, rather than abandoning Empiricism, we should see whether there is any way the Empiricist can explain abstract ideas.

Confidence in Empiricism can actually be restored if you just reflect on how you might go about answering questions about things that are very abstract. Consider justice. One way to understand this lofty idea is by grounding it in very concrete scenarios. There are different kinds of injustice and each can be captured by simulating an event. First, there is inequality. This can be simulated by imaging a situation in which I get two cookies and you get three. Second, there is inequity. For example, you might give me one cookie in exchange for two. Third, there are violations of rights. Suppose I try to eat my cookie and you prevent me from doing so. These simple schoolyard scenarios can be adapted to more complex cases. One might conclude that disparities between rich and poor are unjust by comparing them to the first case (the rich have more cookies); one might infer that heavy taxes are unjust because they are like the second case (giving without getting much in return); one can infer that censorship is unjust because it is like the third case (restricting speech is like preventing someone from speaking, which is analogous to preventing one from eating a cookie). Of course, there is some latitude in how simple schoolyard cases scale up to grand societal issues, but that may explain why issues of justice are often so hotly debated. We learn the concept by means of very simple cases and then need to figure out whether more complicated cases are sufficiently similar to these, and there are no hard and fast rules for doing that. Still, the simple scenarios can give us a very concrete idea of what justice is, and that is sufficient for grounding our understanding of this seemingly abstract concept.

This strategy works for other cases as well. Consider democracy. Democracies don't look like anything special; they have no characteristic shape on the map. But it is easy to grasp what democracies are by simulating democratic procedures. Suppose you want to decide where to eat dinner tonight in a democratic way, so you ask your family to raise their hands: Who wants sushi? Who wants Mexican? The tally dictates where you go. We have a number of procedures like this, and they all involve counting votes. This can ground understanding of the concept. We think of a nation as democratic if they decide things by voting, and we conceptualize voting by imagining the kinds of procedures we learned in primary schools. We know a range of scenarios (raising hands, casting ballots, saying aye and nay) and we can easily imagine others by extension (stamping feet, waving coloured flags, making marks on a blackboard). In each case, people make an opinion known by some display, and one group of opinions is compared quantitatively to another.

This strategy of simulating simple concrete events works for a surprising range of abstract concepts, but it is not the only resource available to the Empiricist. Consider the philosophical thesis that life is pointless. Philosophers get people to understand this thesis by walking them through an introspective exercise. They ask: 'Why do you report to work each day?' When you think about what motivates you, you may answer that you need money. 'Why do you need money?' To buy food and shelter. 'Why buy food and shelter?' To live. Providing these answers involves simulating events (reporting to work, getting a paycheque, buying food), but also introspecting on motivations. If you think about why you report to work, the desire that immediately comes to mind is receiving that paycheque, and so on. But suppose the philosopher now asks, 'Why live?' Up to this point you could answer each question by mentioning another activity, but now those answers may give out. When you are asked why you should live, you may find nothing but an emotional state: a consciously experienced desire to live. Now the agile philosopher asks, does mere desire give an activity 'meaning'? To help you answer this, she may ask you to consider someone who takes pleasure in chewing gum. That is easy to imagine, but is gum-chewing a meaningful activity? This question brings out something about how we comprehend the notion of 'meaningful activity'. We associate this notion with activities that we find commendable. Finding something commendable is an emotional response, a feeling of praise. But we have no temptation to commend mere pleasure. Pleasure is nice if you can get it, but not worthy of being complemented. If our pursuits in life are all motivated by pleasure, and pleasure is not commendable, then maybe our pursuits are not commendable, even if they seem to be at first. And if our pursuits are not commendable, then life is meaningless, or pointless.

This example is designed to show that some extremely lofty, abstract, philosophical ideas can be grasped by introspecting motivations and emotions. If you weren't persuaded by the argument, it is probably because your motivations or emotions departed from the ones I reported here. Maybe you are motivated to live in order to learn or to help others, and maybe when you think about these things, they feel commendable. So your life feels like it has a point. Whatever conclusion you draw about the meaning of life, the examples show that one can reflect on this highly abstract philosophical issue in a concrete way, grounded in felt motives and attitudes.

Not all lofty philosophical concepts can be explained this way, however. Critics of Empiricism often advance logical concepts as a counter-example. Consider such concepts as truth, negation or identity. These have no obvious link to emotions or motivations, and it's hard to imagine how they could be grounded in simple concrete scenarios.

The key to explaining how logical concepts are understood is to remember that logic is a kind of ability. Mastering logic is a matter of knowing how to make certain inferences. Thus, Empiricists do not need to say that people can form an image of truth or negation. That would be impossible. Instead, Empiricists need to account for logical concepts by explaining how they are used in reasoning.

Consider truth. Suppose I tell you that some claim is true. It's true that aardvarks are nocturnal, I submit. If you doubt me, the first thing you'll do is check. To do so, you must first comprehend my claim. On an Empiricist theory, that involves forming a mental image. You imagine aardvarks foraging at night. Next you need to confirm this. The most direct method would be to find some aardvarks and observe them. If you see aardvarks walking about in darkness, this will match your visualization of my assertion that aardvarks are nocturnal, and you will conclude that I was telling the truth. Of course, it might be inconvenient to observe aardvarks, because you don't live near the African savannah. In that case, you can use an indirect method of confirmation. You can use language. Not a language of thought, but plain old English. You can ask other friends or experts whether aardvarks are nocturnal, or, if you are computer savvy, you can do a Google search on 'nocturnal, aardvarks'. If you find other people or texts saying that aardvarks are nocturnal, you will have reason to accept my claim as true.

In short, mastering the concept of truth consists in learning skills for testing claims against the world. We learn to match mental images with reality and sentences with testimony. If we find a match, we increase belief. This simple matching process, which requires nothing other than perceived words and images, grounds our concept of truth. We may later refine the concept or use it in technical ways, but it should be clear from the explanation just offered that one can grasp the basic idea of truth without abandoning the Empiricist conjecture that we think using stored sensory records.

A similar story can be told about negation. Suppose I tell you that there is no wine in the cupboard. There is no such thing as a mental image of non-wine. What would that image be? Everything other than wine is non-wine: beer, lettuce, submarines, the number three and so on. So you cannot grasp a negated concept, such as non-wine, by forming images of what members of that category look like. But you can understand negation if you treat it as a kind of skill. Mastering the concept of negation is a matter of learning how to test negative claims. If I say there is no wine in the cupboard, one thing you can do is look for wine there. If you fail to find any, you conclude that I was right. Here, you are also using mental images. You imagine what wine looks like, search for a match and then, having failed to find one, you report that there is no wine.

Now consider identity. In logic, identity is used to convey that two names or descriptions designate the same person. We say Lewis Carroll is the Oxford logician named Charles Dodgson, Thomas Jefferson is the inventor of the swivel chair, or the butler is the murderer. These are all identity statements, because they identify a person described in one way with a person described in another. In logic they would be expressed using the equals sign and symbols for each referring term. So Carroll is Dodgson might be expressed by $c=d$. For the Empiricist, identity is puzzling because it doesn't have any appearance. When we discover that Carroll is Dodgson, we don't visualize two men morphing into each other, like some science-fiction fusion experiment. Nor can we visualize two men standing next to each other (that would be a case of non-identity), or one man (since that would not express identity at all). There is no good picture of identity.

There is, however, an ability associated with mastery of this concept.

When I tell you that Carroll is Dodgson, you can confirm it by making sure, for example, that they lived at the same time. Once you have confirmed the identity (or if you accept it on my testimony), you merge the things you know about these two men. You are inclined to say Dodgson is the author of children's books, and Carroll is a logician. You might even look for logical puzzles in Carroll's children's books. The concept of identity is not a picture; it's a capacity to integrate information. Likewise, if you learn that the butler is the murderer, you seek his arrest, since that's the appropriate thing to do with a murderer.

Now consider the basic principle of logic that everything is identical to itself. At first, it might seem impossible for the Empiricist to explain how we can comprehend such an abstract principle. But now an explanation suggests itself. When you learn that two things are identical, you merge what you know about each. All of your images and attitudes are combined together. If I tell you that Lewis Carroll is identical to himself, however, there is nothing to merge. If you try to exercise your merger skills, you will realize you don't need to because you already have an integrated representation. You don't need to combine any facts or confirm birthdays when you say Carroll is Carroll. That discovery should make it obvious that skills underlying the concept of identity apply trivially in the case of self-identity. The principle that everything is identical to itself merely expresses that insight.

The stories I've been telling about how people understand abstract concepts are both speculative and incomplete. The abstract concepts we've been looking at are used in many different ways, and there are hundreds of other abstract concepts to consider. There has been very little psychological research on abstract concepts, so the proposals here are highly speculative and incomplete. But the foregoing discussion does teach a crucial lesson. There is a knee-jerk response to Empiricism that says, 'We can't think in mental images, because mental imagery can't account for abstract ideas.' This quick dismissal is taught in textbooks, and it remains the biggest barrier to getting people to take Empiricism seriously. What we've now seen is that the knee-jerk response couldn't be more mistaken. It's actually pretty easy to see how mental imagery could lend itself to thinking about abstract ideas.

We've considered some of the loftiest ideas known to our species: concepts of morality, politics, philosophy and logic. It might seem impossible to explain these things by an Empiricist theory, but a little reflection is all it takes to see that there are many strategies for explaining such lofty abstractions. We grasp some by considering concrete scenarios, some recruit our emotions and attitudes, and some can be mastered by learning skills for comparing mental images to the world. With a little imagination, it's easy to devise stories about how any abstract concept is understood. We can even come up with stories by introspection: just take an abstract concept and think about how you grasp its meaning. The stories we come up with to explain abstract ideas can be tested by doing psychological research. Little research on abstract ideas has been conducted, but, even prior to any experiments, it should already be clear that there is no reason for doubting that Empiricism has resources to explain our capacity for abstract thought. The textbook reason for rejecting Empiricism has no foundation.

Where Does Language Come From?

6

The Gift of the Gab

In the late 1950s, a young linguist named Noam Chomsky developed a series of arguments that revolutionized the science of linguistics. Chomsky's radical conclusion was that human beings are born with an innate 'language faculty'. The basic rules of grammar must be in place from the start, Chomsky claimed, if a child is to ever learn how to speak. This innateness hypothesis quickly spread, and it is now the dominant view in linguistics. Language has been likened to an instinct, a sense modality, or a biological organ, which is hard-wired in our species and absent in others.

Chomsky's theory of language has had a huge influence on psychology more broadly. The hypothesis that infants' minds are organized like little universities, which we encountered in chapter 4, might never have emerged if it were not for Chomsky's influence. His views about language are taken as a model for other psychological domains. In each domain, the psychologist's job is to identify a set of innate universal principles that underlie adult competence.

Now there is another revolution brewing. Some researchers are starting to doubt Chomsky's arguments for innateness. They are seriously exploring the possibility that language is learned by experience. The gift of the gab is given to us by our parents, not our genes. This return to common sense is part of the larger revival of Empiricism that we have been exploring in the last two chapters. Empiricists emphasize learning, rather than hard-wiring, and, as we will see in subsequent chapters, this shift in focus lays a foundation for understanding human diversity.

RULES AND REGULARITIES

By his or her fourth birthday, the typical child has mastered a language. Four-year-olds know thousands of words, and they have mastered the subtle rules that allow them to comprehend and produce fully grammatical sentences. How do they achieve this?

One natural suggestion is that kids learn by imitation. They hear adults speaking all around them and they copy what they hear. This suggestion has some initial plausibility because kids are great imitators, and imitation is a form of learning that human beings engage in more than any other creatures. There is also an obvious sense in which imitation is involved. We do need to hear words to learn them, and kids do repeat the words they hear. But imitation cannot offer a complete theory of language acquisition.

One problem with the imitation theory is that kids make a lot of errors not found in adult speech. For example, English-speaking children tend to over-extend the -ed ending that we add to regular verbs such as waited, cooked and bamboozled. They use the -ed ending with irregulars such as eated, swimmed and taked, instead of ate, swam and took. Sometimes kids combine the correct irregular form with an -ed ending. When I was a child, my parents would ask me every day what I had done in kindergarten, and I would evasively reply, 'I stooded around.' Another problem for the imitation theory is that kids also come up with novel sentences. Recall that languages are endlessly productive: most of what we say has never been said before. If kids simply repeated what they heard, new sentences would never come about. It follows that kids could not be learning by just imitating what they hear. More must be involved.

From these examples, it should be obvious that kids are not just blindly aping what they hear. They are picking up on rules. Kids over-extend the -ed ending because they have learned a rule that says add -ed when you want to talk about an action in the past. Kids learn that they can construct new sentences because they have mastered rules for combining elements together, as opposed to just learning entire sentences as fixed units. But how do kids learn these rules? One might think, they learn as scientists learn: through observation and explicit induction. Scientists trying to discover the laws of nature observe the world around them and then propose laws that could explain what they see and extend to new cases. In this view, a child sees that adults make the -ed sound when talking about events that have transpired, and they form an explicit hypothesis: 'There is an -ed at the end of every past-tense verb'.

This sounds plausible at first, but it doesn't hold up under scrutiny. It's important to remember that we are dealing with kids who are very young. Kids are picking up on grammatical rules shortly after their first birthday, and the idea that they are forming explicit hypotheses seems quite implausible. After all, even trained linguists who have spent their entire careers studying language cannot agree on the rules of grammar. It's extremely difficult to figure out and formulate the rules of language. For example, we all know how to pronounce the plural -s ending in words like ducks, pigeons and fishes, and we can correctly come up with plurals for nonsense words, such as glip, flig and bliz. How do we do this? How do we know when to pronounce the -s ending softly, when to pronounce it like a z and when to make it into a whole new syllable, as in blizzes? Even as thoughtful adults, it takes a lot of work to explain this pattern. In fact, we might not even be explicitly aware of any pattern until it's pointed out to us. Or consider a harder case. It's grammatical to say 'Sally squirted paint on the wall' and 'Sally squirted the wall with paint', inverting the indirect object and direct object. But when we say, 'Peggy poured paint on the floor' we cannot invert; it's ungrammatical to say, 'Peggy poured the floor with paint'. It has taken linguists years to come up with an explanation for this, but we all master the underlying rule in early childhood. Young children effortlessly acquire rules that experienced linguists find completely perplexing. The linguist Ray Jackendoff calls this the Paradox of Language Acquisition. It proves that children cannot be learning language by consciously formulating rules that capture the sentences they hear and then extending those rules to new cases. Even skilled adults can't do that efficiently.

Jackendoff thinks there is only one way to get out of the paradox: assume that language is innate. Jackendoff is a follower of Chomsky. Chomsky revolutionized linguistics by proposing that children are born with an innate universal grammar, or UG. In the most influential version of the theory, Chomsky proposes that UG consists of a collection of unconscious rules, most of which have more than one possible setting. To take a simple example, consider the fact that English-speakers place adjectives before nouns, and Spanish-speakers can put adjectives after. There could be an innate rule that has both of these options, and then the child sets the rule to the appropriate one based on the sentences she hears during development. Thus, children do not learn grammatical rules by experience. They already possess the rules. Experience merely determines which variation of those innate rules to use. This would solve the Paradox of Language Acquisition. The reason children effortlessly arrive at the rules of the languages they speak is that they know those rules innately. Learning merely activates the relevant rule, like a finger flicking a switch. The rules are unconscious, so linguists cannot figure out the rules by simply introspecting.

Chomsky's innateness hypothesis – often called linguistic 'nativism' – is widely accepted. If you've ever read anything about linguistics, you've probably read that language is innate. Nativism is more plausible than the suggestion that we learn language by imitation or explicit induction. But it's not the only possibility. Shortly before Chomsky's revolution, some researchers had been exploring the idea that children might learn language statistically, by unconsciously tabulating patterns in the sentences they hear and using these to generalize to new cases. Statistical learning is an attractive avenue to explore because we know that it is widespread in nature; animals learn many things by keeping track of statistical regularities, such as where to forage and how to recognize edible objects. We also know that the nervous system is naturally very adept at statistical learning. Neurons are linked together in layered networks, which obey a simple principle: if you fire together, wire together. If a stimulus in the environment causes a pattern of neural activation, the connections between those neurons will get stronger, making them more likely to fire together in the future. Over time, neural connections will reflect the frequency of the stimuli that an organism encounters, with stronger neural connections corresponding to stimuli that have been encountered numerous times. This has been known for decades, and researchers before Chomsky began to suspect that children might learn language by picking up on statistical regularities in adult speech.

The initial efforts to come up with statistical learning models were unsuccessful. The early models could not account for some basic facts about how languages are learned. They tended to treat sentences as linear strings of symbols with no underlying structure, but we know that sentences can be broken down into parts. 'The dentist gave a lollypop to her patient' has a subject, a direct object and an indirect object, each represented by a different noun phrase: 'the dentist', 'lollypop' and 'her patient.' The last of these phrases has a possessive pronoun, 'her', that refers back to 'the dentist', even though these words are not consecutive in the sentence. It wasn't until the 1980s that researchers figured out how statistical learning could discern structure in strings of words, and, over the last two decades, statistical learning theories have become even more powerful and sophisticated.

Defenders of statistical approaches to language learning usually reject Chomsky's nativism. They deny that there is an innate faculty that evolved specially for language acquisition. Unlike UG, statistical learning mechanisms are not specialized for language. They can be used for other purposes, such as pattern recognition or learning the sequences of muscle movements required for physical skills. It just so happens that these learning mechanisms also allow us to learn language. Kids are bombarded by sentences that have statistical regularities. For example, nouns tend to sound different from verbs because they have different endings and different positions in sentences. Nouns and verbs are also used in different conversational contexts; 'dog' is used when dogs are present regardless of what they are doing, and 'run' is used when running occurs, regardless of who is running. Children automatically and unconsciously keep track of such things, and the observed patterns generate unconscious statistical predictions. For example, English-speakers predict that a word that has been used in the presence of objects will be followed by a word that has been used in the presence of actions. Defenders of statistical approaches to language acquisition speculate that language acquisition will one day be fully explained by statistical features of the sentences children hear and a general capacity to keep track of these statistics. Linguistic rules are just unconsciously extrapolated from statistical regularities.

The statistical approach to language acquisition is more plausible than the imitation theory. Kids don't just ape what they have heard.

They discover underlying rules and use these to generate new sentences. After being bombarded by -ed endings, kids start using that ending whenever they produce past-tense verbs, because, statistically speaking, that's what follows. Regularities are turned into regulations. The approach is also more plausible than the induction theory. Kids do not form explicit hypotheses about the regularities they observe. All this is done unconsciously. Kids unconsciously pick up on which sounds are followed by each of the three ways of pronouncing the English plural. Kids can also statistically discern when verbs are used to refer to actions and when they are also used to refer to outcomes. 'Pour' is used to refer to an action, whereas 'squirt' is used to refer to both an action and an outcome (there can be a squirt of paint on a canvas, but not a pour of paint). Outcome verbs can be used immediately before words referring to the object used in the action or the object affected by the action: 'squirt the paint' or 'squirt the wall'; 'pack the clothes' or 'pack the suitcase'; 'pluck the feathers' or 'pluck the chicken'. But verbs that do not designate outcomes cannot be used this way: 'pour the paint', but not 'pour the floor with paint'; 'squeeze the clothes', but not 'squeeze the suitcase with clothes'; 'count the feathers', but not 'count the chicken with feathers'. Language-learners may unconsciously pick up on such regularities.

The statistical approach handles the Paradox of Language Acquisition in a way that shares something in common with the UG approach. Both appeal to unconscious knowledge. Professional linguists have great difficulty figuring out the rules of grammar because the rules cannot be consciously accessed. But statistical learning theorists and Chomskyans part company there. Chomskyans say that children acquire language effortlessly because they possess grammatical rules innately. Statistical learning theorists say that children acquire language effortlessly because statistical learning is something we all do incessantly and automatically. The brain is designed to pick up on patterns of all kinds.

At this stage, the statistical approach to language learning is speculative. No one has come up with an actual model that explains how every feature of language can be learned statistically using the same resources by which we pick up on regularities outside of language. Computer simulations of statistical learning have met with some success in this domain, but only on learning very specific features of language or learning simple artificial languages invented by researchers. We don't have a statistical learning machine that can be given a bunch of sentences that children hear and output flawlessly in English. We don't really have a Chomskyan model that can do that either. No one has taken the rules that Chomsky or his followers have proposed for UG, programmed them in a computer and then presented the computer with batteries of English sentences. If they did, the computer would not generate correct English. We are far away from that point. But, with the innateness hypothesis, one can at least see that some version of the approach could explain how languages are acquired. If all the world's grammars are innate, then there isn't much learning the child has to do. So, we have reason to think that the UG approach could ultimately explain how acquisition is possible. We have no comparable guarantee with statistical learning. We may be decades away from having an understanding of what statistics children might use learning language and how they integrate all the information that comes in when they hear sentences (word sounds, intonation, facial expressions, gestures, nearby objects, recent context and so on).

It is best to think of the statistical learning hypothesis as a promising research programme that is still in the early stages of development. We don't yet know if it will succeed. But, it has certain obvious advantages over Chomsky's UG theory. In science, less is more. It's bad policy to postulate things that are not necessary. If we can explain the acquisition of language by appeal to general-purpose pattern recognition capacities that keep track of statistical regularities in the environment, we don't need to postulate a specialized psychological mechanism for language acquisition. The statistical learning theory would be easier to explain as a gradual evolutionary outgrowth of simpler capacities, it would be easier to implement in the nervous system and it would be less costly for the genome. In science, simpler theories are preferable, and they should be abandoned only if evidence forces researchers to adopt more complex explanations. Most linguists today think that is precisely the situation we are in. There are some very powerful arguments that suggest we need an innate capacity dedicated to language acquisition. If these arguments go through, the statistical learning theory is hopeless. But those arguments may not go through. The most famous arguments for an innate language faculty may be seriously flawed. New research is casting doubt on assumptions that have been cherished by a generation of linguists, and the statistical learning account may be vindicated in the end.

CHOMSKY'S POVERTY

Chomsky and his followers have devised a number of influential arguments for an innate language faculty. The most influential of these are called arguments from the 'poverty of the stimulus'. The word 'stimulus' refers to the sentences children are exposed to while learning a language. Chomsky sometimes calls this the primary linguistic data. The word 'poverty' refers to the fact that children don't hear enough linguistic data to explain how they correctly discover the rules of grammar using their general-purpose learning abilities. This may sound surprising, because children do hear a lot of sentences. Before Chomsky, it was assumed that kids get enough exposure to language to learn how to speak correctly. The poverty of the stimulus arguments were designed to undermine this assumption. Each argument points to some limitation in what children are exposed to before mastering a language. I will consider these limitations in turn, in an effort to rebut Chomsky's claims.

Early Acquisition

The simplest poverty of the stimulus argument emphasizes the fact that kids acquire language very early in life. At this age, the argument goes, they simply haven't heard enough sentences to explain their success.

So formulated, this argument isn't very impressive. First of all, four years is a pretty long time. Kids show a gradual increase in their linguistic skills in this time, suggesting that they are incrementally building on what they have learned. The build-up is from single words, to two words, to simple sentences, and finally master sentences with several clauses. They don't do this overnight. It takes a year before they utter their first words, and years more before they can produce sentences that sound like adult speech. Statistical learning theory predicts this kind of incremental development. If language were innate, it's not clear why it should develop so slowly.

Statistical learning theory also has a straightforward explanation of why language learning can get off the ground at the start of life, before children have mastered many other skills. On this approach, language is learned perceptually, by unconsciously tabulating regularities in perceived speech. Perceptual systems are in place at birth, so learning can start right away. Of course, some children are born without the capacity to hear. These children do not benefit from the same early exposure to language as other children. They can observe lip movements and gestures, but most don't begin language acquisition until they have ample exposure to people who have mastered sign language. This can result in considerable delays in language learning.

Degraded Sample

A more powerful poverty of the stimulus argument draws attention to the fact that kids are often exposed to grammatically incorrect sentences while growing up, yet they end up producing correct grammar. If they were extrapolating from what they heard, this should not happen. The incorrect sentences that Chomskyans have in mind are usually the ones produced by the children themselves. At early stages of development, kids often say things that are perfectly intelligible, but grammatically erroneous. A child might point to his mother and say, 'He a girl.' The sentence omits the verb, uses the wrong gendered pronoun and over-extends 'girl' to adults. Yet, it's perfectly clear what the child meant to convey, and rather than correcting these mistakes, adult onlookers will likely say, 'Yes, that's right!' This example comes from a pioneering study by Roger Brown and Camille Hanlon, which showed that parents frequently approve of ungrammatical sentences.[1] If children learn by observation, such approval should reinforce errors and prevent accurate language acquisition. But kids soon stop making these mistakes, and that suggests they are not learning by observation.

There are two problems with this argument. First, it is self-refuting. The very fact that children produce such woefully ungrammatical sentences favours a learning story, rather than an innateness story, and adults' failure to correct such speech may in fact have an impact on the duration of these errors. Second, statistical learning theory says that children extrapolate language from regularities in what they hear. If they listen to adults speaking regularly, then correct speech should eventually flood these errors out statistically, and that would explain why mistakes of this kind don't last for ever. The argument could succeed only if kids heard ungrammatical sentences most of the time, but that is not the case.

In fact, there is evidence that children usually hear sentences that are unusually grammatical and clear. When adults talk to children they speak clearly and simply, avoiding fragments, and articulating speech sounds with extra care. This clear pattern of speech has been given the politically incorrect name Motherese. Hearing the kinds of sentences that mothers stereotypically produce may give children an especially helpful data set to work with when learning a language. The Motherese hypothesis remains controversial, and some researchers have argued that many children are not spoken to in this special way, but the current balance of evidence suggests that this is a standard practice. Thus, the primary linguistic data may be far less degraded than some proponents of innateness allege.

Insufficient Negative Data

To learn a language, children must generalize from the sentences they hear to new cases. Otherwise they'd be stuck repeating familiar sentences, and they would never come up with anything new. Learning in this way is tricky, because it is very easy to over-generalize. Any pattern can be taken too far and used to generate sentences that are not grammatical. For example, if you hear the sentences 'The dog ate the doll' and 'The dog ate' you might generalize the pattern and conclude that it is always possible to drop direct objects after verbs. This would allow you to go from 'The dog took the doll' to 'The dog took', but the latter is not a proper sentence of English. Such over-generalizations can be avoided if children receive what linguists call negative data: if they are informed which sentences are ungrammatical in addition to being told which sentences are grammatical. The problem is children don't get enough negative data. Negative data could come in the form of correction, when children make errors, but it turns out that errors are rarely corrected. As we saw from Brown and Hanlon's example, 'He a girl', parents usually don't bother to correct mistakes, as long as they can understand what a child is trying to convey. Moreover, children often seem impervious to corrections on the rare occasions when negative feedback is received. If someone corrects the child, and says, 'You mean, she is a girl', the child may quickly retort, 'Yeah, he a girl'.

Given the dearth of effective negative data, children should continue making mistakes for ever. But they do not. They eventually get it right. Moreover, there are certain mistakes kids never seem to make. For example, kids hear the irregular verb 'eat' and its past form 'ate', but they don't infer from this that past tense of the verbs 'beat' and 'meet' should be 'bate' and 'mate'. Also, kids may hear an active-voice sentence like 'Hilda expects the train to come' and a passive rendition, 'The train is expected to come'; from this pair they should infer that 'Hilda expects the train *will* come' can be converted into 'The train is expected will come'. But kids don't make this inference. If children can avoid over-generalizations without negative data, then they must have hard-wired constraints that determine which generalizations are permissible and which are not. This points to an innate language faculty.

This is one of the most influential arguments in modern linguistics, but it hasn't persuaded everyone. The argument depends on several assumptions that can be challenged. First, the fact that children rarely make certain mistakes is surprising only if we assume that children are prone to make risky generalizations on scant data. If language learning is statistically driven, children may be very conservative, awaiting considerable evidence before making generalizations. They may over-extend -ed endings because they hear numerous examples of the regular past, but they don't over-extend irregular endings, because irregulars are infrequent. They don't say 'The train is expected will come' because there isn't enough statistical support. It may look like 'The train is expected to come', but the word 'to' and the word 'will' behave very differently in the language, so it is not a conservative substitution.

The argument also assumes that children rarely receive effective explicit correction. But this assumption is somewhat exaggerated. Kathryn Hirsh-Pasek and her collaborators discovered that adults are twice as likely to repeat what a child says when that sentence is ungrammatical as opposed to grammatical, and, in repeating the sentence, adults routinely fix the grammar. This feedback may be effective even if children do not pick up on the correction right away. The fact that children are initially impervious to correction is consistent with the statistical learning theory. A single episode of feedback may be insufficient to correct the statistics from which the error initially arose, but parental corrections along with subsequent observations can gradually push the child towards correct performance.

More importantly, explicit correction is not the only form of negative data. Children can get extensive negative feedback *implicitly*. One source of correction comes from prediction. Statistical learning often works by assigning probabilities. For every sound, one can assign a probability to what sound will come next. In English, for example, an *r* is usually followed by a vowel. For each word, there are probabilities for what the next word will be, or what type of word it will be (a noun or a verb, for example). Phrases also have more and less probable successors in a sentence. By unconsciously tabulating such probabilities, children may form predictions about what will follow what when listening to speech. If a prediction fails, that will produce an error signal that reduces the probability assignment. This is a form of negative data.

Other forms of implicit correction are available as well. When kids make mistakes, they may get negative feedback from facial expressions (adults may grimace or giggle at grammatical errors). Mistakes can also result in communication failures, which children may find frustrating. Interestingly, even positive data can serve a negative role. Brian MacWhinney proposes that children don't so much correct errors as replace them.[2] If a child produces an incorrect sentence and subsequently hears a way of expressing the same thought, the child can use the new sentence as evidence against the permissibility of the old.

Recently, researchers have been creating computer models that use statistical learning and comparing their performance to how children learn language. These models make very specific predictions about what kinds of over-generalizations children will make and how those errors will improve over time. Michael Ramscar has been using this method to study how kids learn regular and irregular plural endings.[3] Kids famously over-extend the plural -s. They say foots and mouses instead of feet and mice. Curiously, they often get the irregular plurals right initially and then start to make errors. In a statistical learning model this is explained by the fact that early learners still haven't heard the regular ending often enough to extrapolate the statistical rule. As they hear more regulars, they start to over-extend the -s. But over time, kids also start to hear more and more irregulars, and this reinforces the statistical probability that an irregular ending heard earlier is correct. Ramscar's computer model predicts that kids will correct their own over-extensions without any explicit negative feedback, as they get exposed to more regular and irregular verbs. His model also makes the highly counter-intuitive prediction that older kids' performance on irregular plurals will improve after playing a game in which they need to generate regular plurals. After they talk about dogs and frogs, the model predicts, older kids will be more likely to recall that it's mice and not mouses. For younger kids, the model predicts that using the regulars will tend to lead to more mistakes with irregulars. The reason for this is that, in statistical learning systems, a regular pattern will interfere with an irregular pattern if the latter is very unfamiliar, but, as irregulars gain familiarity, regular patterns reinforce them because such learning systems seek to differentiate familiar forms. All these predictions are borne out in studies of children's errors, suggesting that they learn the past tense statistically, and that this method can eventuate in accurate performance without any explicit correction.

The upshot of all this is that one cannot establish that language is innate by appeal to the fact that adults don't correct every mistake that children make. Children have many implicit sources of negative data. Statistical learning models provide a promising strategy for explaining why kids over-generalize some forms and not others and how they recover from their mistakes. These models make specific predictions, which are not made on nativist models, and research is beginning to suggest that these predictions are right.

Insufficient Positive Data

The argument from insufficient negative data says kids don't get enough correction to explain how they learn the rules of their language. We

have just seen that kids do in fact get plenty of correction, just implicitly rather than explicitly. But there is a related argument for innateness
that emphasizes the lack of positive data rather than negative data.
Kids learn how to form sentences that are very different from any they
have heard. A form can be totally absent from the primary linguistic
data, yet applied correctly. Statistical models allow for some innovation, provided the new forms can be extrapolated from examples
that children hear. But many linguistic rules are difficult to extrapolate from the collections of sentences that kids hear. In fact, that limited
sample should lead kids to extrapolate incorrectly. But they don't.
Kids arrive at grammatical rules that are not simple generalizations of
the statistical input. That has been taken as evidence for the conclusion that kids are relying on innate principles.

The most famous example of this involves rules that convert assertions into questions. Consider the sentence 'The puppy is barking'. To
turn this into a question, you move the verb to the front and get 'Is the
puppy barking?' That's easy enough, and kids may hear lots of pairs
like this. But now consider a more complex sentence: 'The puppy that
is barking is angry'. How would you convert this into a question?
After reflecting a moment, you probably came up with: 'Is the puppy
that is barking angry?' This is what linguists call a complex polar
interrogative (where polar just means you answer yes or no). As adults,
we have little difficulty generating complex polar interrogatives, and
it turns out that kids can form them too. That is an astonishing fact.
If kids learn language by observation, then it's natural to assume they
learn how to form complex interrogatives by extrapolating from
simple ones, which are much more likely to come up in casual speech.
But it doesn't look like such extrapolation is possible. In the sentence,
'The puppy is barking' there is one verb, and a question is formed by
moving it to the front. But 'The puppy that is barking is angry' has
two verbs. Which one should move? Statistically, there is no clear way
to decide. The simplest policy might be to move the first verb in the
sentence. But that rule would result in an ungrammatical monstrosity,
'Is the puppy that barking is angry?' Kids never produce sentences
like that. Somehow, they figure out the right rule.

We could account for this success if there were complex polar interrogatives in the primary linguistic data. But that seems unlikely. Nativists

claim that such sentences are rare enough that learners might never hear them, especially if we remember that adults tend to simplify language when talking to kids. So kids seem to be figuring out the right rule without the positive data they need, and that suggests that the rule is dictated by an innate language faculty.

This case of complex polar interrogatives appears more frequently in the literature on poverty of the stimulus arguments than any other. It is the canonical example used by nativists to argue that kids can't learn language by observation. It is also one of the few examples that nativists have investigated empirically. Sometimes nativists claim that children are capable of producing certain kinds of sentences without actually testing to confirm that this is true. Not so with complex polar interrogatives. In a clever study, Stephen Crain and Mineharu Nakayama devised a game in which children were told to get information from the *Star Wars* character Jabba the Hut.[4] The experimenters said: 'Ask Jabba if the boy who is watching Mickey Mouse is happy.' Amazingly, even three-year-olds came up with the correct complex polar interrogative in response. Crain and Nakayama conclude that they must be using an innate grammar.

The argument looks like a deathblow to statistical learning, but it fails on closer examination. Some critics have challenged the assumption that kids never hear complex polar interrogatives. Linguists test such claims by searching through records of actual sentences and conversations that have been recorded. These are called speech corpora. Barbara Scholz and Geoffrey Pullum searched two corpora containing sentences directed towards children and discovered that complex polar interrogatives do occur, albeit infrequently (about 1 per cent of questions have the relevant form).[5] This suggests that children may actually be repeating patterns that they have heard before.

Nativists will reply that that 1 per cent is too infrequent to guarantee that every child is exposed to enough examples to learn the correct rule. They will also point out that some corpora of child-directed speech contain no complex polar interrogatives. So the argument for innateness retains some force. Fortunately, there has been a more decisive rebuttal. Children may be able to learn the rule *indirectly* by statistically analysing simpler sentences and extrapolating from those. This may sound far-fetched, but recent research by Florencia

Reali and Morten Christiansen provides resounding proof of this conjecture.

Reali and Christiansen did a statistical analysis of a speech corpus that contains sentences from nine different mothers speaking to their children over several months, spanning from the age of 1 to 1.9 years.[6] They picked this corpus because it is representative of what typical English learners hear early in life, and because it doesn't contain any complex polar interrogatives. They wanted to know, can kids learn to produce such questions if they never hear them. As a first step in their study, Reali and Christiansen simply computed how frequently every combination of two and three words co-occurs in the corpus. This is just the kind of statistical information that learners can unconsciously tabulate. The phrase 'the puppy' might occur more frequently, for example, than 'puppy a' (as in, 'Give the puppy a ball'). Using these statistics one can take any sentence that is not in the corpus and compute how closely it conforms to statistical patterns in the corpus. For example, 'Give the doll a kiss' might be very statistically probable, even if it doesn't exist in the corpus, because every pair of words in the sentence frequently co-occurs. With their statistics in hand, Reali and Christiansen set out to see whether grammatical or ungrammatical questions would seem equally probable to a child who had been exposed to sentences in the corpus. To do this, they randomly generated a series of complex polar interrogatives that were either grammatical or ungrammatical – questions like our earlier examples: 'Is the puppy that is barking angry?' and 'Is the puppy that barking is angry?' They then computed the probabilities for these, relative to the corpus and they discovered that almost every one of the grammatical interrogatives had higher probability than the ungrammatical interrogative (94 per cent). If children construct sentences by selecting the option that has higher statistical probability, they would almost always get it right.

This is an extremely important finding. By their second birthday, children have already heard enough sentences to select between grammatical and ungrammatical questions even when they are more complex than the questions they have heard. Complex polar interrogatives are the parade example used to argue for the poverty of the stimulus. They play a central role in arguments for innateness, and the standard assumption has been that no statistical explanation can explain how children form these questions correctly. But now we have an explanation. If a child encounters an assertion with two verbs and wants to make it into a question, there are two options favoured by prior experience: move the first verb to the front or move the second. But these two options are not equally consistent with prior statistics. Moving the second verb is almost always more consistent with other sentences a child has heard. The fact that kids are accurate in complex question-formation can be explained by appeal to statistical learning, and therefore this most celebrated argument for innate rules is undermined.

Grammar *Ex Nihilo*

The argument that we've just been looking at tries to establish that children master rules that are not exemplified in their primary linguistic data. In response, we saw that the data are less impoverished than Chomskyans assume. The most famous example of children going beyond what they've heard can be handled by appeal to statistical learning. We are still far away from showing that statistical learning can explain every linguistic achievement, but there is reason for optimism. After all, adults do speak correct English, and children hear accurate sentences for years. The rules kids need are being used by their teachers, and that raises the possibility that kids might be able to infer these rules by some kind of statistical extrapolation.

But there is one thing in the poverty of the stimulus argument that cannot be dismissed with such simple optimism. In all the cases we've been looking at, kids learn grammatical rules from linguistically competent speakers. Astonishingly, there is also evidence that kids can learn grammatical rules even when the speakers they listen to are incompetent. Learning tends to obey the garbage-in/garbage-out principle: bad teachers produce bad students. But in the case of language, this principle seems to be violated. There are cases where kids acquire languages that are much more grammatically complex than what they hear from their teachers. Since nothing comes *ex nihilo*, this is evidence for an innate language faculty. Perhaps students outperform their teachers because they possess linguistic rules innately.

There are three examples of this phenomenon that crop up regularly in the literature. The first is a case study involving a boy named Simon, who was born deaf and learned sign language.[7] Sign language is just like spoken language in its complexity. For example, just as spoken words can have different endings (what linguists call inflections), gestures can have subtle variations that express things such as quantity and tense. Simon is remarkable, however, because he learned to sign from his parents, who are also deaf, but learned to sign as teenagers and are therefore not very accurate signers. They make numerous errors – omitting signs, using the wrong signs or expressing things in overly complex ways. Simon's sign language is comparatively consistent and accurate. He does not make the errors that were pervasive in his parental input. He follows rules, rather than just repeating the inconsistent patterns that he observed during learning.

The second case involves a language called Hawaiian Creole. Creoles are hybrid languages that combine various different sources. They tend to emerge when groups of immigrants from different countries are brought together in the same isolated place. Immigrants often need to communicate with each other, and if they have no common language, they will improvise by combining elements of the various languages they speak. The resulting system of communication is called a pidgin language. Pidgins are quite different from creoles in that they lack complexity. Sentences in pidgins tend to be much simpler and they lack fixed word order and inflections. Creoles are spoken by descendants of pidgin speakers. Over time, improved systems of communication become more elaborate, systematic and structured. What's remarkable is this sometimes seems to happen in the space of a single generation. Derek Bickerton has argued that Hawaiian Creole was spontaneously created by the children of people who spoke Hawaiian Pidgin.[8] The children who first spoke Hawaiian Creole were introducing structure not present in their parents' language. Therefore, they could not have simply been picking up on the rules governing their primary linguistic data. The rules were imposed on anarchic inputs. They must have come from within.

The third example integrates elements of the first two. There has been some controversy about how quickly Hawaiian Creole emerged, but recently a new language was created in the space of about ten years. The new language is not a hybrid spoken by the offspring of immigrants. It is not spoken at all. It is a language called Nicaraguan Sign. Until recently, deaf people in Nicaragua were not being taught a systematic sign language. But that changed when the country created centralized schools for the deaf in Managua. Children and adults came together and began communicating with each other using improvised methods of signing that they had each developed independently. But in a short period of time, these disparate systems of communication were merged and perfected, resulting in a full sign language governed by systematic rules. The result was documented by Judy Kegl, a colleague of Chomsky. She argues that rules of Nicaraguan Sign were not statistically extrapolated, but rather they were imposed by a universal grammar innate in each of the learners.[9]

These fascinating case studies shed light on how language is learned, but Chomskyans are wrong to think they provide decisive proof that language is innate. Far from it. The case of Simon is especially interesting in this regard. Chomskyans sometimes imply that Simon acquired his highly systematic sign language from parents who could barely sign a sentence. But the truth is that Simon's parents were pretty good. They made errors frequently, but got things right most of the time. Jenny Singleton and Elissa Newport, the linguists who worked with Simon, discovered that Simon's accuracy could be directly predicted on the basis of his parents' performance. When his parents consistently applied a rule 60 per cent of the time, Simon mastered it, but when their performance dipped lower, Simon's skills suffered as a result. The really important discovery here is not that Simon created a structured, inflected language *ex nihilo*, but rather that Simon turned statistical regularities into rules. This is exactly what statistical learning theory predicts. Suppose that a foraging squirrel finds nuts under a particular tree on six out of ten visits. You can be sure the squirrel will start checking that tree every time she passes by. Now consider the kind of information Simon had to work with. Simon was learning American Sign Language (ASL). In ASL, verbs of motion are inflected to indicate such things as the direction and manner of movement. If the signer wants to say that a car is going up a hill, the usual sign for driving is altered to indicate the upward movement. But Simon's parents often forgot to add such inflections.

They expressed verbs of movement without always altering the gesture to convey the nature of the movement. Now imagine that Simon wants to express the fact that a car drove up a hill. He has seen his parents modify that motion sign with an upward gesture about 60 per cent of the time. This looks like a pretty solid pattern, and it clearly helps express the thought that Simon is eager to convey. It's unsurprising, then, that Simon's motion signs regularly include the information that his parents left out. They were good enough to teach him how to make such gestural inflections, even though they didn't always follow the rules themselves. Simon's statistical learning mechanisms convert regularities in his parents' speech into reliable rules, but this is not evidence for innateness; it simply reveals an important fact about how statistical learning works.

The same principle may be at work in Hawaiian Creole and Nicaraguan Sign. Despite advertisements to the contrary, there are not cases of highly structured languages appearing without any highly structured inputs. Hawaiian Creole emerged out of Hawaiian Pidgin, and, while this pidgin was limited in its degree of structural complexity and consistency, its speakers all also spoke other languages, such as Cantonese, Japanese, Portuguese and Hawaiian. In addition, the pidgin emerged in a colonial context, where settlers were highly motivated to communicate with English speakers who controlled the labour market. Thus, the pidgin speakers and their offspring were exposed to English, from which most of the vocabulary derives. The grammatical rules of Hawaiian Creole reflect these roots, borrowing from various languages, but disproportionately borrowing rules from English. So it is not a case of structure *ex nihilo*. Contrary to nativists' advertisements, Hawaiian Creole did not emerge in a single generation. It is not a case of children hearing unstructured inputs and then instinctively imposing innate rules. Rather it's a case of a community slowly perfecting a communication system so people can express their thoughts effectively and function in English-dominated workplaces.

Nicaraguan Sign also emerged more slowly than is sometimes implied by casual commentators, but it did develop with remarkable speed, and some of the deaf signers who first came to Managua and influenced its development had no knowledge of any highly structured language. So it's a powerful case in the nativist arsenal. But it is not a case of structure emerging from nothing. In the school setting in which Nicaraguan Sign emerged, there was some limited exposure to fully developed sign languages and considerable exposure to written Spanish and Spanish lip reading and the mouthing of Spanish words. There were concerted efforts to teach Spanish, which were unsuccessful, but may have exerted an influence. Those who have studied the emergence of Nicaraguan Sign have not fully documented the communicative inputs, and as a result the case has not been presented with enough detail to assess whether it supports the innateness hypothesis or not. When *The New York Times* published an article describing Nicaraguan Sign as proof that language is innate, William Stokoe, the linguist who pioneered the scientific study of sign language, wrote an irate letter rebuking the paper for biased journalism. Many researchers have documented ways in which sign languages become more refined through social interaction, energy-conserving streamlining and intentional innovations by signers. For example, Ivani Fusellier-Souza has studied emerging sign languages in Brazil, and she finds that signers sometimes dramatically simplify a newly innovated sign in the course of a short conversation, making it easier to convey in one simple gesture. This is precisely the kind of innovation that has been noticed in Nicaragua. But it does not imply the workings of an innate language faculty. The fact that signers streamline their signs and use them more consistently in a social context is easily explained by the demands of efficient communication.

The examples we've been considering illustrate the poverty of the stimulus argument in its most radical form. They have been presented by nativists as evidence for the claim that children will end up with a grammatically rich language even if they are never exposed to one. But none of the examples show that. In each case, the learners have extensive contact with people who have mastered grammatically rich languages, and, in cases where the details of the primary linguistic data are available, there is a predictable correlation between what children produce and what they are exposed to. These examples do show that children do not merely copy the statistical distributions of their inputs. They turn regularities into rules, they innovate and they streamline. But these improvements, which usually emerge incrementally over time, are not evidence for an innate language faculty. They are consistent with a statistical learning story and also reflect the effects of extensive practice and efforts to improve efficiency and successful communication.

APES AND ANOMALIES

Poverty of the stimulus arguments are the most common reasons offered in support of the hypothesis that there is an innate language faculty. As we have just seen, these arguments are far from decisive. But nativists have another family of arguments that they often use. There are cases where learners fail to acquire fully developed language skills. Such failures are instructive because they help identify the prerequisites for language acquisition. Nativists argue that the prerequisites include an innate language faculty. But statistical learning theorists think this conclusion is hasty.

Language Deficits

In arguing for the innateness of language, Chomsky has often emphasized the fact that language skills do not vary with intelligence. Clever or clueless, we all learn to speak. In fact, people who are mentally retarded can have perfectly intact language skills, and individuals with Williams Syndrome (a rare genetic disorder) are linguistically precocious, despite very low IQ scores. These findings have been slightly exaggerated – vocabulary size and sentence complexity can increase with intelligence – but it is incontrovertible that you don't need to be a rocket scientist to master inflections and embedded clauses. This casts doubt on the idea that language learning requires conscious reflection, but it comes as no surprise to the defender of statistical learning. It is plausible that our capacity to unconsciously tabulate statistical frequencies is independent of general intelligence, just like our capacities to perceive objects, recognize patterns and hone physical skills. But there is another kind of dissociation between intelligence and language that puts more pressure on the opponent of innateness. Some people who are perfectly smart have profound difficulties with language.

First consider aphasias, which are deficits in linguistic abilities caused by brain injury. There are various kinds of aphasia, but the most famous are Broca's aphasia, which is an impairment in language production, and Wernicke's aphasia, which is an impairment in language comprehension. These conditions are named for the people who discovered them, and those names have often been assigned to the brain areas associated with the disorders. Broca's area is located around the base of a fold on the side of the left frontal cortex. Wernicke's area is located towards the back of a bulge in the temporal cortex. These have been called 'language areas' because they are thought to be specialized for language skills. Individuals with aphasia are often highly functional and intelligent. They can carry out ordinary activities, pursue goals and otherwise exhibit normal intelligence. This is hard to square with the idea that language is learned using general-purpose mechanisms. If that were the case, linguistic impairments should be caused by brain areas that serve multiple functions, and they should co-occur with other deficits. Nativists claim this is not the case, and they have used aphasia to argue for the conclusion that we have evolved brain structures that are dedicated to language.

A closer look at aphasias tells a different story. While it's certainly true that people with aphasia can retain a high level of intelligence, they usually have problems that go beyond language. Elizabeth Bates and her colleagues found that aphasic individuals have other deficits, including problems with the recognition of non-linguistic sounds and the interpretation of pantomimed actions.[10] Bates also found that these other deficits are associated with injuries to Wernicke's area and Broca's area respectively. That is unsurprising. In healthy individuals, Wernicke's area is right next to the brain structures that specialize in hearing, and Broca's area is next to premotor cortex and implicated in action perception. Bates concludes that there is no such thing as a language area in the brain, because the areas that contribute to language do other things as well. Her research suggests that aphasia is not a selective language deficit, and it supports the conclusion that language is learned using psychological mechanisms that may have evolved for more general sensory and motor skills.

Aphasia is not the only disorder that has been described as a selective deficit in language skills. There is also a disorder called Specific

Language Impairment, or SLI. Aphasias are caused by brain injury, but SLI is inherited. In fact, it has been linked to a specific gene, which is sometimes described as the 'language gene' by over-zealous nativists. People with SLI can communicate linguistically, but they make systematic errors. They tend to omit inflections (like the -s or -ed endings in plurals and past-tense verbs), and they also leave out function words (such as 'the' and 'he'). These errors are impervious to correction, suggesting that people with SLI are genetically incapable of mastering certain aspects of language. At the same time, they can have normal levels of intelligence. Nativists conclude that language skills are genetically based and independent of more general cognitive capacities.

This conclusion may be too swift. People with SLI have other subtle deficits. For example, studies have found that they make errors when repeating sequences of sounds, and they have difficulty controlling their facial muscles. There is also some evidence that the disorder has to do with a general processing deficit that affects memory access to linguistic information rather than innate linguistic rules. For example, they tend to make more errors when they are constructing structurally complex sentences, as opposed to simpler sentences. If they were incapable of learning certain linguistic rules, they should make errors regardless of sentence complexity. Further support for this interpretation comes from research on people with normal language skills. Arshavir Blackwell and Elizabeth Bates have shown that we all make grammatical errors when our minds are occupied, and these errors look a lot like what we find in SLI. If you hold six digits in your mind and try to speak, you will leave off verb endings. In other words, you will temporarily perform like someone with SLI or aphasia. If linguistic deficits resulted from impairments in cognitive mechanisms that were exclusively dedicated to language, then it should not be so easy to induce such impairments in healthy individuals.

At this stage, we don't know exactly what causes SLI, and we don't have a full understanding of what gets impaired in aphasia. We do know that these conditions co-occur with other deficits, and that blocks the argument from language disorders to innateness. These disorders may arise because of malfunctions in mechanisms that are used for pattern recognition, or memory-processing or muscle control. We don't currently know how such mechanisms contribute to language, and research on language deficits may help us find out. But that research may ultimately confirm that language skills derive from more general abilities, rather than refuting that possibility as nativists would have us believe.

The Critical Period

People with aphasias and SLI have brains that function differently than the rest of us. Their impairments owe to something unusual about how their minds work. But sometimes people with perfectly ordinary minds end up with language impairments. This is what happens when people do not get exposure to language before puberty. It takes extraordinary circumstances for that to happen, since most of us are bombarded by language wherever we go. But it happens occasionally, often under tragic circumstances.

In 1797, a twelve-year-old boy was found near a forest in Aveyron, France. He was naked, had uncut hair and walked on all fours. It was presumed that he had been abandoned early in life and managed to raise himself in the woods. The boy, dubbed Victor, soon escaped, but three years later he emerged again, and this time he was taken in by a physician named Itard, who set out to civilize him and teach him how to speak. Victor adapted to life in a house, to table manners, and to clothing, but efforts to teach him language failed dismally. After intense training, his only words were *lait* (milk) and *o Dieu* (oh God).

No one could figure out why Victor was incapable of learning to speak, but an explanation emerged 170 years after his discovery, when the linguist Eric Lenneberg proposed that language has a 'critical period' for acquisition. That means if a person does not get exposed to language by a certain age, acquisition will be impossible. Lenneberg borrowed the idea of a critical period from research on animals. Konrad Lorenz had claimed that there is a critical period for imprinting in ducks. Lorenz showed that ducks will bond with the first moving object they encounter and treat it as their mother. He thought this instinct was limited to a brief period after birth, though we now know this to be false. The Nobel-prize-winning neuroscientists David Hubel and Torsten Wiesel had also argued for a critical period in the acquisition of visual abilities.[11] They took a cat and stitched one of its eyes closed for the early weeks of life. As a result, the visual brain areas linked to that eye didn't develop, and when the stitches were removed, the cat was blind in that eye and incapable of stereoscopic vision. More recent research has shown that vision can be restored, to some degree, after congenital blindness, but, when Lenneberg advanced his proposal in 1967, critical periods were regarded as a hallmark of innateness. The guiding idea was that innate capacities, such as imprinting or vision, are easy to exercise early in life, but if they don't get used, the neural resources that have been reserved for that capacity get co-opted for other purposes. In a similar spirit, Lenneberg suggested that language is easy to acquire in childhood, gradually becoming more difficult, and, after puberty, impossible. The case of Victor seemed to corroborate this conjecture, as did the fact that immigrants who leave their native land after puberty tend to have more difficulty learning a new language and losing an old accent than immigrants who move earlier in life. Lenneberg himself had fled his native Germany as a young man to escape Nazi persecution and knew first hand how difficult it is to learn a language if you come to it too late. He took this as evidence for the thesis that language, like imprinting and seeing, is an instinct.

Lenneberg's hypothesis caught on with the linguists of the 1960s, who were increasingly under the spell of Chomsky's nativism. The idea of a critical period also entered the popular imagination, and three years after Lenneberg published his influential book, François Truffaut released his film *The Wild Child*, which dramatizes the story of Victor of Aveyron. But evidence for the critical period was still limited. Knowledge of Victor was limited to Itard's short memoir, and anecdotes about immigrants could be used to prove that it is difficult to learn a second language late in life, but not that it is impossible to acquire a first language after the critical period. Of course, scientists could not test the critical period hypothesis without depriving a child of exposure to language for a dozen years, which would be grossly unethical. But tragic events sometimes provide evidence for scientific theories. That is what happened for the critical period.

In 1970, the year that Truffaut's film about Victor was released, social workers rescued a girl named Genie, who had been the victim of unimaginable abuse. Genie's psychotic father and ineffectual mother had kept her locked up in isolation from the age of twenty months to the time she was thirteen, when she was discovered. During her isolation, Genie's parents did not speak to her, and they punished her when she made any noise. When she emerged, she could not speak. Intense efforts were made to teach her language skills, but they met with limited success. With several years of training, Genie got to the level of a 2.5-year-old child. She could form sentences, but they tended to contain two or three words, and she omitted inflections and articles, like 'a' and 'the'. Beyond that, progress seemed impossible. After four years, research funding ran out, and Genie was bounced from foster home to foster home. Some years later, Genie's mother sued the linguists who had worked with her for excessive and abusive testing. The case was settled out of court, and Genie now lives in an undisclosed care centre.

Genie is a complicated case because she was horrifically abused, and there is evidence that she was mildly retarded. But there is no evidence that child abuse or retardation prevent language acquisition. So it is likely that her linguistic limitations arose because she had passed a critical period. Subsequent cases have added further support. The most compelling is Chelsea, who, as a child, was incorrectly diagnosed as profoundly retarded because of a hearing deficit, and, consequently, she was not exposed to language. The diagnosis was ultimately corrected, and her hearing was restored at the age of thirty-one. But it was too late for her language skills to recover. After a dozen years of training, she remained linguistically stunted. Like Genie, she only achieved the skill level of a 2.5-year-old. Nativists conclude that this level of ability may not require an innate language faculty, but anything more sophisticated does. There seems to be a critical period in which we are able to master inflections, complex sentence structure, fixed word order and function words. Beyond that, this may be impossible. That has been taken as evidence that these aspects of language depend on an innate UG.

The evidence for a critical period is now pretty strong, but nativists are mistaken when they take this as evidence for an innate language faculty. The inability to master a first language after puberty may result from a change in some more general cognitive capacity. A proposal of this kind has been put forward by Elissa Newport.[12] One of the challenges facing young language learners is that linguistic inputs are often long and complex. Parents speak to children in whole sentences, and that's a lot of information to take in all at once. Confronted with a five-word sentence, a statistical learning mechanism would have to record the co-occurrence of each adjacent word pair in the sentence (there are four of those), each triad (there are three of those), and each group of four words (there are two of those), as well as more complex combinations, such as the co-occurrence of pairs with triads (another two). That's a lot of co-occurrences to keep track of all at once. It might outstrip the book-keeping abilities of any ordinary human being. Even an adult. But suppose for a moment that we can only keep track of pairs. That's a manageable task, but it has limited utility. Language is complex, and some of the most important patterns and relationships involve whole phrases made up of many more than two words. So statistical learner mechanisms that track very local co-occurrences are not very useful to language learning, and if they try to track everything that co-occurs in a sentence, they will be overwhelmed. Newport noticed an elegant solution to this problem. Imagine a mechanism that begins with a very limited capacity and then grows over time. Initially, such a mechanism would be able to track co-occurring pairs of words, but nothing more. Then, over time, it would improve. It could track pairs of pairs, and so on. If such incremental growth occurs in a statistical learner, intractably complex input patterns become learnable. But suppose you present sentences to such a learning mechanism after it has grown to its full capacity. Without having built up slowly, it will be overwhelmed by typical English sentences and it will fail to learn. The critical period could arise because, by puberty, we are capable of tracking multiple things (but not everything) at once, and that makes linguistic inputs too overwhelming to master.

This proposal gains support from computer simulations. Jeffrey Elman programmed a statistical learning model with an adjustable short-term memory capacity from tracking regularities in linguistic inputs.[13] He was interested in testing whether these models could learn one important feature of English grammar – the idea that a subject and verb in a sentence must agree in number even if there are words that come in between them. This requires what linguists call long-distance binding: coordinating words that are not adjacent in a sentence. Elman tried to train his computer models to master long-distance binding statistically, by presenting English sentences as inputs and seeing whether they could generalize to new cases. When he made the short-term memory capacity large, the models failed because they were overwhelmed by the complexity of the inputs, and when he made the short-term memory capacity small, they failed because they could not track relationships between non-adjacent words. But his models solved the statistical task successfully if they began with a small short-term memory, which then increased incrementally.

This pattern of growth is exactly what we find in human children. Very young children can only hold a couple of items in their heads at once, but over time that capacity increases, and ultimately, by chunking smaller bits of information together, we can hold as many as seven items in our heads at once. This incremental growth may be exactly what we need to master the complex statistics in language. It also explains why there is a critical period for language acquisition. If our memory capacities gradually increase, then by puberty we may already have a fully mature capacity capable of tracking multiple items at once. But that large capacity gets swamped by language. Complex sentences, inflections and function words that are not essential for communication are too much to keep track of. Full linguistic mastery is impossible.

This elegant explanation of the critical period contrasts with Lenneberg's account. He proposes an innate language capacity that stops working, for some mysterious reason, in adolescents. Newport proposes a general memory capacity that grows slowly and allows for gradual improvement in linguistic performance. If the capacity matures before exposure to language, mastery is impossible. This takes the mystery out of the critical period. Moreover, there is demonstrative evidence that short-term memory grows over time, so Newport's conjecture is consistent with established facts. The bottom line is that the critical period can be explained without postulating an innate language faculty.

Species Specificity

You might have heard that apes can learn to use language. There is some truth to this. Efforts to teach rudimentary linguistic abilities to our closest living relatives have met with some success. But there are limits. Certain aspects of language seem to be uniquely human. This fact has been taken as support for innateness. After all, apes are a lot like us. There is certainly little reason to think their perceptual systems differ dramatically from ours, and, like any foraging animals, they must be reasonably good at tracking statistical regularities. If language acquisition relied on such general resources, there should be no limit to what apes can learn. The fact that human language is unique to us might mean that we evolved a language organ some time after we split from the ancestors we share with our hairy cousins.

Efforts to teach language to apes have been underway for decades now. Initially researchers tried to teach apes to speak, but their vocal abilities are not up to the task. A breakthrough came when researchers switched from speech to sign language. Apes turned out to be pretty good at expressing themselves through gesture. A chimpanzee named Washoe learned to use about 250 signs, and a gorilla named Koko is alleged to have command of 1,000 signs. The most celebrated language-using ape is a bonobo named Kanzi. Bonobos are a subspecies of chimp known as the hippies of the ape world; they are far more peaceful than their common chimp cousins, and they seem to have an insatiable appetite for casual sex. Kanzi uses lexigrams instead of signs. These are symbols on a large grid that he can point to when he wants to express a thought. His trainer, Sue Savage-Rumbaugh, had initially planned to teach these lexigrams to Kanzi's adoptive mother, Matata. Those efforts failed, but Kanzi, who had been observing these lessons from infancy, began to pick up the signs. He is now estimated to know about 500 signs and he responds to verbal commands. Kanzi's success has been more carefully documented than Koko's, and his comprehension of basic syntactic skills has been carefully tested. For example, Kanzi can use word order to comprehend sentences: he can differentiate the commands 'Put the ball on the toothpaste' and 'Put the toothpaste on the ball'.

For all this success, ape language is limited. One difference is size. While apes reach vocabularies in the hundreds, humans master tens of thousands of words. Another difference is function. Herb Terrace, who trained a chimp named Nim Chimpsky to sign, later noted that apes don't use language to describe the world around them or express thoughts; rather they use language to manipulate others, by making requests and commands. The most important difference in the present context, however, has to do with the limitations of ape grammar. Their sentences are short, with no complex clauses, and they use neither function words nor inflections. This pattern should sound very familiar. Apes achieve the same kind of skills that Genie attained, and their limitations resemble what we observe in acquired and genetic language deficits. This is the level of achievement we find in 2.5-year-olds. Nativists argue that abilities of this kind can be acquired using general-purpose cognitive resources. The fact that normally developing human beings progress past this stage and acquire highly structured inflected languages suggests that we have mechanisms that are absent in people with language impairments, people who come to language after puberty and apes. According to nativists, our comparative success derives from an innate language faculty.

This is a seductive argument. It looks like there is a qualitative divide between what apes can learn and what healthy humans can learn, and that could be explained by positing a Chomskyan UG in our species. But we have already seen that there are explanations of human language deficits that do not postulate an innate grammar. It is especially instructive to recall how statistical learning theorists explain the critical period. According to that story, our capacity to learn a language depends on a slowly developing short-term memory capacity. Because our capacity starts small, we don't get overwhelmed by linguistic inputs and can attend to co-occurrence patterns in pairs of words rather than whole sentences. Gradually, short-term memory increases, and we can master more complex statistical relationships, building up to a mature grammar in all its complexity. Now suppose that apes begin with a small short-term memory capacity that never grows significantly larger. That would mean they would be stuck at the developmental stage we see early in human life. They could master rudimentary sentences with two or three uninflected words, but little more. That is precisely the pattern that research on ape language has revealed.

This explanation of ape language limitations is not mere speculation. The anthropologist Dwight Read has documented the short-term memory differences between humans and apes in great detail.[14] Apes, it turns out, can hold about two items in their heads at any given time. That limit is exactly what we find in human beings until they are about two and a half. At that age their memory capacities continue to grow incrementally. By puberty, human beings can hold about seven items in their head. Limitations in ape memory capacity align with what we see in humans at exactly the age-level that apes can master language. They cannot move beyond that level because their memory capacities stop developing. This explains why humans have more sophisticated language skills without postulating any innate grammatical rules.

THE STATISTICAL UPSHOT

This ends our tour of arguments for the hypothesis that there is an innate language faculty. Despite their tremendous influence and initial appeal, none of those arguments is decisive. It is very possible that language is acquired using statistical learning mechanisms that did not evolve for the purposes of communication. This may sound like a weak conclusion – a mere possibility. But it's actually a heretical suggestion. Nativism is deeply entrenched, and its defenders are passionate about their beliefs. Debates about the innateness of language are some of the most heated in all of science. The discovery that there are chinks in the Chomskyan armour is a bit like the discovery that the Earth might not be at the centre of the universe. It forces us to reconsider fundamental assumptions about human nature. The ultimate success of the statistical learning approach remains to be seen. We don't yet know if statistical learning is powerful enough to explain language acquisition, and it may be many years before the issue is settled. But we are already in a position to see that the arguments for nativism are inconclusive. The statistical alternative is a serious option.

The statistical learning approach has a great advantage over the innateness hypothesis: simplicity. All sides to the debate must admit that humans automatically engage in statistical learning. Our ability to notice patterns, recognize familiar objects and make decisions based on prior decisions depends on this. All sides to the debate must also acknowledge that human short-term memory grows over time, and that incremental growth considerably increases the power of statistical learning. These are established results. If language acquisition can be explained by appeal to general cognitive resources that human beings are known to possess, then there is no reason to posit an innate language faculty. No reason, that is, unless there are strong arguments for thinking that language is innate. By casting doubt on these arguments, we have removed the main reasons for preferring the nativist approach to the statistical approach. In one sense, this is a stalemate. Nativist approaches clearly can explain language acquisition (they can build in as much innate language as the evidence requires), but there is no decisive evidence that these approaches are on the right track. Statistical approaches have not yet proven that they can explain language acquisition, but there are no decisive arguments for the conclusion that these approaches will fail. So which should we pick? One answer is that we should actively pursue both possibilities. It's good for science to explore all serious options. But, there is also a sense in which the statistical approach wins out in a stalemate. As the simpler theory, it should be the default explanation. We should not posit an innate language faculty unless we are forced to by decisive arguments, and we are not in that position yet. So even if we grant that both approaches should be actively pursued, we should also feel confident enough to regard the statistical approach as more likely. That is not a weak conclusion. It's the first step in a scientific revolution.

7

Words and Worlds

Before the 1950s, professional linguists were primarily concerned with variation. They were struck by how different the world's languages are and they believed that these differences were extremely important. Language, they surmised, can play a role in shaping how people think and experience the world. If so, English, Russian and Spanish are not just ways of speaking; they are ways of thinking as well. This 'linguistic relativity' hypothesis was once widely accepted, and it was one of the things that made the study of language seem exciting and worthwhile. Things changed when Chomsky came along. Chomsky's hypothesis that language is innate led many to believe that linguistic differences are really superficial. At their core, all languages are alike. With this new conception of language, linguists became preoccupied with uncovering hidden universals, rather than documenting differences, and they soon abandoned the idea of linguistic relativity. Learning a language is just a matter of learning how to express the universal grammar.

In the last chapter we saw that the innateness hypothesis has been called into question. Language may actually be learned. This heretical hypothesis has been leading to a new perspective on the very nature of language. Language is increasingly seen as a kind of tool that humans invented, and this raises a question of what the invention might do for us. One answer is that language may influence they way people think. That possibility has triggered renewed interest in the questions that got linguistics off the ground in the late nineteenth century. If language can influence thought, and there is no innate universal grammar, then maybe distinct languages influence thought in different ways. Linguistic relativity is having a comeback.

INVENTING LANGUAGE

If language is not the result of biological evolution, then where did it come from? The obvious answer is that we created it. Language is a human invention. That does not mean there was an inventor. There may have been no linguistic Thomas Edison who patented the first subordinate clause or filled each household with prepositions. It is more likely that language came about through a long cooperative effort involving many generations of speakers. It may even have been invented several times by isolated groups in different parts of the world. That is a hotly contested question in historical linguistics.

The precise details of how language emerged are irretrievably lost in the past, but it's not difficult to imagine some of the relevant factors. Like all animals, human beings are good associative learners, and this may be a precondition to linguistic development. Pavlov's dog can associate a bell with a bowl of food and, in so doing, treats the bell as a *sign* that food is coming. In nature, this associative ability is invaluable. Squirrels associate large trees with nuts because smaller trees are too young to produce nuts. An association of that kind can be regarded as a *natural* sign. Large trees naturally signify nuts. When it comes to signs, human beings have a distinct advantage: we can invent signs. We don't simply exploit naturally occurring associations. We invent sounds or gestures and associate them with things. I say sounds or gestures, since those are the two things we can most easily produce. Our ability to devise signs probably comes from the fact that we have a greater capacity to control our behaviour than other creatures, and we also have the cognitive ability to recognize that, in wilfully producing a sound, we can influence the behaviour of other people. The easiest words were probably not mere labels but could also play the role of commands, requests, warnings and so on.

Once we got into the practice of using signs, the next step would be producing sentences. As beings smart enough to devise a vocalization that signifies lions and a pointing gesture that means over to the right, we would have been in a position to combine signs: vocalizing lion while pointing to the right. This combined sign signifies that there is a lion to the right and is thus a primitive kind of sentence. Sentences can also be produced by combining two vocalizations or two gestures. This level of linguistic achievement may have already been available to our ancestors and cousin species, such as the Neanderthals, but *Homo sapiens* had an extra advantage. Through incremental learning, we could make our sentences increasingly complex.

The necessary biological breakthrough came when we evolved large short-term memory capacities that emerge slowly over development. As we have seen, this trait allows us to learn languages that have complex sentence structures and inflected words. But why did we evolve short-term memory capacities of this kind in the first place? The answer is probably that an incrementally increasing working memory facilitates the acquisition of complex skills, such as tool construction. Dwight Read, who has documented memory differences between humans and apes, has shown that human tool-making requires a number of steps and a temporally protracted learning period. It may be hard for a child to comprehend all the steps at once, but children who can only process a few things at a time may gradually build up the capacity to memorize a multi-step procedure. This confers a significant survival advantage. Human beings are the most versatile and inventive tool-users in nature, and all our major advances, from arrowheads to agriculture, have come through technological innovation.

The cognitive resources that made us good incremental learners also made us good linguists. We acquired the ability to learn complicated grammatical rules as a byproduct of the ability to learn complicated skills. At this point the stage was set for modern language. It's not clear how or when modern language appeared, but it's not surprising that it did. If our ancestors were already using primitive two- or three-word sentences to communicate, their descendants could build on this ability. The first generation of children born with incrementally increasing short-term memory capacities could already start constructing sentences that were more complex than their parents' by combining simple sentences together. Then, over time, linguistic complexity could group, and rules could be passed on from generation to generation. And each generation could add new innovations, increasing vocabulary and adding ways to express tense, quantity, aspect and so on. These additions might have depended on the insights of some clever individuals, but they could not have gained traction without wide adoption, and it is more likely that innovations came about through collaborative efforts.

It has sometimes been suggested that modern language emerged about 50,000 years ago, because that is the time at which the archaeological record begins to show use of diverse materials in making tools, stylistic variations across geographical areas, arts and other products of culture. Some people surmise that this explosion of productivity occurred because of a biological mutation that gave humans the capacity to use language. That would be very surprising, however. Anatomically modern humans have been around for almost 200,000 years, and there is no physical evidence that our minds radically changed 50,000 years ago. Plus, it's not clear why modern language would lead to a revolution in tool design. Some contemporary hunter-gatherer groups still use very simple stone tools, despite having complex languages. The increase in cultural artefacts may have more to do with innovations in technology, climate change or some other factor that allowed humans to have more leisure time and diversified labour. Moreover, the anthropologists Sally McBrearty and Alison Brooks argue that cultural products emerged more slowly than initial interpretations of the archaeological record would suggest, so there is no reason to posit a sudden change in human cognition.[1] It is more likely that language has been with us for the better part of the last 200 millennia.

VOICES IN THE HEAD

It is unlikely that the invention of language triggered a cultural revolution, but language may have had an influence on the way we think. However, the extent and nature of that influence is a matter of controversy, and this controversy hinges on a broader dispute about the nature of language. Let's consider three options.

The most modest option can be called the *expressive* theory of language. According to expressive theory, language primarily serves to express thought. We think, in this view, using something other than the languages we speak. For example, we might think in a language of thought or in imagery – the two options discussed in chapter 5. Spoken languages, like Hindi and Hungarian, simply allow us to translate those thoughts into symbols that can be shared with others. This view was put forward by the Empiricist John Locke, and it was shared by many of his Rationalist rivals, including Chomsky. It's Chomsky's contention that the language faculty is independent of other cognitive faculties. Its main function is to communicate, not to think.

The expressive theory denies that language is a tool for thought, but it is perfectly compatible with the view that language has an indirect influence on thought. For example, you can influence my thoughts by telling me things I didn't know. When that happens, you express your ideas in words, which I then hear and decode. You can also use language to point out things ('Have a look at that!') and to keep records ('Dear diary . . .'). The ability to share ideas and store information has a tremendous impact on what we are capable of as a species. We can accumulate knowledge this way, building on what others have learned. But language does not change *how* we think on the expressive approach; it only has an impact on *what* we think, by increasing our access to the ideas of others.

One problem with the expressive theory of language is that we often seem to think using language. Many of us experience an endless stream of sentences in our heads as we make our way through life. It is as if there were a little narrator in there, describing what we do, evaluating our actions or telling us what to do next. It could be that these narratives are just enabling us to express our thoughts to others, but it often seems as if they are actually contributing to how we think. This is especially clear in cases where we find ourselves using inner speech while trying to solve technical problems. When doing arithmetic in our heads, the numbers we mutter silently to ourselves help us arrive at the solution.

The intuition that language helps us solve certain kinds of problems can be confirmed experimentally. Psychologists test for the role of language in thinking by devising tasks in which they prevent people from verbally describing things to themselves. The preferred method of doing this is called shadowing. In this technique, participants in the experiment listen to words being spoken though a pair of headphones and repeat whatever they hear. They might be asked to repeat a series of words or sentences or a talk radio broadcast. This prevents them from engaging in silent speech. While shadowing what they hear, the participants in the experiment must solve a problem, and psychologists measure whether shadowing interferes with their ability to do so.

Using this technique, psychologists have confirmed that repeating words makes it difficult to keep track of numbers. For example, Michael Frank and his collaborators found that people had difficulty matching an array of spools with exactly the same number of deflated balloons if they were repeating the words in a radio broadcast at the same time.[2] Language clearly helps us count.

Language helps with other problem-solving tasks as well. Ashley Newton and Jill de Villiers had people repeat sentences while observing a cartoon in which a cat moves a rabbit's carrot when the rabbit isn't looking.[3] Afterwards people were asked whether the rabbit would look for its carrot in the original location or the new location. Normally this task is easy for adults, but, when they repeated sentences, performance dropped off to chance levels. This suggests that language plays an important role in keeping track of other individuals' beliefs. In another study, Linda Hermer-Vazquez and her colleagues showed that repeating sentences prevents people from correctly recalling spatial locations.[4] In their study, they placed an object in a location that could be recalled by combining geometrical information (e.g., on the left) with colour information (e.g., by the blue wall). People who were shadowing sentences had difficulty remembering where those objects were located. This suggests that we use language to integrate different kinds of spatial cues.

In light of such findings, some researchers have concluded that the expressive theory of language underestimates the role of language in thought. Language is not only a tool for expressing our thoughts, they say, it can also help us think. Some researchers have gone so far as to adopt the view that most or all thought is carried out in one of the languages we speak. This can be called the *cognitive* theory of language, because it implies that the primary function of language is not expression, but cognition. The cognitive theory is a dramatic departure from both the traditional Empiricism of Locke and the Rationalism of Chomsky. Those authors claim that the languages we speak are separate from the mechanisms of thought.

The cognitive theory has a clear advantage over the expressive theory favoured by Locke and Chomsky: it explains the results of shadowing experiments. But it may go too far. The claim that language is the primary medium of thought is hard to reconcile with several established facts. First, animals that lack language can clearly think. They learn, classify, solve problems and make decisions. There is a measurable continuity between the way animals solve problems and the way that we humans do, suggesting that we think in similar ways. But animals clearly don't think using spoken languages. Second, human beings who suffer from language impairments seem to be capable of thought. There are people who suffer from transient or short-lived aphasias, during which their language capacity is profoundly impaired. However, during this time their behaviour is organized and intelligent, and afterwards they report that they were able to think without language. Finally, as noted in chapter 5, we could not learn a language if we did not have the ability to assign meanings to the words in its vocabulary. Words are arbitrary signs, and, to understand them, we must be able to think about what they mean in some non-linguistic way. Therefore, it is very unlikely that thought depends on language in the way that the cognitive theory implies.

If the cognitive theory is too radical, how can we explain the results of shadowing tasks? One answer is that language is used as a kind of shorthand in thought, but it isn't essential. Consider Hermer-Vazquez's suggestion that we need language to integrate information about geometry with information about colour. This could not be true, because even dogs and chickens can solve the tasks that she uses in her experiment. A more likely suggestion is that we store facts linguistically when doing so increases efficiency. We could integrate colour and geometry by storing a visual image of a scene, but it is easier to just describe a location in words. When shadowing prevents us from using a verbal strategy that we have come to depend on, we make mistakes. Likewise for the study in which we have to attribute beliefs to a character who has been deceived. We are capable of attributing false beliefs without language. I can imagine what it is like for children to believe in the tooth fairy by simply visualizing what they think happens when their baby teeth are replaced by gifts during the night. But language is a very efficient way to store information about false beliefs. It is difficult to simultaneously imagine how misguided people think of the world while tracking how the world actually is. We make this easier by representing their beliefs linguistically.

Maths provides a stronger case for the cognitive theory of language. There is a limit on how many items we can keep in our heads at once. So linguistic counting may be necessary for counting large quantities. But even so, our basic understanding of what numbers are is grounded in perception. We can visually distinguish a group of two apples from a group of three. We learn numbers by labelling these perceivable differences and then mastering words for quantities that exceed the number of objects we can attend to at once. For these large quantities, the cognitive theory of language is right, but the practice of counting would be unintelligible if we couldn't keep track of smaller quantities in a non-linguistic way. Thus, even with numbers, there is only a very limited sense in which thought depends on language. Here too, it would be equally accurate that non-linguistic thought grounds our comprehension of language.

The cognitive theory overestimates the role language plays in thought, and the expressive theory underestimates it. We need a compromise. The most plausible view can be called the *interactive* theory of language. According to this view, comprehension of language is grounded in non-linguistic ways of understanding the world, which is what expressive theorists propose. The interactive theory parts ways with the expressive theory in saying that we sometimes use language as a medium of thought because it can be more efficient than constructing images. This is a fairly weak concession to the cognitive theory, because it implies that the silent sentences in our heads are kind of like cheques that ultimately get cashed in for the hard currency of mental imagery. But the interactive theory says something more. Language is not just a temporary placeholder for non-linguistic thoughts, it can also influence the way non-linguistic thoughts are formed.

To see how this might work, think of what happens when you read the caption on a picture in a gallery. There is a picture by Edvard Munch that shows a faceless couple in the background, and a close-up portrait of a man with an ambiguous expression occupying much of the foreground. At first, one might think the couple represents one of the man's memories or perhaps a desire, but the title in the caption corrects these interpretations: the painting is called *Jealousy*. When we read the title, the man's face suddenly looks darkly disturbed, his eyes look crooked and sad, his shoulders seem slumped, and the woman in the background, who looked alluring at first, now looks like she is vulgarly exposing herself to seduce another man. Knowing the title disambiguates the work and changes our experience of it profoundly. For another example, consider Andrew Wyeth's masterpiece *Christina's World*, which depicts a young woman in a pink dress sitting in a grassy field and staring at a house on the horizon. If you read the description next to the painting you discover that Christina was paralysed from the waist down, and that she could get back home only by crawling. Suddenly, Christina's posture looks twisted, and we reconstrue relaxation as desperate determination. Psychological research shows that the distance between Christina and the house looks greater when we know the story behind the painting;[5] the distance expands as we imagine her dragging herself, with slender arms, across a massive, lifeless field. In these examples, words alter our perception, because they draw our attention to features that might otherwise be ignored or misinterpreted.

Psychologists have recognized the effects of words for some time. In the late nineteenth century, the psychologist Joseph Jastrow produced an ambiguous drawing that is now known as the duck-rabbit; it looks like a duck's head when you focus attention on the left, and like a rabbit's head when you focus on the right. The duck's beak gets reinterpreted as the rabbit's ears. This reversal is perfectly possible without language, but words can have an impact on which interpretation we arrive at first. If the picture is labelled *Duck*, you might never notice that it can be seen as a rabbit. In the 1930s, Leonard Carmichael and his colleagues discovered that people misremember pictures depending on how they are labelled. In the experiment, if you see an image like this O-O labelled 'dumbbells' you will recall it having a thicker and longer connecting line. If the same image is labelled 'eye-glasses', you will recall a shorter connecting line, curved slightly upward in the centre. In the 1960s, Sam Glucksberg and Robert Weisberg investigated the role of language in physical problem-solving.[6] They gave participants a candle, a box of thumbtacks and a book of matches and told them to affix the candle to the wall so that it didn't drip on the floor when lit. Think for yourself how you would solve this problem. For most of us, it is very difficult. We try to tack the candle to the wall, the candle falls, and even if we could get it to stick, wax would drip on the floor. But the problem is solved effortlessly if the box containing the thumbtacks is labelled 'box' rather than 'tacks'. Now it becomes clear that the box can be tacked to the wall, and the candle tacked inside the box, which will also serve to catch the falling wax.

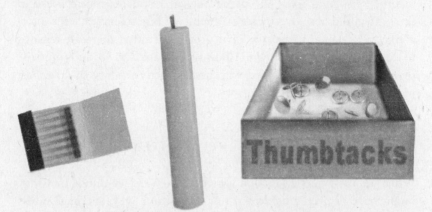

Figure 3. Use these items to affix the candle to the wall so that it can light without dripping on the floor.

These findings do not imply that we think in language, as defenders of the cognitive theory would have it. We interpret a duck-rabbit by focusing visual attention, we recall a picture by storing a mental image and we solve spatial problems by visualizing how the parts can be used together. But language is actively influencing how we construct our mental images. A title on the duck-rabbit makes us search for one set of features rather than another. A caption on a picture can distort visual memory by bringing associated images to mind. A label on a box can lead us to see how that object would help solve a physical problem. Words alter perceptual experiences in all these cases. This departs from a merely expressive theory of language. Language can directly interact with the representations that we use in thought.

These examples indicate that Locke was wrong to think that language plays no role in cognition. But it is important to notice that this does not undermine the central tenet of his Empiricist theory: the claim that we think in imagery. Locke failed to realize that an imagistic account of thought is fully compatible with the view that language plays an active role. We think in images, but those images can be influenced by verbal labels. Historically, Empiricists have overlooked this possibility, but they can embrace it without giving up their theory of how the mind works. A Rationalist like Chomsky is in a more difficult position. It is crucial to his innateness hypothesis that the language faculty is isolated from the psychological resources that we use in thought, and that makes it more difficult for him to embrace the interactive theory. In any case, we have already seen that there are reasons to doubt both Rationalism and Chomskyan nativism. So, an Empiricist theory of thinking, combined with an interactive theory of language, may be the best way to go.

WORDS FOR SNOW

When an artist puts a label on a picture, viewers see things that they might have neglected before. In much the same way, language labels reality and draws our attention to things that we might not have seen without it. In this way, language can have an enduring impact on thought. One of the most fascinating and controversial claims in modern psychology concerns the nature of this impact. Do all languages influence thought in the same way, or are there linguistic differences that impact thought in different ways?

Chomskyans tend to think that differences between languages are relatively superficial. We all have the same innate language faculty, which constrains the languages that we acquire and minimizes cross-linguistic differences. But what if the Chomskyan programme is wrong? If there is no innate language faculty, then there may be more room for variation, and that variation may promote differences in how people think.

This idea was advanced most influentially by Edward Sapir, one of the leading linguists in the generation before Chomsky. Sapir was a student of Franz Boas, the founder of modern anthropology, and he regarded language as fundamentally infused by culture. Each language builds in the beliefs of a cultural group and serves to transmit those beliefs. When you learn a language you are learning a culturally specific way of thinking. Language shapes the way we think.

Sapir's view of language was soon taken up by an amateur linguist named Benjamin Lee Whorf. Whorf earned a living as a fire insurance adjuster; he'd catch the people who burned down their homes to claim insurance money. But his real passion was linguistics, and he attended Sapir's lectures and wrote both scholarly and popular essays on language in his spare time. Whorf observed that we are profoundly unaware of the ways in which language influences thought. The way we think about things seems perfectly natural, even inevitable, but, Whorf surmises, there is no privileged way to understand categories in the world and no uniquely correct way of drawing inferences. The categories and inference rules that we use are built into the languages we speak. We think we have direct mental access to how things are in the world, when actually we are thinking through a verbal veil that imposes an order that is not actually out there. Each language is a worldview – a way of organizing reality.

Sapir and Whorf endorse two dramatic claims: languages organize the world differently, and languages shape the way we think. Together, these claims entail that differences in the languages we speak result in different ways of thinking. This is known as the Sapir–Whorf hypothesis. We have already seen that there is reason to believe that language can influence thought. But is it really true that languages influence thought in different ways? And, if so, how profound are these differences?

Whorf offers a number of examples in his writings.[7] He says that the Nootka language, spoken by indigenous people of the Pacific North-west, contains nothing but verbs. So instead of saying 'There is a house', they say something that would be literally translated as 'It houses'. Whorf invites us to infer that this language forces its speakers to see the world as a collection of events or processes, rather than things. He also describes the way tense is represented in the language of the Hopi Indians. Rather than conjugating verbs to convey when an action took place in objective time, Hopi expresses tense by referring to speakers' attitudes. To speak about the present, speakers use a word that indicates that they are conveying a belief report; the past is expressed by indicating that what's being reported is a memory; statements about the future are presented as expectations. In Whorf's view, this means that Hopi speakers are forced to construe time as completely subjective. Whorf also gives the example of Eskimo words for snow. By his count, they distinguish about seven kinds of snow in their language, and therefore they do not see snow the way we do, as a uniform substance.

The Sapir–Whorf hypothesis caught on in the popular imagination, but it also came under scientific attack. First, there is a chicken-and-egg problem. Sapir and Whorf assume that language shapes thought, but it might also be that thought shapes language. Perhaps the cultural beliefs of these groups influence their worldviews and then get encoded in the language. I don't think this is a very damaging concern, since Sapir and Whorf need not deny that thought can be affected by cultural factors outside of language; they need only say that language is one of the ways culturally shaped beliefs are spread. Perhaps culture affects thought, thought affects language, and then language sustains patterns of thought that were first arrived at non-linguistically. A more damaging criticism is that Sapir and Whorf are often relying on hearsay. Sapir had a Hopi informant, but neither author had much direct knowledge of languages they were describing. As a result, they are often guilty of gross mischaracterizations. The Eskimo snow case is a dramatic example.

You've undoubtedly heard that there are many Eskimo words for snow. How many words are there? Boas, from whom Whorf filched the example, says there are three. Whorf more than doubles that number, and subsequent reports in textbooks and newspapers have systematically inflated it over the years. *The New York Times* reports that there are fifty Inuit words for snow in one article, and 100 in another. None of these numbers is accurate. For one thing, there are numerous Eskimo languages (the word Eskimo is actually an ethnic slur, but it is still used to refer to the language group). Once you pick a language, you need to decide how to count words. Eskimo languages allow speakers to combine multiple words into one phrase-size unit. Do we count all of these? If we avoid such compound phrases and just pick the words that might wind up in a dictionary, the count in representative Eskimo language comes out around fifteen. That might look like a resounding confirmation of Whorf, but we have at least that many snow-related words in English: snow, sleet, slush, flurry, hardpack, dusting, hail, blizzard, frost, ice, crust, snowflake, slope, avalanche and whiteout. Of course, we shouldn't be surprised to find that experts have augmented vocabularies – wine buffs have words for wine that would be lost on the typical carton quaffer – but Whorf and Sapir can't base their bold conjecture on junk data.

The final problem is the most fatal. Sapir and Whorf spend pages describing odd features of obscure languages, but they never offer any evidence that these features have psychological effects. Does a language full of verbs make speakers see the world in endless flux? Who knows? Sapir and Whorf never bother to check. They seem to think it's sufficient to document a linguistic difference, but it begs the question at issue to assume that these differences have any impact on thought. Consider tense. Whorf would have us believe that Hopi speakers are subjectivist about time because they express tense by referring to speakers' attitudes. But we can find equally odd features in English. For example, we can express past and present by simply conjugating verbs (eat and ate), but we express the future by adding auxiliary verbs (will eat). Does that mean we conceive of the future as fundamentally different from the past? If so, that's news to this native speaker.

To find evidence for the Sapir–Whorf hypothesis, we need three things: a well-established linguistic difference between two languages, a correlation between that difference and a psychological difference between speakers of those languages, and a reason for thinking the latter is caused by the former. Early efforts to satisfy this burden of proof were unsuccessful. For example, a group called the Dani in Papua New Guinea were found to have only two colour words, but when they were tested on colour perception, they seemed to distinguish the same colours as English speakers, and they shared our judgements about which colours are similar. Chinese-speakers were hypothesized to have difficulty with counterfactual thinking because there is no subjective tense in their language, but tests failed to support this prediction when idiomatic Chinese was used. Navaho children are more likely than some English children to classify objects by shape, as opposed to colour, because Navaho sentences explicitly reference shape, but this difference disappears by adulthood.

THE RELATIVIST'S REVENGE

By the 1990s, it was widely believed that the Sapir–Whorf hypothesis was without merit. The received view, bolstered by Chomksy's universal grammar and a generation of failed tests, was that linguistic differences have no significant impact on thought. If people in different cultures have different worldviews, language has nothing to do with it. But that conclusion turned out to be too hasty. In the last twenty years, there has been a resurgence of interest in linguistic relativity, and new carefully designed experiments lend support to the idea that the way we think may be influenced by the languages we speak.

First, there is evidence that language influences the way we think about space. Imagine you are sitting at a table in a café, and I line up three cards in front of you: a ten on the left, followed by a two and a queen. Then I ask you to rotate your chair 180 degrees around and repeat this arrangement of cards on the table that was behind you a moment ago. If you are a typical English-speaker, you will reproduce the series, ten-two-queen, beginning with the ten on your left. Stephen Levinson tried this kind of game with speakers of Tzeltal, a language spoken by indigenous people in the south of Mexico. When Tzeltal-speakers rotated 180 degrees and reproduced the sequence, they reversed the left-right order. Ten-two-queen would be reproduced as queen-two-ten, with the queen on the left. The reason for this is that Tzeltal doesn't have words for left and right. Those are called relative spatial terms, since left and right are relative to the speaker. Tzeltal has only absolute spatial terms, which locate objects relative to fixed points in space, such as north and south, or uphill and downhill. Notice that if I place the cards in front of you, with the ten on your left, that card is also located towards some fixed geographical point. Perhaps it's closer to the north than the two and the queen. If you rotate 180 degrees and have to reproduce the sequence, one option is to keep the ten closest to the north. If you do that, you will get the sequence queen-two-ten, which is what Tzeltal-speakers do. What seems counter-intuitive and difficult to us is the default option for them. We tend to think of space in relative terms by default, and they think in absolutes. Both English-speakers and Tzeltal-speakers could be trained using each other's

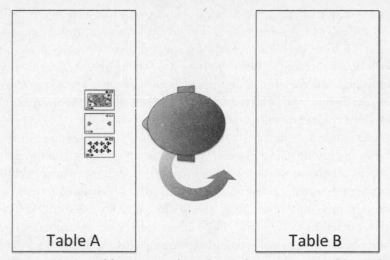

Figure 4. How would you reproduce the card sequence on Table A after turning around to Table B?

preferred methods, but language seems to determine which method seems most natural or obvious.

Here's another example. English has two kinds of nouns: count nouns and mass nouns. A count noun is one that can be preceded by a number: one potato, a dozen biscuits, six pence and so on. They usually refer to objects. Mass nouns usually refer to non-solid substances that cannot be counted. We can say some clay, much moisture, abundant smoke, but not one clay, two moistures and three smokes. Grammar makes the difference between object and stuff very salient to us. Yucatec Mayan, another Mexican language, does not make this distinction. It uses mass nouns for objects that we would count. They would say 'There is much pig over there' when pointing to a crowded pen. John Lucy has shown that this has an impact on thought.[8] He showed English- and Yucatec Mayan speakers images of several bucolic scenes and then gave them a memory test, asking which of two scenes matched each original. The test scenes either varied in the amount of some stuff (e.g., a pile of grain would shrink) or in the number of objects (e.g., the number of pigs would shrink). English-speakers were quick to note when the number of objects changed, but they were easily duped when the amount of stuff changed. The reason for this is clear. If you take one pile of grain, we'd call it some grain. Cut that pile in half, and it's still called some grain. Now take four pigs. Remove one, and we call it three pigs. Yucatec Mayan speakers had difficulty detecting both kinds of changes. When they see a group of pigs, it is encoded as some pig, and that encoding applies even when the group grows or shrinks. Swipe a pig, and they may not notice, because their language does not compel them to count.

Some languages differ even more dramatically from English when it comes to counting. The Pirahã people, a group of Amazonian hunter-gatherers, have no words for exact numbers. They have two words that correspond roughly to few and many. Peter Gordon wanted to find out whether this limitation affects how Pirahã speakers think.[9] He gave them numerous tasks where they had to keep track of quantities. For example, an experimenter would tap several times and ask the Pirahã speakers to repeat the taps. Or the experimenter would place some fruit on the floor and ask Pirahã speakers to put up the same number of fingers. Their accuracy was pretty good for two or three items, but dropped off precipitously after four. If the experimenter tapped four times, the Pirahã-speaker might echo with only three taps. What is effortless for inveterate counters like us is seriously challenging for them.

These examples show that language can influence what we notice, what we recall and what we can keep track of. But what about perception? It is possible that speakers of Tzeltal, Yucatec Mayan and Pirahã see the world just like us but classify things differently. Early studies on colour perception suggested that vocabulary does not affect how colours look to us (recall the Dani). But subsequent research has revised that conclusion. Consider the Tarahumara, another indigenous group from Mexico. They have one word that subsumes both blue and green. We see those colours as categorically different, but the Tarahumara may not. To test this, Paul Kay and Willett Kempton took three colour chips that we would classify as one shade of blue and two shades of green.[10] But they selected the chips so that one of the greens is actually more physically similar to the blue than it is to the other green. The Tarahumara had no difficulty discerning this fact. When asked which two chips were more similar, they got the answer right. But when English-speakers look as these three colours, we can't help but see the greens as more alike. Stare as long as you like, and they just look more similar.

The presence of a linguistic colour boundary between blue and green makes it impossible for English-speakers to perceive colour distances objectively. Likewise, we see pink as very different from red, and give it different social significance. A man who is uncomfortable with his masculinity might sport a red shirt, but would hesitate to wear pink. Other cultures do not linguistically mark this difference and treat pink as just a light shade of red. In Russian and Hebrew, blue and light blue are given different words, and, like our pink, light blue is considered feminine. In Russian, the word for light blue is slang for homosexual. Jonathan Winawer and his colleagues found that Russians could discriminate a light blue and a medium blue faster than two distinct light blues or two distinct medium blues; English-speakers do not show this speed advantage.[11]

The case of colour is striking because we humans have colour-sensitive cells built into our eyes and brains. Colour perception is biologically rooted if anything is. Unsurprisingly, then, there are many similarities in how languages label colours. But there are also differences. Lots of them. The number of named colours and the location of colour boundaries varies from language to language. The Himba of Namibia, like the Tarahumara, do not distinguish green and blue; in Bermino, a language of Papua New Guinea, there is no divide between blue and purple; for the Japanese, green is a shade of blue (they refer to blue apples). It is likely that all of these differences affect colour experience, as well as discrimination abilities and resulting memory.

We don't know exactly how language influences perception, but here is one possibility. During language acquisition, competent users teach colour terms to children by labelling examples: 'Those apples are red', 'That cotton candy is pink'. Children gradually store a set of examples for each term, and these are averaged together in memory to create a colour prototype that corresponds to an ideal instance of each category. We know that such prototypes are used in object recognition, so this account readily explains how language can affect how we categorize colours. But it is less often noticed that they may also affect experience. To categorize an object as red, we must compare it to a category representation stored in memory, and doing that involves matching a current perception against a stored image of the prototype. When we activate the prototype, it must be visualized using the brain mechanisms that are involved in colour perception. That means the prototype and the perceived colour must be active concurrently. If so, the experience of the perceived colour may be blended with the prototype, shifting it away from its objective position in colour space. Every colour may look slightly more like the prototype than it actually is. This would explain why colours look different to speakers of different languages, and discrimination across labelled boundaries is easier.

Notice that this explanation does not assume that language is literally part of the experience. We don't think in language and we don't see in language. Language simply establishes habitual ways of thinking and seeing. To illustrate, consider one more example. Some languages, including most in western Europe, assign gendered articles to every noun. In German, the word for key is masculine (der Schlüssel) and bridge is feminine (die Brücke). These gender assignments are arbitrary. In Spanish, key is feminine (la llave) and bridge is masculine (el puente). In the days when people doubted the Sapir–Whorf hypothesis, no one would have believed that these linguistic differences have any impact on thought. Now that attitude seems naive. People have deeply entrenched gender stereotypes, and gendered articles could easily call these to mind by association. For this reason, the neo-Whorfian psychologist Lera Boroditsky surmised that gender stereotypes may infect how speakers of gendered languages think about ordinary objects. She also made a bolder prediction. These associations may become so deeply entrenched, so habitual, that they will continue to exert an influence even when speakers of gendered languages drop their native tongue and pick up a gender-neutral language like English.

To test this, Boroditsky brought native speakers of German and Spanish to her lab and asked them, in English, to simply describe keys, bridges and other familiar things.[12] Amazingly, gender stereotypes were clearly at work, even though the test was conducted in English. Native German speakers described bridges as beautiful, elegant, fragile, peaceful, pretty and slender, whereas native Spanish speakers described them as big, dangerous, long, strong, sturdy and towering. In contrast, Germans described keys as hard, heavy, jagged, metal, serrated and useful, while Spanish speakers said they were golden, intricate, little, lovely, shiny and tiny. Clearly gender stereotypes are at work. Gendered languages may lead people to construe every object in a way that conforms to these stereotypes, and those influences may have enduring effects, even when speakers become fluent in another tongue.

Boroditsky's findings make it clear that language can influence thought. She identifies a linguistic difference, correlates it with a psychological difference and makes it clear that the former causes the latter. The gender effect must be driven by language because there is no other explanation. Languages assign gender in arbitrary ways. The fact that speakers associate masculine stereotypes with one object and feminine with another cannot be explained by appeal to some fact about the objects, but must instead be pinned on language. This, by the way, shows that we must be cautious when using gender-specific language. English has no gendered articles, but we do sometime infuse language with gender. For example, it is customary to use a masculine pronoun when speaking about people generically. We say, 'If a person is oppressed, he will seek freedom'. The male pronoun here represents the typical person in a gender specific way and that may influence thought. For example, when we think about what is involved in seeking freedom, we might imagine something aggressive when the male pronoun is used – a violent struggle for liberation. But suppose we said, 'If a person is oppressed, she will fight for freedom'. Now the image that comes to mind might be less violent, and, say, more verbal, like the consciousness-raising tactics of the women's liberation movement. Steven Pinker, a dedicated Chomskyan who rejects the Sapir–Whorf hypothesis, says we should avoid using masculine pronouns in generic sentences, because they are offensive, not because they influence thought.[13] I think we should resist them for both reasons. Pronoun choice can almost certainly influence how our words are construed.

The research we've been looking at provides the evidence that Sapir and Whorf were missing. They claimed that languages influence thought in different ways, but they had no proof. Now we have proof. We can predict how people will perform on various psychological tests by knowing something about the languages they speak. People raised with different languages think differently, and these effects endure even when they are not using those languages. Language inculcates habits of thought, just as Whorf and Sapir proposed.

Still, one might wonder how deep these effects are. Whorf claims that each language embodies a worldview. He implies that all our categories and inference rules are linguistically imposed. This may be an exaggeration. It is certainly a mistake to conclude that we cannot think without language, and it is equally implausible that a speaker of one language cannot come to think in ways that are similar to a speaker of another. Still, there is a sense in which Whorf was probably right. If you look at a scene, you immediately and automatically label the salient objects. Each label we use is an artefact that we learned from others by being presented with examples. In some cases, labels also come packaged with other culturally specific associations, such as we get with gendered articles, or the gender roles associated with the colour pink. That means we are automatically associating a lot of linguistically conveyed information with the categories we encounter. The decisions we make (Do I buy a pink shirt?), the inferences we draw (Is that bridge sturdy or is it pretty?), the things we recall (How many pigs were there?), the actions we perform (How do I reproduce this array?) and the qualities we experience (Which two look more alike?) can all be affected by language. Languages may not embody fundamentally and irresolvably different metaphysical frameworks, and they are certainly not essential for thinking, but they do have significant influence. Linguistic variation is not superficial. It is a powerful example of how something we learn through experience can shape our understanding of the world.

Steven Pinker argues that language is a window into human nature. Pinker is a Chomskyan, so he thinks the lesson we learn from language is that the mind is a constellation of highly specialized innate mechanisms, and that these are so heavily constrained by evolution that there is little variation in the way people think. I have argued that quite the opposite is true. Language is an invention, not an instinct, and it is a conduit for human variation, rather than an inflexible universal. If language teaches us about who we are, the lesson is that we are fundamentally flexible. The way we divide categories and experience the world is not fixed by what's out there or by what is innately specified within. Learning, including linguistic mastery, can impose a structure on reality that is not biologically inevitable.

Where Does Thinking Come From?

8

The Tao of Thought

Aristotle defined human beings as rational animals. What distinguishes us from other species is that we are especially good at thinking. Other animals can make decisions and solve problems, of course, but we do it better and more compulsively. Rather than reacting in predictable ways to the stimuli impinging on our sense, we reason about what we should do and bring past knowledge to bear in adaptive ways. If anything deserved to be called a feature of human nature, it is our distinctive capacity for thinking. More than our bipedal gait and our hairless bodies, it is thinking that puts us on the map.

The fact that humans think in a way that is unique to our species does not mean that all humans think alike. It is a genuine possibility that different people think differently. Indeed, there are obvious examples. Brilliant scientists, engineers and artists often innovate by approaching familiar problems in new ways. It is possible that some of the differences in how people think are the result of genetic differences. Einstein may have been born with a brain that deviated in marvellous ways from the standard assembly-line model that you and I received. In other cases, however, cognitive differences may result from experience. Evidence for this comes from the discovery that members of different cultures are conditioned to think in subtly different ways. These differences have important implications. They bear on international relations and, more generally, testify to the malleability of the human mind.

'I' CULTURES AND 'WE' CULTURES

Would you like your elderly parents to live with you, or would you rather they were cared for in a nursing home? Would you consult your uncle and aunt when considering a career change? Do you like to talk to your neighbours daily, or do you feel more comfortable with anonymity? Would you like your boss to know about your personal life? Do you feel honoured by the accomplishments of relatives, or are you simply pleased for them? Questions like these are used to distinguish two kinds of people: individualists and collectivists.[1] Individualists are concerned with personal achievement, and they value autonomy. They do not regard themselves as dependent on, or subordinate to, others. An individualist would not feel honoured by the accomplishments of another person: individualists think that people deserve credit for our own actions, not for the actions of others. And an individualist would not want her employer meddling in her personal life. Individualists think people should respect each other's 'personal space'. They also think it is important to tolerate variation, saying, 'To each his own' and 'Different strokes for different folks'. Individualist cultures also tend to be loose: they allow considerable variability in how people behave, and social roles are not very strictly enforced. It is not unusual, for example, for individualists to have schools without dress codes.

Collectivists have a very different outlook. They tend to focus less on individual achievement and more on the groups to which they belong. A sign in a Singapore restruarant reads:

Essential Words to Remember

The most damaging one letter word: avoid it.

I

The most satisfying two letter word. Use it.

We

The most poisonous three letter word. Kill it.

Ego

Collectivists have a strong sense of duty to others, and find deep value in interdependence. A collectivist would be very reluctant to marry someone without parental approval. Indeed, collectivists often have close ties to their extended families, and when a collectivist accomplishes something, it may bring honour to parents, grandparents and even second cousins, great uncles and other more distant kin. Misdeeds bring shame. Collectivists also feel a close sense of connection to non-relatives – such as members of the same creed, corporation or community. Conformity is valued, and humility is a cherished emotion. For collectivists, 'blood is thicker than water', and some collectivists warn: 'the nail that sticks out will get hammered' (a Japanese proverb). Collectivist cultures tend to be tight: behaviour is very strictly regulated. People are expected to play very specific social roles, and, in some cases, there are rules prescribing everything from etiquette and attire, to posture and facial expressions.

The contrast between individualism and collectivism can be discerned in the arts. In painting, European individualists historically liked portraits, while in China, landscapes were preferred. When collectivists paint people, they are often stylized, whereas individualist artists like to create exact likenesses, so they can capture the features that make each person unique. In literature, the quintessential individualist fable is Daniel Defoe's *Robinson Crusoe*, which celebrates our capacity to survive in total isolation from other people. Collectivists are more likely to appreciate the Japanese story of the forty-seven Ronin (or masterless samurai) who avenged their master after he was wrongfully sentenced to death and then took their own lives. The forty-seven Ronin function as part of a uniform whole, and they live and die for their master. When individualists write about teamwork, as in the *Three Musketeers* or the Arthurian legends, we hear more about the personality of each individual than about any deep sense of connection they might feel for each other. The most celebrated film in America is Orson Welles's *Citizen Kane*, which describes one man's quest for wealth and power. Collectivist filmmakers are more likely to make movies that deal with people's obligations to family, such as Satyajit Ray's celebrated *Apu* trilogy and Yasujiro Ozu's *Tokyo Story*, which is regarded by some as the best film from Japan. Collectivists think in terms of the first person plural, 'we', and individualists think in terms of the singular 'I'. It is no coincidence, then, that the languages spoken in many collectivist societies do not require use of a subject pronoun; consequently, when a person speaks about himself, the offending 'I' can be discreetly left out of the sentence.

No nation could be described as entirely individualist or entirely collectivist. There can be considerable variation between individuals and between subcultures. Still, generalizations can be helpful. When researchers average over a large number of people, they find that some nations are predominantly individualistic, and some are predominantly collectivist. Individualism is the dominant ethos in western Europe and in countries that are dominated by people of European descent, including Australia, Canada and the United States. Collectivism prevails in Asia, the Middle East, Latin America and much of sub-Saharan Africa.

Predominantly individualist nations are sometimes collectivist in certain respects. Germany is a good example. If you try to jaywalk in Germany, a complete stranger might reprimand you. Reprimanding strangers for minor offences is characteristic of collectivist cultures, where people value conformity to local norms. In purely individualistic cultures, the prevailing attitude is 'mind your own business'. Germany has an individualistic culture with some collectivist tendencies. There are also collectivist cultures with some individualist tendencies. This is the case in India, which was once a British colony and has a long tradition of celebrating individual accomplishment in science and culture. Consequently it is best to think of individualism and collectivism coming in degrees. Some countries, like the United States, are extremely individualistic, and some, like China, are extremely collectivist. Other countries, such as Israel and Argentina, fall closer to the middle.

A variety of factors influence the degree to which a culture is individualist or collectivist. One factor is wealth. Affluent nations are more likely to be individualistic, as are affluent people within nations that are not affluent over all. Wealth gives people greater mobility and freedom, so it tends to promote an ethos of self-reliance. Wealthy people are also better educated, which is another factor that promotes individualism. Education brings exposure to multiple customs, and that tends to promote a degree of looseness. People are less likely to obey a rigid set of social rules if they know that other people live and thrive without those rules. Of course, education does not promote individualism if it is non-secular; religious schooling tends to promote collectivism, because it encourages a uniform set of values and a close connection with other members of the religious group. A third factor is pluralism. Homogeneous societies are more likely to be collectivist, because it is easier to subordinate oneself to a group if one sees the members of that group as very similar to oneself. Pluralistic societies tend to be more individualistic, because the distance between self and other can be considerable, and members of pluralistic societies are exposed to many ways of living. Given the impact of pluralism, it's not surprising that urbanization is also a factor that tends to promote individualism; urban settings tend to be more culturally varied. In urban environments, it can also be difficult to maintain close ties to the community. Rural communities tend to be more collectivist; people are more likely to know each other and to get involved with each other's business. A final factor is occupation. People who make their living doing things that require close collaboration are more likely to develop a collectivist ethos. Farming is an example of such an occupation. Societies that are predominantly agricultural tend to promote collectivist thinking, because collaboration is so essential for survival. Occupations that allow for greater self-sufficiency tend to promote individualism. This is the case in traditional hunter-gatherer societies where each individual learns the skills necessary to acquire food without the aid of others. Individualism also tends to be high in cultures that subsist through herding animals or fishing, both of which can be done by a single individual. Industrialization tends to promote individualism, because it promotes urbanization, and it also introduces a wide range of jobs in which people have very different responsibilities, unlike in agricultural economies, where many people make very similar contributions to subsistence. With the rise of the information age, some researchers think that individualism will become the norm around the globe. Through information technologies, people in homogeneous societies can learn about how people in other societies live, and they gain access to goods and job opportunities that were once off limits because of geographical distance.

It's quite clear that individualists and collectivists have different values. Individualists like to stand out and collectivists strive for harmony. What is surprising is that these differences in values have a measurable impact on perception and cognition. Individualists and collectivists think differently, and they may even see the world in different ways. It is as if collectivists and individualists were running different mental programmes.

THINKING HOLISTICALLY

Hunters and Farmers

One of the first to explore this proposal systematically was Herman Witkin, who argued that people from different cultures have different 'cognitive styles'.[2] Suppose you enter a room with slanted walls. You might mistakenly believe that the room is straight and you are standing at an angle; if so, you will adjust your posture. Alternatively, you might sense that the position of your body is perfectly straight and discern that the room must be slanted. Witkin used the term 'field-dependent' to describe the cognitive style of people who attribute the tilt to their posture. Such people process information in a way that is very sensitive to the surrounding context. Rather than ignoring the visual appearance of the room, they factor that in when they assess the position of their bodies. They use all the information available to compute an answer. Field-independent people process information differently. They decontextualize, or abstract away from context, to pinpoint and consider each bit of evidence independently. They can easily ignore the appearance of the room when assessing their posture. Witkin used other tests for distinguishing people who are field-dependent and field-independent. He developed something called the rod and frame test, in which subjects view a rod surrounded by a crooked frame. Their job is to adjust the rod so that it's perpendicular to the ground. Field-dependent thinkers make a lot of errors on this test, because they find it hard to ignore the frame. Witkin also developed an embedded figure test, in which subjects are shown a complex shape with many overlapping lines, and they are asked to determine whether a simpler shape, such as a triangle, can be found in

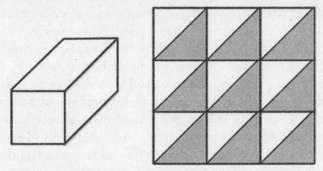

Figure 5. The embedded figure test. Find the figure on the left in the pattern on the right.

the complex shape. Again, field-dependent people make more errors, because they have difficulty decontextualizing.

Witkin did not use the terms 'individualism' and 'collectivism', but he did assume that a person raised in a culture that emphasizes the interdependencies between people is likely to have a field-dependent thinking style, and a person raised in a culture that emphasizes personal autonomy is more likely to be field-independent. Witkin was especially interested in the connections between ecology, culture and thought. Ecology – the environment in which a group of people lives – can influence the form of subsistence they choose, and subsistence methods can influence thought.

In small-scale societies that have been isolated from the modern world, two forms of subsistence are especially commonplace. In some small-scale societies people survive by hunting animals and foraging for insects, fruits, edible plants and other naturally available foodstuffs. These hunter-gatherers tend to be nomadic, because, once they have depleted food supplies in one area, they need to move on. Hunter-gatherers also tend to have a comparatively individualist orientation, because an individual person can hunt or gather successfully without depending heavily on others. To survive, hunter-gatherers must be self-reliant. In this respect, hunter-gatherers are quite different from people who survive by farming. Farmers, or agriculturalists, are typically sedentary; they remain in one place, because they can reharvest the land each year. They also tend to be collectivists, because farming is most effective when done collaboratively, and collaboration is required to build permanent houses and irrigation systems. Agriculturalists encourage conformity, because collaborative efforts require that people play set roles, respect each other and work towards a common end.

Witkin's hypothesis was that hunter-gatherers and agriculturalists would have different cognitive styles because of their different styles of life. He predicted that hunter-gatherers would have a field-independent cognitive style because they are enculturated to think individualistically, and agriculturalists would have a field-dependent cognitive style, because they are enculturated to appreciate the value of interdependence. This conjecture has been most systematically investigated by a follower of Witkin named John Berry. In one study, Berry compared the Inuits of Baffin Island and the Temne of Sierra Leone.[3] The Inuits (more popularly known as Eskimos, which, as already stated, is seen as a term of slander) live in arctic conditions, so they cannot cultivate the land; instead, they subsist by hunting sea mammals and fishing for char. The Temne live in forests and grasslands, where they plant rice, nuts and cassava root. The Inuits are nomadic, and the Temne are sedentary. The Inuits emphasize autonomy in their culture, and they raise children to be very independent. The Temne raise children with strict discipline, and they enforce rigid social roles. Berry wanted to know whether these differences in lifestyle have any impact on cognition. To find out, he used Witkin's methods of measuring field-dependence. As predicted, the Temne were less capable of seeing embedded figures, and they were more likely to make errors on the rod and frame test. In short, the Temne are less likely to decontextualize.

These results have been replicated in other populations. For example, in India, there are some relatively isolated small-scale societies with different forms of subsistence. The groups that continue to rely on hunting have been shown to have a more field-independent cognitive style than the groups that harvest the land. Similar results have been found among contrasting groups of North American Indians, aborigines of Australia and indigenes of New Guinea. In all these diverse locales, different forms of subsistence correlated with differences in cognitive tests. And subsistence is not the only variable that matters. Any lifestyle that promotes a collectivist orientation can engender a cognitive style that is field-dependent. For example, among orthodox

Jews, there is a strong spirit of collectivism; members of this close-knit community abide by strict rules, and they regard themselves as part of an interconnected whole. Reform Jews have a much less cohesive community and much greater emphasis on individualism. Consequently, reform Jews do better than orthodox Jews on Witkin's tests for field-independence even when they have the same IQ.

East and West

Much of the research on field-dependence and -independence was carried out in the 1960s and early 1970s, and investigators focused on small-scale societies. In recent years, there has been a resurgence of interest in the topic fuelled by an acclaimed social psychologist named Richard Nisbett and his collaborators. Nisbett is primarily interested in showing that people in Western nations have a different thinking style to that of people in the Far East. It has been known for a very long time that people in the West tend to be individualists, and people in the Far East tend to be collectivists, but the cognitive implications of that contrast in value systems has not been systematically examined until recently.

Before looking at that research, it's worth asking about the origins of the East/West contrast. Why are people from east Asia so often collectivists? Why are Europeans and Americans so often individualists? The answer may have to do with economics. East Asian countries owe their deepest cultural debts to ancient China. Historically, China was a nation of farmers. They grew millet, wheat and rice. These crops depended on irrigation systems along China's rivers, which were controlled by warlords. Under the Zhou dynasty, which began in the twelfth century BCE, social organization was feudalistic: peasant farmers worked for vassals, who paid tribute to the land-owning nobility. Each person had a set place in society and was expected to play specific social roles. During this period, China made a transition from ancestor worship to organized religion, and great philosophical traditions, most notably Confucianism, were born. China also perfected farming techniques under the Zhou, and these led to an exponential population growth. Chinese farming techniques spread throughout Asia, along with Chinese culture. The prevalence of farming and the rigid social organization promoted collectivist value systems, which became dominant throughout Asia.

In contrast, the West owes its biggest cultural debt to Classical Greece. Farming began in Greece much later than it began in China, and it was a less central part of the economy, because Greece is too mountainous to support very large-scale agriculture. Greek farmers often had small-scale operations, and, unlike the Chinese, they worked double-duty as traders, selling the goods that they farmed. Greek ecology was more amenable to other forms of subsistence, especially herding animals and fishing. Both of these activities can be performed by individuals working on their own without assistance from others. With herding and fishing as mainstays, the Greeks had little need for massive farms, complex irrigation systems or large-scale cooperative labour. This promoted a spirit of individualism, which helped lay the seeds for the democratic government that blossomed in Athens at the end of the sixth century BCE. In this climate, new schools of philosophy were born, and new emphasis was placed on individual achievement. Greek individualism influenced the ethos of ancient Rome, but it did not endure. The Roman Empire was culturally pluralistic, which fuelled individualism, but, after the fall of Rome, things began to change. The Church increased cultural homogeneity, and power shifted from Rome northward into territories that had great agricultural resources. The Church was unable to directly control these regions, and, as a result, there was a decentralization of power resulting in a feudal economy. As in China, there were local fiefdoms controlled by vassals and supported by a large underclass of farmers and artisans. Consequently, medieval Europe had a collectivist orientation at the time. Studies of medieval biographies suggest that people did not conceive of themselves as unique individuals with distinctive traits and abilities, but rather, they saw themselves as dependent parts of a larger whole, with each person playing a preordained social role.

Individualism re-emerged in the late Middle Ages as feudalism waned. To escape the drudgery of agrarian fiefdoms, people started moving to cities, and with urbanization came an increase in trade. Whereas people had once bought food and other goods directly from farmers and artisans, they now bought them at an inflated price from merchants, who served as middlemen between producers and consumers.

Merchants transported goods across considerable geographical distances, and this led to greater mobility, greater communication and greater wealth. People in Europe became far more autonomous than they had been under feudalism, and individualism became the prevailing ethos. Biographies began describing people as unique individuals, and artists began adopting distinctive styles. These changes laid the groundwork for the Renaissance, the Enlightenment, capitalism and democratization. To the individualist, each of these developments is regarded as great progress.

If these historical accounts are right, then Eastern collectivism was sown on the farmlands of ancient China, and Western individualism was shepherded in by the Greeks, and then repackaged, after a long hiatus, by the merchants of medieval Europe. Economic systems have determined the extent to which people depend on each other for subsistence, and variations in patterns of interdependency have shaped the cultural ethos. If Witkin and Berry are right, cultural ethos influences cognitive style. The collectivists of the East should have a field-dependent style, and the individualists of the West should have a field-independent style. This is exactly what research has shown. For example, American college students do better than Taiwanese college students on Witkin's rod and frame test; the Taiwanese students have a harder time rotating a rod to a vertical position when it is surrounded by a tilted frame. People from some east Asian countries also make more errors than Americans on the embedded figure test. Surprisingly, when this experiment was first done with Chinese subjects, they performed better than Americans; they were more capable of finding a shape that was hidden in a complex figure, which is not what we would expect if people from China have a field-dependent cognitive style. But the experimenters realized that there was an explanation: Chinese subjects have years of experience reading Chinese, and Chinese letters are essentially complex shapes with embedded parts. Consequently, Chinese subjects become very adept at discerning embedded shapes. To avoid this source of exposure, a group of researchers gave the embedded picture task to a group of Malaysian subjects. Malaysia is an east Asian collectivist culture, but they use the Roman alphabet in their writing system. As predicted, Malaysians make more errors than Westerners on the embedded picture test.

The embedded picture test and the rod and frame test give the misleading impression that east Asians are less competent than Westerners. This is a gross distortion. East Asians are, on average, more sensitive to context. Sometimes this leads them to make errors, but it also leads them to notice things that Westerners miss. For example, if we tested a Westerner's memory of the frame position in the rod and frame test, or their memory of the relation between the frame and the rod, they might make more errors than Easterners. This is nicely demonstrated in a study by Takahiko Masuda and Richard Nisbett.[4] They presented Japanese and American subjects with an animated sequence of fish swimming around in an underwater scene. When asked to describe the film, Japanese subjects usually began by talking about the setting, whereas Americans usually began by talking about the largest fish. Japanese subjects made 70 per cent more comments about the background than American subjects, and Japanese subjects were twice as likely to mention relations between fish and the objects in their environment. After viewing the first film, subjects were shown a picture of a fish and asked whether it appeared in the original scene; in some

Figure 6. Look at this fish tank and describe what you see.[6]

cases the fish was presented against the original background and in some cases it was presented against a novel background. Changing the background had no significant effect on American subjects, but it caused Japanese subjects to make more errors. When the fish were presented against the original background, Japanese subjects made fewer recognition errors than American subjects. In another study, Masuda and Nisbett also showed Japanese and American subjects street scenes and then tested recall by swapping the original images with slightly altered images.[5] In some cases, they changed a focal object, such as a car, and in others, they changed a background object, such as a building. American subjects were more oblivious than Japanese subjects to changes in the background, and Japanese subjects were more oblivious to changes in focal objects.

This research confirms that Western and Eastern subjects process information differently, and there are advantages to both cognitive styles. If you are enculturated in the East, you may have some difficulty abstracting away from contextual information and you may fail to notice features of the objects that you encounter. If you are enculturated in the West, you may be good at abstracting away from background information and focusing on objects, but you will thereby fail to notice features of the background and relations between attended objects and their surrounding context. In Witkin's terms, Easterners are more field-dependent than Westerners. Nisbett and his colleagues use different terminology: Easterners tend to process information holistically, while Westerners process information analytically. Easterners focus on the *whole*, so they see relations between objects and background, and they recall those relations later on. Westerners *analyse* a scene into parts, and focus on foreground objects, seeing their distinctive features and recalling those features.

The tasks that I have been describing involve perception, attention and memory. Nisbett and his colleagues have shown that the analytic/holistic contrast crops up in other domains as well. Consider categorization. In one study, Chinese and American children were presented with a picture of a man, a woman and a baby, and asked which two go together.[7] Most American children said the man and woman go together because they are both adults, while the Chinese children said the woman and the baby go together because women take care of babies. In other words, American kids group things together on the basis of intrinsic similarities, while Chinese kids were more likely to group things on the basis of how they relate to each other.

East Asians are also likely to think holistically when they explain behaviour. Westerners typically explain behaviour by appeal to psychological factors, such as motives or character traits. Easterners are more sensitive than Westerners to the environmental influences on behaviour. Nisbett and colleagues asked Chinese and American subjects to explain why a mass murderer had committed his crimes. American subjects focused on the killer's mental instability, while Chinese subjects often cited societal factors. Sometimes, the tendency to focus on psychological traits leads Westerners to make errors when they are explaining behaviour. Westerners tend to think motives and character traits drive behaviour even when the environment is to blame. Social psychologists call this the fundamental attribution error. Nisbett and his colleagues have shown that Easterners are less likely to make mistakes of this kind. In one study, Americans and Koreans were asked to read essays either defending or opposing atomic testing. They were then told that the authors had no choice about what position to adopt when writing those essays. This led Koreans to withdraw their statements about whether the authors actually held the views expressed. Americans ignored the fact that the authors had no choice, and continued to attribute to the authors the opinions expressed. In fact, the American tendency to ignore external influences may lie behind the fact that Western psychologists tend to assume all human behaviour is driven by the genes or other internal causes, while ignoring the effects of experience and context.

The research summarized so far suggests that Easterners and Westerners have different ways of thinking about the objects they encounter. Easterners tend to focus on how objects are related to their environments, and Westerners tend to abstract away from the environment. This pattern supports Witkin's conjecture that people who have a strong sense of social interdependency will be more sensitive than others to interdependencies between objects in the world. Nisbett's research has shown that this is true of collectivist cultures in the Far East. In conducting his research, he has also uncovered another East/ West contrast that is even more surprising. There is a difference in how Easterners and Westerners reason. Westerners are very committed to formal logic. If a conclusion follows logically from a set of premises, then the inference must be a good one, according to Westerners, even if the conclusion is at odds with experience. Easterners are more willing to reject logically valid inferences when the conclusions seem peculiar. In one study, Koreans and Americans were asked to consider the following argument:

Premise 1: All things that are made of plants are good for the health

Premise 2: Cigarettes are things that are made of plants

Conclusion: Cigarettes are good for the health

When asked if the conclusion follows from the premises, Americans were considerably more likely than Koreans to say yes. For Koreans, the fact that the conclusion is at odds with prior knowledge makes them less likely to recognize that the premises of the argument logically entail the conclusion; the argument is logically valid, but it happens to have a false premise. Nisbett speculates that such reliance on logical rules is a direct consequence of individualist culture. The ancient Greeks, who developed the most influential systems of formal logic, were not afraid of arguments. As individualists, they felt no pressing need for social harmony, and would engage in debates frequently. They probably also had field-independent thinking styles, which means they would have been adept at abstracting information out from its surrounding context. Argument and abstraction are the essence of logic. The rules of logic abstract away from content and specify the form that a premise must have in order to support a conclusion. For example, a logical rule might tell us: if all As are B, and some As are C, then some Bs are C. This kind of rule is likely to be discovered by enthusiastic debaters with a knack for abstract thought.

One consequence of the Western enthusiasm for logic is that Westerners tend to abide by the principle of non-contradiction: if two claims are conflicting, one of them must be false. People from the Far East tend to have a different view. They are prone to think dialectically, recognizing that two opposing sides may both have some truth to them. This principle is expressed in the idea of Yin and Yang, which are conceptualized as opposing forces existing harmoniously in the universe. The idea of Yin and Yang is central to Taoism, and it also plays a role in Confucianism and in the *I Ching*, a classical work of ancient Chinese philosophy. Having been raised with these traditions, east Asians are more willing than Westerners to accept contradictions when they reason. In one demonstration of this, Nisbett and Kaiping Peng asked Chinese and American subjects to consider everyday conflicts, such as the desire to conform to one's parents' wishes and the desire to follow one's own wishes.[8] When asked to assess and resolve these conflicts, about three-quarters of the Chinese subjects adopted a dialectical position, granting merit to both sides. Three-quarters of the American subjects took a single side, in accordance with the principle of non-contradiction. Similarly, when Chinese subjects are presented with an argument against a position that they endorse, they weaken their initial endorsement, giving credit to both sides. When American subjects are presented with an argument against a position that they endorse, their initial endorsement becomes stronger.

Lessons from the East

All of these findings demonstrate that people in different cultures think differently. There is not a single way of thinking shared by all human beings – a universal logic of thought. Rather, there are different thinking styles that emerge through enculturation. Ecological factors affect how people make their livelihood; methods of subsistence affect values; and values affect how people process information. All aspects of information processing seem to be amenable to cultural influence. We have just seen that reasoning varies across cultural boundaries, and studies using the embedded picture test and related methods suggest that there may even be cultural differences in perception. It is plausible that when collectivists view a scene, they see a network of interconnected parts, while individualists see a collection of interacting objects. These differences probably aren't fixed. With training and guidance individualists can perform like collectivists on cognitive tasks, and collectivists can perform like individualists. It's a gross exaggeration to say that members of other cultures think in ways that are entirely alien and inaccessible. Rather, we must recognize that our way of thinking is not the only way, and experience can inculcate different ways of thinking.

The discovery of cultural differences in thinking styles has important implications. Let me mention just three. One implication has to do with marketing. If you are trying to promote something (a product, a theory or a policy) across cultural boundaries, then you should recognize that consumers process information differently. To make this point, a group of marketing researchers conducted the following study.[9] They told a group of Chinese and American consumers that product A is better than B and C in certain respects, and C is better than A and B in other respects. When asked which product they would buy, Chinese respondents tended to go for B, the product that falls between the other two. American subjects express a preference for either A or C, rejecting the compromise. As with logic, Westerners like to pick between opposing extremes, and Easterners like to find a middle way.

A second implication concerns health care. In the East, there has been an emphasis on holistic medicine, which treats mind and body as a unified whole. In the West, medicine involves the analysis of symptoms, and little attention is paid to the relationship between illnesses and the context in which they arise, which can include things like diet and mental state. Westerners have resisted Eastern medicine for a long time, and conversely. The tendency to adopt one approach to health care over the other may stem directly from the respective cognitive biases promoted by Eastern and Western culture. Medicine will surely improve if we heed lessons from both traditions.

A third implication concerns international relations. When dealing with other countries, it's important to recognize that they may see the world differently. Americans tend to assume that people in all countries have the same interests as they do. The effort to bring freedom and democracy to the rest of the world is symptomatic of this. As individualists, Americans value choice, and they think everyone wants to express their views in selecting a government. In countries with a more collectivist orientation, social harmony may be more important than self-expression; authorities that warrant respect may be more valuable than the right to vote. This is especially true in culturally homogeneous countries. When dealing with the United States, other countries should know that Americans tend to dig in their heels when faced with opposition, rather than attempting to see both sides of an issue. Diplomats who want to be more successful brokering deals across cultural boundaries should learn how members of other cultures are likely to respond to conflicting opinions.

BEYOND EAST AND WEST

We have been looking at the contrast between the Far East and the West, but this contrast is just a simple illustration of a much broader phenomenon. We should resist the temptation to think that there are just two kinds of minds in the world, oriental and occidental. This way of looking at it promotes racial reification and stereotypes. There is a long dark history of Westerners treating the people of east Asia as the 'exotic other'. The East/West contrast is just one of many that can be drawn, and it brushes over many finer distinctions that deserve to be studied.

One thing to bear in mind is that the collectivism/individualism distinction cuts across the continental divide. Some western Europeans score high on tests for individualism, but collectivist tendencies can be found in southern Europe and northern Europe. The culture of honour in Sicily and the social welfare programmes of Scandinavia show signs of this. There are also collectivist tendencies in central and eastern Europe, and Nisbett has recently been involved in research that replicates the East/West findings using eastern and western Europe as the contrast.

We must also recognize that individualism and collectivism may come in many forms. We find collectivist tendencies in the regions of Europe just mentioned, in east Asia, in south Asia, in north Africa and in South America. People in all of these places may score higher on tests for field-dependence than people in Australia or North America, but there are undoubtedly numerous differences between collectivist cultures. We should not assume that collectivists in China think in the same style as collectivists in Japan, India, Saudi Arabia or Mexico.

One variable that has not been adequately explored is social organization. Some collectivist cultures, such as Japan and India, are hierarchically organized. But others, including China and the kibbutzim of Israel, are egalitarian. Likewise, some individualist cultures are more hierarchical than others, as in the case of Great Britain, which retains a clearly articulated class structure. Americans are a little less class conscious, but there is strong cultural emphasis on achievement and movement up the economic hierarchy. In contrast, Scandinavians tend to emphasize equality. Americans score more than twice as high as Danes on a scale that measures this inegalitarian orientation.[10] It is possible that this difference affects cognitive styles in ways that crosscut the individualism/collectivism distinction. Also some cultures are more religious than others, some are located in hotter climates, and some are more frequently engaged in warfare. These and many other variables need to be systematically explored.

We must also recognize that thinking styles can vary within a culture. Most obviously, cultures can contain subcultures, and subcultures may educate their children in different ways. Thinking styles can also vary across any groups within a culture that are treated differently. There are cultural boundaries of wealth, occupation, geography, ethnicity, sexual orientation, political affiliation and religion. There are also umpteen subcultures divided along boundaries of taste, recreational interests and lifestyles. There are hippies and hipsters, preppies and punks, jocks and junkies. For all we know, there are systematic differences in the way these groups think.

This may sound far-fetched, but we must remember that subcultures have different values and pastimes, and these are sources of social influence. If a group affiliation can influence a person's wardrobe and record collection, why not their cognitive style? Isn't it possible that goths are smarter than jocks? It is important that such questions be addressed using science, not stereotypes. Empirical studies can both confirm and correct our prior assumptions. Herbert Marsh and Sabina Kleitman struck out the dumb jock cliché by showing that involvement in athletics is positively correlated with academic achievement.[11] Jocks tend to get good grades, do their homework and perform well on standardized tests.

These are early days for research on culture and cognition. There are a myriad cultures and subcultures that can be compared. The main lesson of the East/West research is that culture can have an impact.

Cognitive styles are affected by group membership. The effects are sometimes modest and almost always reversible, but they are important. They show that human beings have different default thinking styles and these defaults can be set in place by experience. This casts doubt on the idea that there is a set of universal laws of thought that we all use to the same degree in the same contexts. Theories of how the mind works that are taught in most psychology textbooks might be better described as theories of how the Western mind works, since most of the research reported has been done on American college students. The study of the mind cannot be separated from anthropology, since the mind is always informed by culture.

9

Gender and Geometry

When we think about cultural differences, we tend to think about groups who live in different places, speak different languages and worship different gods. But cultural differences can be very local, as when urban subcultures live side by side in the same town. The most local cultural divide of all, however, is the gender divide. Men and women work the same fields, worship in the same churches and sleep in the same beds, but they reside in different cultures. Men and women are treated differently, they often do different things with their leisure time, and they are subject to very different cultural expectations. Of course, men and women are also biologically different. And this raises a puzzle for science. If men and women perform differently on tests of intellectual ability, should the difference be pinned on nature or nurture or both?

DIFFERENCE AND DISCRIMINATION

The Summers Debacle

On 14 January 2005, Lawrence Summers, the president of Harvard University, sparked a media frenzy by suggesting that innate cognitive differences are a leading cause of the fact that women are under-represented in the science and engineering faculties of elite universities. He voiced this opinion while speaking at a private conference at the National Bureau of Economic Research, but soon his assessment was being reported by newspapers across the globe. Critics argue that Summers's remarks were uninformed and irresponsible. In his speech, Summers claimed that discrimination and socialization play little role in gender inequity within the academy. There is a considerable body of research to the contrary. Summers also implied that women are biologically inferior to men, in that they are genetically less likely to attain the levels of aptitude demanded by prestigious programmes in science, maths and engineering. This, we will see, is also at odds with the evidence. Biology may make some contribution to cognitive differences between men and women, but differences in academic achievement may owe more to socialization.

The same people who presume that the cognitive differences between men and women are primarily biological also tend to conclude that these differences are inalterable. If this conclusion is combined with the view that women are cognitively inferior to men, then the inevitable upshot is that they are incapable of achieving the same standards. This is exactly what Summers implied, and that is why his speech was offensive to so many. The offence was compounded by the fact that Harvard has had a depressingly bad record when it comes to hiring women. During Summers's reign as president, only 12 per cent of the new tenured faculty appointments went to women. Summers was not in charge of selecting new faculty – departments do that – but he participated in tenure decisions, and he could have encouraged departments to recruit women more actively. Instead, female appointments declined appreciably during his time at the helm. When Summers raised the spectre of biological differences, his detractors inferred that he might be guilty of gender bias, falsely believing that men are more likely than women to be naturally brilliant.

Before presenting the evidence against this conjecture, we should note that it is nothing new. In 1873, a respected Harvard medical professor named Edward Clarke published a book called *Sex in Education, or a Fair Chance for the Girls*, in which he warns that women who attend college risk becoming infertile and hysterical. He conjectured that when women tried to use their underdeveloped cognitive capacities to learn, blood would be diverted to the brain from the uterus, which would then atrophy. In 1889, C. C. Coleman, an American physician, issued a similar warning:

Women beware. You are on the brink of destruction: You have hitherto been engaged in crushing your waists; now you are attempting to cultivate your mind . . . you are exerting your understanding to learn Greek, and solve propositions in Euclid. Beware!! Science pronounces that the woman who studies is lost.

The French psychologist Gustave Le Bon went even further:

[T]here are a large number of women whose brains are closer in size to gorillas' than to the most developed male brains. This inferiority is so obvious that no one can contest it for a moment . . . [Women] represent the most inferior forms of human evolution and . . . they are closer to children and savages than to an adult, civilized man . . . A desire to give them the same education, and, as a consequence, to propose the same goals for them, is a dangerous chimera.

Such attitudes were not esoteric or anachronistic. Clarke's book went through seventeen printings, and the scientific community widely believed Le Bon's contention that women are no smarter than children. The fact that women have more youthful proportions than men was taken as incontrovertible physiological evidence for the conclusion that their intellectual development did not advance beyond childhood. The prevailing view throughout the nineteenth century was that women are intellectually inferior to men.

This prejudice had a measurable impact. Most obviously, women were not allowed to vote. Women's suffrage came to Great Britain and Germany in 1918, to the United States in 1920 and to France in 1944. Women were also excluded from many professions. At one time, women were deemed incapable of working as stenographers or secretaries, two fields they came to dominate. The presumption of inequality seriously delimited women's access to education. Women were generally excluded from college education until the nineteenth century. In 1837, Oberlin College in Ohio became the first college to admit female students, but they were assigned a special curriculum, which included cooking and cleaning rather than Latin and Greek. Even the feminist reformers of this period were happy to admit that women could never equal men. In 1823, Harriet Martineau argued that women should be given access to higher education in England, so that they could become 'companions to men, instead of playthings or servants'.[1] This may sound like a plea for equality, but Martineau was also quick to concede that 'the acquirements of women can seldom equal those of men, and it is not desirable that they should'. Accordingly, women were often educated in separate schools, and they were discouraged or prevented from pursuing graduate degrees, especially in maths and science. Sofia Kovalevskaya was the first woman to earn a mathematics doctorate in Europe, in 1874. In 1895, Caroline Baldwin Morrison became the first woman in the United States to receive a doctorate in science. The first European woman to receive a doctorate in science was Marie Curie, in 1902; she went on to win two Nobel Prizes. For the majority of women, graduate education was not an option, and, though almost half of all college students were women in the early twentieth century, many went to women's schools that were not always equal to their male counterparts. Widespread coeducation is a recent development. Princeton and Yale opened their doors to women in 1969. Harvard beat them to the punch by conferring degrees to women in 1964, but those women had to be enrolled in Radcliffe Women's College, which did not officially merge with Harvard until 1999.

Summers struck a nerve against this background. His remarks were especially wounding to women in academia who have extensive firsthand knowledge of inequitable treatment. Women are routinely ignored, talked down to and hit on by male college professors. They are often not encouraged in their academic pursuits and not believed in. Women in academia also know that the struggle for equal treatment is a slow one. Most had many more professional opportunities than their mothers, and it seems implausible that bias would simply evaporate in the space of a single generation.

The Science of Difference

By the 1970s, few people would openly suggest that women are less intelligent than men, but the same period saw an increase in scientific testing of gender difference. Flagrant claims of male superiority were replaced by the rhetoric of separate but equal. Scientists began broadcasting evidence that women think differently, and, more often than not, they assumed these differences were biologically based.

There is now a considerable body of evidence showing that men perform better on some tasks, while women perform better on others.

The male advantage shows up most frequently in two areas: spatial reasoning and maths. In spatial reasoning, men are on average better at imaging geometrical objects at different orientations ('mental rotation'), finding an object that has been embedded in a complex picture and orienting a rod so that it is perpendicular with the floor of a room. When it comes to spatial navigation, men are more likely than women to use their sense of compass directions and geometrical information. In maths, male scores on standardized tests tend to be higher. In 2004, male high school students scored 7 per cent higher on the maths portion of the Scholastic Assessment Test (SAT), and, in earlier years, those numbers have been as high as 15 per cent. Moreover, 78 per cent of the students who got perfect scores on the maths SAT were male.

Women's strengths tend to lie elsewhere. They outperform men on certain verbal tasks and on tasks that involve recognition of fine details and contextual information. In terms of language, women do better than men at coming up with words that begin with a particular letter, and they are better at recalling words from lists; they also use considerably more words than men in the course of a day. In visual memory tasks, women also have some advantages over men. They are better at recalling where an object was located in an array. Unlike men, women tend to navigate using landmarks rather than compass directions. Where a man might recall that the bank is three blocks north, a woman might recall that the bank is just past the post office.

Women tend to be less efficient than men when it comes to spatial tasks that involve understanding three-dimensional configurations of objects or object parts. Some of the largest gender differences have been reported in studies of mental rotation. In mental rotation tasks, subjects are presented with a picture of two objects at different orientations, and they are asked whether the two objects are the same. To answer, subjects must mentally rotate one object to see if it aligns with the other. Women make more errors than men, and there is some evidence that they tend to use a different strategy. One way to see how a person solves a problem is by giving them two tasks at the same time and seeing if one interferes with the other. Men do badly at mental rotation tasks if they are doing another spatial task at the same time, such as keeping an arrangement of dots in their minds. Women are not impaired at mental rotation while they are memorizing arrangements of dots, but they are impaired if they are trying to hold a list of words in their minds. This suggests that women may be relying on their language skills when they mentally rotate objects. Perhaps they are labelling each part of the object and reasoning about how it would change when rotated.

Some naturists have advanced evolutionary explanations of gender differences. Differences in maths and language are difficult to explain in evolutionary terms, because sophisticated maths and language skills appear recently in human evolution, and it's far from obvious why either sex would have greater use for capacities than the other. Are women with greater vocabulary and men who excel in algebra really more likely to procreate? Most evolutionary speculation has centred around spatial skills. According to one popular view, men are better at spatial cognition because male ancestors were hunters, and hunting requires a high degree of spatial precision. This hypothesis is not really plausible, however. First of all, it's not clear what specific skills such as mental rotation have to do with hunting. Second, some spatial skills, such as finding embedded objects in a complex scene, are equally useful for both hunting and gathering, which is believed to have been dominated by women. Third, the male advantage in spatial cognition has been reported in species that don't hunt, such as rats, who are scavengers by nature. If gender differences in cognition are at all based in biology, we have no good explanation of why they evolved. It is possible that such differences are just a freak by product of how male and female brains happen to be wired.

The differences between men and women are often small, and some people perform in ways that are atypical for their sex. But, however

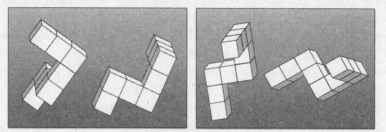

Figure 7. Mental rotation task. Which pairs are the same?[2]

small, the differences do show up reliably on a variety of tests, and they often ring true anecdotally. For example, it's something of a cliché that men have a better sense of direction, and women have a better eye for details. Men refuse to ask for directions because they feel confident about where they are going. Women may be more likely to remember where the car keys are, and they may be more likely to notice an interesting building or odd looking person as they drive along the road.

Gendered Jobs?

In his speech, Summers suggested that gender differences in thinking might be used to explain why women are under-represented in certain university departments. In particular, it might explain why there are comparatively few women in maths, engineering and scientific fields that are highly quantitative, such as physics. Summers also implied that the cognitive differences are biologically determined. Both of these conjectures are misguided. Biology contributes to cognitive differences between men and women, but there are important cultural factors as well, and cultural factors may be the primary cause of academic hiring inequity.

The under-representation of women in university departments may owe something to cognitive differences, but it owes much more to discrimination. The disproportion of men to women in the academy is far greater than the extent of the alleged cognitive differences. Based on data from 2001, the National Science Foundation reports that, in American maths and physics departments, male full professors outnumber female full professors by a ratio of 10 to 1. In engineering departments, the ratio is about 36 to 1. If faculty employment ratios were driven entirely by statistical differences in thinking styles, we might expect women to outnumber men in fields that rely heavily on language skill, such as English and philosophy. This is not the case. In Harvard's English department, 20 of 53 faculty members are women, and in the philosophy department, 5 of 18 are women. For all the rhetoric about women being better than men in some cognitive domains, there is little evidence that their superior aptitude ever affords greater opportunities for women than for men. Up until very recently, men have dominated in all areas of the academy. We mustn't forget that, one hundred years ago, there were virtually no women teaching in universities. It would have been absurd to think this was due to differences in cognitive style. Women just weren't given the opportunity. The current numbers suggest that there has been exponential progress in women's educational equity, but they also suggest that discrimination remains a serious problem.

In fact, there is direct evidence for discrimination against women in hiring. Rhea Steinpreis and her colleagues at the University of Wisconsin in Milwaukee sent out a CV to a large number of psychology professors and asked them to assess whether the person named on the CV was worthy of hiring.[3] Each professor received a CV with exactly the same content, but in half the cases, the name on the CV was Brian Miller, and in the other half it was Karen Miller. Despite the fact that the imaginary job applicants were equally qualified, those who received the male applicant's CV were about 50 per cent more likely to say that he should be hired than those who received the female applicant's CV. The majority of professors evaluating the female applicant said she should not be hired, and the overwhelming majority of professors evaluating the male applicant said that he should be hired. It must be noted that all the professors who participated in this study probably believe that it is wrong to show preferential treatment to a man, yet that is exactly what they did unwittingly. Female professors were as likely as male professors to show this form of bias. This is direct and powerful evidence for the existence of discrimination in academic hiring. Similar studies have shown that the very same paper is rated as superior if it has a male author's name on it rather than a female name. There is also evidence showing that female professors receive less mentoring than their male counterparts when they are starting out, they are given lower salaries and they are regarded more negatively when they are assertive. Each of these factors can negatively impact prospects for women in academia.

Given the evidence for discrimination, it is possible that employment inequity has very little to do with cognitive differences. If graduate admissions committees, hiring committees and tenure committees are unconsciously biased against women, then we have a perfectly good explanation of why men outnumber women in the academy. The fact

that inequity is greater in some fields than in others ι
of residual stereotypes about women's capabilities.
very few women in law and medicine, and now wc
up with men rapidly. Our conception of what wor
tinually shifting. Given this history, and evidence for ιυ..
in hiring, there is no reason to think that cognitive differences are a
major factor in the current distribution of university jobs.

EXPLAINING GENDER DIFFERENCES

In blaming academic employment inequity on discrimination, I don't
mean to deny that there are cognitive differences between men and
women. There may be. As we have seen, men and women tend to perform differently on certain tests. Men do better with maths, mental
rotation and embedded pictures, and women do better with verbal
memory and fluency and with recalling where objects were located.
These differences need to be explained. There are three possibilities.
One possibility is that men and women are equally good at the skills
in question, but they just perform differently on tests. Another possibility is that there are biological differences that have an impact on
cognition. A third possibility is that cultural variables lead men and
women to think somewhat differently. It turns out that each of these
variables is partially right.

Testing Troubles

Let's begin with the possibility that gender differences in thinking are,
in part, an illusion generated by misleading performance, on tests.
There is some strong evidence for the suggestion that differences
between male and female math scores can be partially explained this
way. The primary evidence for male superiority in maths comes from
the fact that men do better on the maths portion of standardized tests,
such as the SAT and the Graduate Record Exam (GRE). But these
results actually conflict with records of classroom performance. In
American high schools, girls take about the same number of mathematics classes as boys, and they get better grades. Women also comprise

.ost half of the maths majors in American colleges, and they do just
.s well as men. These indicators suggest that women and men have
comparable aptitude for mathematics. Why, then, do men do better
on standardized tests? One possibility is that women under-perform
because they believe that they are less capable than men. In a simple
experiment Claude Steele and his colleagues gave a maths test to a
group of male and female college students, after telling them in
advance that men tend to do better than women.[4] Lo and behold, the
women did worse. Then the experimenters gave the same test to
another group of male and female college students without saying
anything in advance, and their scores came out the same. This phenomenon, which has been replicated many times, is called stereotype
threat; if you make a negative stereotype salient to people, they will
inadvertently conform to it. These effects are widespread. If you tell
people of colour that they do not generally perform as well on a test
as whites, their scores will drop, and if you tell white men that their
scores are usually lower than Asians', there will also be a significant
decline in performance.

A negative stereotype can become salient without even mentioning
it. To demonstrate this, Michael Inzlicht and Talia Ben-Zeev gave
female college students maths portions from the GRE.[5] The women
took the test in a room with other women, with men, or with a combination of men and women. When women were placed in a room
with men, their performance declined, and the extent of the decline
was proportionate to the number of men in the room. Male performance was not affected by the presence of women. Apparently, when
women take standardized tests in the presence of men, they unconsciously recall the stereotype that men are better at maths, and their
performance drops off. If the presence of men adversely affects female
performance in maths, one might wonder why women do as well as
men in their maths courses. One possible explanation is that a male
presence has an adverse affect only when women are taking standardized tests. Unlike the ordinary tests that students take for their maths
courses, standardized tests are overtly comparative. Everyone knows
that the SAT and GRE are used to make college and university admissions decisions, and scores are given as a percentile in comparison to
other students. With ordinary classroom tests, students are less likely

to see themselves as competing with other students, so stereotypes pertaining to comparative performance (women are worse than men) are less likely to come to mind.

This research suggests that the cognitive differences between men and women may be exaggerated. Some of the variation in test performance may result from the unconscious fulfilment of negative stereotypes. But this probably isn't the whole story. First, it's not clear that there are negative gender stereotypes associated with every cognitive test that shows gender differences. Why, for example, are women better at verbal memory and recalling how objects are arranged? Why are women worse than men at embedded picture tasks? Second, there is a nagging question of where the stereotypes come from. Many stereotypes have no basis in reality; they are merely used to denigrate. We can instil fear by saying that Jews are greedy, and we can justify economic disparity by saying that blacks are lazy. The claim that women are worse than men at mental rotation is potentially insulting, in that it implies that men are more intelligent, but it's hard to see how this particular claim would have been deliberately devised to hold women back. Third, there is evidence that biology has a role in the cognitive differences between men and women.

Biological Factors

The evidence for a link between biology and cognitive differences comes from several sources. First, there is evidence from neuroscience. Compared with women, men seem to have more grey matter, the pinkish grey tissue comprised of cell bodies covering the surface of the brain. But women have more white matter, the connective tissue just below the surface that allows cells to communicate. Male brains are larger overall, but women have faster brains, and some studies suggest that women's brains have more cells and larger areas dedicated to language. There are also differences in how male and female brains function. For men, IQ scores correlate best with activity in the frontal cortex and parietal cortex. In women, IQ scores are correlated with different areas of the frontal cortex, including language areas, and there is little correlation with the parietal cortex.[6] Differences in brain function could explain male and female performance on spatial tasks.

For both men and women, mentally rotating an object involves many of the same brain areas, but in women there is greater activity in areas associated with object recognition, and in men there is greater activity in areas associated with motor control. One possibility is that women try to mentally rotate an object by visually analysing its parts, whereas men are more likely to imagine physically moving the object around.

These findings are intriguing, but difficult to interpret. We often don't know the significant brain differences; for example, we have a very limited understanding of the link between brain size and brain function. Moreover, different labs report different results, and some alleged contrasts between male and female brains have been called into question. For example, it was widely reported that female brains are more symmetrical than male brains, with certain language functions actively involving both hemispheres rather than being predominantly located in the left. But a recent analysis of multiple brain scanning studies suggests that this isn't the case; both male and female brains seem to be equally asymmetrical. Studies of the brain also raise a difficult chicken and egg problem. If male and female brains function differently, those differences could result from differences in socialization. Differences in how the sexes are educated could affect brain function. We know, for example, that trained musicians, mathematicians and taxi drivers have brains that function somewhat differently from those of the rest of us.

Even if we take current findings from brain science as tentative, there is some reliable evidence for the conjecture that biology contributes to gender differences in cognition. For one thing, gender differences show up in other creatures. Male rhesus monkeys outperform females on a spatial memory task, in which they have to find a food reward that changes locations on each round. The male spatial advantage can even be found in rodents. In rats, mice and meadow voles, males often outperform females when learning the location of food in mazes. In rats, the pattern changes when the food can be located by memorizing landmarks; females do better than males at using such information. These findings are intriguing, because they confirm the pattern we see in humans: males in many mammalian species are more adept than females at spatial tasks.

Research on animals suggests that some of the cognitive differences between men and women may be deeply rooted in biology, but there are reasons to exercise caution when drawing this conclusion. First, gender differences have not been found in all animals. Male meadow voles are maze masters, but male prairie voles are not. Second, when gender differences are found, they are often ephemeral. In some species, the gender differences disappear with age, training or at different stages in the reproductive cycle. Third, in some species, the gender differences we find in animals may actually contrast with the human case on closer analysis. As I mentioned, male rhesus monkeys outperform females in remembering the location of objects, but in human beings spatial memory is often better in women. Finally, there is also some risk in drawing inferences from one species to another, because each has its own evolutionary history. Consider an analogous case. Members of polygynous societies might find comfort in knowing that male gorillas keep a harem of females, but it would be a mistake to infer from this comparison that polygyny is the natural arrangement for human beings; a wide range of sexual arrangements can be observed in the animal kingdom.

The best evidence for an interaction between biology, gender and thought comes from hormone studies. Consider testosterone, the principal male hormone. It turns out that fluctuations in testosterone correlate with fluctuations in cognitive performance. As men age, testosterone levels drop, and, when those levels drop, there are correlative drops in performance in maths and spatial skills. Hormone replacement therapy can improve performance. Similar effects have also been observed in women. Women naturally produce some testosterone, but only about a seventh of the amount that men produce. Studies have shown women with comparatively high testosterone levels outperform women with low testosterone levels on spatial tasks and maths tasks. Giving women a single dose of testosterone improves their performance on mental rotation tasks. That doesn't mean we should all take mega-doses of testosterone. Optimal performance is associated with moderate levels of the hormone. Women with high testosterone levels and men with low levels perform better than people with too much or too little.

Cognitive effects have also been associated with the principal female hormone, oestrogen. In particular, oestrogen is positively correlated with verbal memory and verbal fluency, two skills that tend to be better in women than in men. Two to three weeks after menstruation, when oestrogen levels are high, women score better on verbal tests. When oestrogen levels decline in menopause, there are correlated drops in verbal skills. When men with prostate cancer take oestrogen, their verbal memory improves.

These findings suggest that cognitive differences between men and women are influenced by hormonal differences. Testosterone and oestrogen can change the way we think. But we should not get too carried away. For one thing, the correlations between hormone levels and cognitive abilities are ceiling high. A maths wiz can have low testosterone, and a verbal savant can have low oestrogen. In fact, there is little reason to think that individuals who make great achievements in these domains have impressive levels of the corresponding hormones. For example, it was recently discovered that people working in the hard sciences tend to have low testosterone levels, or at least levels that were low during crucial periods of early development. This underscores the point that women are not being excluded from science because of inadequate biology.

A second reason to doubt the importance of hormones comes from the fact that gender gaps are closing. A few decades ago, the performance gap between men and women was twice as large, but the hormonal differences were, we can presume, just as great as they are today. If hormones were the primary source of cognitive differences between the sexes, we should see greater stability over time.

A third reason for caution is that hormone differences may have social causes. Suppose that hormones levels were perfectly correlated with cognitive skills. It still would not follow that hormones are the *ultimate* cause of gender differences, because variation in hormone levels can be affected by environmental factors including socialization. For example, depression causes testosterone levels to drop. It also happens to be the case that women are twice as likely to be depressed as men. Why? Perhaps it is because women are socialized into feeling inadequate, subordinate or limited in their opportunities. Thus, socialization can cause depression, depression lowers testosterone, and low testosterone levels in women diminish performance on maths and spatial tasks. Perhaps women do worse than men statistically because societal factors make women more depressed. Hormone levels might be the proximate cause of sex differences, and not the ultimate causes.

Finally, hormones cannot explain all the data on cognitive differences, because there are demonstrable interactions between hormone levels and environmental factors. Here's a case in point. As we have seen, women are more likely to do badly on maths tests when they are reminded of the stereotype that women are less numerically competent than men. It turns out that the effects of stereotype threat are magnified for women who have high levels of testosterone. Remember, these are the women who are ordinarily likely to do best on maths tests. When women who are talented in maths are reminded of negative stereotypes, their performance plummets. Women with low testosterone do not show the same effect. Consequently, when stereotypes are primed, naturally talented women actually perform worse than women who have not had a helping hand from biology.

Learned Limitations

The impact of stereotypes on cognitive performance suggests that social environment plays a role in the cognitive differences between men and women. If drawing attention to a negative stereotype can affect a woman's performance while she is taking a test, imagine what a lifetime of exposure to negative stereotypes can do. There is overwhelming evidence that women are treated differently from men, and these differences begin from the earliest days of life. In the face of such obvious and overt differences in socialization, it is remarkable that researchers ever looked to biology as the primary source of differences in cognition. In the nineteenth century, scientists thought that women's child-like appearance, such as their lack of facial hair, explained the fact that women were more ignorant than men. The more obvious explanation was that women were prevented from having equal education, employment opportunity, government involvement and personal autonomy. If women were less informed than the men who controlled their lives in the nineteenth century, it was a consequence of the fact that women were treated like children. With 20/20 hindsight we know that biological differences are too small and too ephemeral to explain the gross inequality that existed at that time.

History teaches a sobering lesson. Our contemporary attempt to explain gender differences by appeal to biology alone may look preposterous fifty years from now, when women have had more time to prove themselves in every branch of the academy. Biological differences exist, but they can be swamped, exaggerated and shaped by culture. In response, a well-intentioned naturist might argue that men and women now enjoy equal education and equal opportunity. In the nineteenth century, women were denied equal access to higher education, but now they are not. So, the naturist will say, any residual cognitive differences must be due to biology. This line of argument is sheer folly. Women and men may attend the same schools, but they are not treated the same way. Attending the same classes does not entail having the same education.

Studies have shown a pervasive and systematic pattern of unequal treatment in contemporary classrooms. Dedicated and well-intentioned teachers fall prey to societal gender bias and treat female students inequitably. Here are some unsettling facts. Male students are given more praise and criticism than female students; teachers call on male students more often; male students are given more time to answer questions when called on; male students are asked harder questions; female students are more frequently asked to report matters of fact, rather than matters of opinion or analysis; teachers generally give male students more feedback; in that feedback, teachers are more likely to give male students advice that helps them arrive at correct answers the next time around rather than just telling them the correct response. This pattern of preferential treatment starts early and has an impact. By the time women are in college, their style of academic engagement differs markedly from their male classmates. Where men make assertions in the classroom, thereby advertising their intelligence, women are much more likely to ask questions, advertising their ignorance. In fact, women who make assertions in classrooms tend to make them with the same intonation as a question, inadvertently playing dumb when they know the answer.[7] Men speak four times as often and shout out answers eight times as often.[8] Teachers are more receptive to these male interruptions, they direct more questions at men, are more likely to develop remarks made by men, and they offer men more encouragement.

A die-hard naturist or an unrepentant sexist might argue that all these classroom differences are the result of biological differences and not the other way around. Perhaps teachers treat males as more intelligent because they are more intelligent. The problem with that explanation is that female performance on aptitude tests suggests that they are as capable as men in most areas, and better than men in some areas. Women's aptitude for science is comparable to men's during teenage years, but social factors are working against them. A recent study shows that teenage boys and girls have comparable interest in, and aptitude for, science, but parents systematically report that their daughters have less interest and talent.[9] In the same study, fathers were shown to be significantly more demanding when helping their sons with science projects. The biases at home reinforce the pattern at school. Ultimately, girls lose confidence in their ability to become good scientists. There is some evidence that these deleterious effects can be mitigated by sending girls to single-sex schools. Girls who graduate from single-sex schools have higher educational aspirations than their coed counterparts, they are more likely to attend top universities and they are more likely to pursue graduate degrees. These girls are also more confident, and they are considerably less likely to fall into the stereotypically female pattern of turning assertions into questions when they contribute to classroom discussion. Girls in single-sex schools are also more likely to profess an interest in maths.[10] These girls are not biologically different from those who attend coed schools; they have just been socialized differently. With boys around, girls become second-class citizens.

Socialization is not restricted to educational settings. Children are exposed to an endless barrage of images showing men and women playing gender-specific sex roles. In movies, television shows, magazines and pop music, kids learn gender-specific attitudes and behaviours. These differences even show up in sources of entertainment designed for young people. In children's books, male characters are five times more likely than female characters to be portrayed as aggressive, and more than three times as likely to be portrayed as competitive. Girls are more than twice as likely to be portrayed as emotionally expressive, and almost four times as likely to be portrayed as passive.[11]

Gender socialization begins at birth. Girls and boys are named differently, dressed differently and put in differently decorated rooms. These overt differences cue care-givers into different patterns of socialization. This has been nicely demonstrated by a series of studies in which adults are presented with a baby wearing either pink clothes and bearing a female name or wearing blue clothes and bearing a male name.[12] In these studies, the same baby is used, but some adults think it's a boy and others think it's a girl, and that makes all the difference. For example, when a six-month-old baby is labelled 'Beth', adults described 'her' as soft, nice and delicate. When the same baby is introduced as 'Adam', adults describe 'him' as strong, active and intelligent. If adults see a video clip of a baby reacting to a jack-in-the-box, they will describe that baby as frightened if they think it's a girl and as angry if they think it's a boy. Adults will also play different games with male and female babies. If several toys are present, adults will hand 'Beth' a doll. If they think the same baby is named 'Adam', they will hand over a toy hammer or truck. Adults also give positive feedback to babies when they pick up toys that fit the gender stereotype. This is not just fun and games. Studies suggest that, regardless of sex, children who play more with stereotypically three-dimensional toys, such as construction sets, do better on maths tests than children who play with dolls. Socialization encourages girls to be less active than boys, more timid and more preoccupied with beauty than brains or brawn. Such socialization could easily affect thinking styles. A typical boy may spend hours making models and building go-carts, and a typical girl may spend hours imagining dialogues between Barbie dolls. The boy gets extensive training in spatial reasoning, and the girl becomes a master of language. Later, when they are given tests as teenagers, the boy will do better at mental rotation, and the girl will show greater verbal fluency.

A Trucker Instinct?

In his speech about gender differences, Summers said children's toy preferences have led him to think that gender differences are driven by nature, rather than nurture:

While I would prefer to believe otherwise, I guess my experience with my two-and-a-half-year-old twin daughters, who were not given dolls and who were given trucks, and found themselves saying to each other, look, daddy truck is carrying the baby truck, tells me something.

This anecdote is striking, but it should not be taken as evidence for nature over nurture. After all, these girls had already undergone two and a half years of socialization. By six months, kids are getting heavy non-verbal cues telling them what sort of toys they should like.

This is not to say that nature is irrelevant. Biology may make a small contribution to toy preferences in childhood. Gerianne Alexander and Melissa Hines set out to prove this by studying toy preferences among vervet monkeys.[13] They observed forty-four male and forty-four female monkeys as they played with a variety of toys. There were two stereotypically male toys (a ball and a truck), two stereotypically female toys (a doll and a cooking pan) and two neutral toys (a stuffed dog and a book). They found that male monkeys showed greater preference than females for the male toys, females showed greater preference for female toys, and males and females showed equal preference for neutral toys. Striking evidence indeed. But, on closer analysis, it's not entirely clear what to make of the study. It would be preposterous to propose that monkeys have an innate gender-linked interest in trucks and pans, since these are human inventions. It's also preposterous to say the females liked the doll because of a mothering instinct, because the doll portrays a human baby. In this context, it's noteworthy that there were no gender differences in how much the monkeys played with the stuffed dog, which arguably resembled a baby vervet more than the human baby doll. Alexander and Hines suggest that the male vervets might like the truck and the ball more than the females do because those toys allow for more active play. This is possible, but the data suggest that the males liked the pan at least as much as the active toys, and they liked the stuffed dog more than anything. So males did not show a general preference for toys that move around. Indeed, the authors concede that the results in their study may be largely driven by colour preferences. The authors used a red pan and a pink doll, and there is evidence from monkeys and humans to suggest that females prefer warm colours. This is a major problem with the experimental design. Another problem concerns the authors' way of assessing the monkeys' preferences. They imply that females preferred pans and dolls to males, but they actually show only that females contact these toys more frequently than males relative to their total number of contacts with toys. But males' contact toys more overall, so males actually contacted dolls and pans more than females. Thus, it is misleading to say that females like these toys more than the males do. At best the study shows that males like moving toys more than females do.

Let's suppose that biological factors make males more likely than females to play with active toys. Let's also suppose that this is true in humans. That natural preference could give males more opportunities to hone their spatial skills. But nature is not working alone in the human case. There is overt pressure on boys and girls to play with gender-specific toys. Small biological differences can be dramatically magnified by socialization. By discouraging girls from playing with active toys, we prevent them from acquiring the skills that they need to perform on a par with boys. Socialization could be used as an equalizer, but instead it is used to exaggerate differences.

The naturist might scoff at this. Naturists tend to be biological determinists. They tend to think that gender differences are indelibly etched in our genetic building blocks. But, as we will now see, that assumption turns out to be false.

DEFLATING DIFFERENCES

Wherever they come from, differences between the sexes are not immutable. Socialization, experience and training can all have an impact on cognition, and such factors can augment gender differences, as we have seen, but they can also diminish those differences. It is even possible that cognitive differences between the sexes can be eliminated or reversed.

Gender across Cultures

Before considering cognitive traits, consider an example of the link between gender roles and social psychology. In the 1970s, a Harvard anthropology student named Carol Ember did her Ph.D. research on the Luo, a Nilotic society in Kenya. In this group, there are clearly demarcated gender roles, but, when there aren't enough women around, some boys are called on to do stereotypically feminine chores, such as childcare and housework. Ember compared boys who had done a lot of feminine work with those who had done relatively little, and she found significant differences in their social behaviour. The boys who had done a lot of feminine work were significantly less aggressive, less likely to try to dominate others and more likely to engage in pro-social behaviour. In all these measures, their behaviour was more like that of the girls whom Ember tested in the study. Simply doing more domestic work shifted male social behaviour towards the female stereotype. This is striking evidence for the effects of socialization.

Ember did not investigate cognitive differences, so we don't know if stereotypically feminine work affects spatial cognition and other abilities that vary as a function of gender. But the circumstantial case for such effects is extremely strong. In the discussion of field-independence and field-dependence in the previous chapter, we saw that socialization can affect thinking styles. In fact, many of the cognitive tests that are used to distinguish individualists and collectivists are also used to distinguish men and women. Women make more errors than men on embedded picture tasks (in which subjects find a hidden picture) and rod and frame task (in which subjects orient a rod perpendicular to the ground when it is surrounded by a frame that is not perpendicular). Collectivists make more errors on these tasks than individualists. This is not merely a coincidence. Women are socialized to be dependent on other people and to downplay their own individuality. In other words, women are socialized to be more like collectivists. If socialization can make an entire culture have a collectivist orientation, and, if such an orientation has cognitive effects, then we should conclude that the socialization of women can have cognitive effects. And, conversely, men are socialized to be more independent than women, and we know that socialization that emphasizes independence can lead to improved performance on certain cognitive tasks.

The research on individualism and collectivism suggests that we should make two predictions about gender differences. First, we should predict that in cultures that are highly collectivist, gender differences should be smaller, because men in such cultures will be more likely to have cognitive styles that emphasize dependence, and that cognitive style will be shared by women in those cultures. This prediction finds confirmation in a study by Li-Jun Ji, Kaiping Peng and Richard Nisbett.[14] They gave Chinese and American subjects a rod and frame test. Among American subjects, the men did considerably better than the women, but among Chinese subjects, gender differences were negligible. Collectivist socialization promotes field-dependent, or context-sensitive, cognitive processing in both men and women, so gender differences diminish.

The second prediction that we can distil from research on individualism and collectivism is that cognitive differences between men and women should diminish in individualist cultures when women are socialized to be very independent. In these cultures, women are socialized to have a more field-independent cognitive style, and that improves their performance on spatial tasks. For confirmation of this prediction, consider the Inuits. Earlier we encountered John Berry's discovery that, in comparison to the Temne agriculturalists of Sierra Leone, Inuit people perform considerably better on embedded picture tests. Inuits are hunters and gatherers, and each individual makes contributions to subsistence without depending on joint collaboration with others. Temne farmers work collaboratively, and they enforce strict social roles. Berry also investigated gender differences in these cultures. Inuit women are given considerable independence, and Temne women live very restricted lives. Unsurprisingly, Temne men outperformed women on spatial tasks. Among Inuits, however, Berry found no sex differences in spatial tasks: women were just as good as the men, and both were better than the Temne. Similar patterns have been found in other societies. For example, Durganand Sinha compared a highly stratified urban population in India to less stratified tribal groups.[15] In the urban population, boys of all ages outperformed girls on an embedded picture test, but gender differences were negligible among those raised in tribes. In another study, Anneliese Pontius looked for gender differences among the Auca, a hunter-gatherer group in the Ecuadorian rain forest, and she found that Auca women actually outperformed men on some spatial tasks.[16]

These variations suggest that gender differences are not immutably fixed by biology. In cultures where women are more independent and contribute more to subsistence, they tend to perform very well on spatial tasks. In cultures where men are more collectivist in orientation, both male and female performance is equally prone to error. Our own society is individualistic, so men tend to perform well, but men and women are socialized differently, so gender differences are found. The fact that European and American women do not score as highly as European and American men on spatial tasks suggests that women here are less individualistic than men, and that may reflect profound differences in socialization.

Training

Cross-cultural comparisons show that gender differences are not fixed, and that has implications for how we should understand and address such differences here at home. Let's suppose that women perform less efficiently than men on some task. Let's even suppose that this difference has roots in biology. It might be possible to bring women up to male levels of performance by simply providing supplementary training (and conversely to improve male performance on tasks where they are outperformed by woman).

The claim that training can improve performance is hardly revolutionary. We know that human abilities get better with practice. If Bob is a naturally gifted musician with little training, and Ben is less gifted but intensively trained, Ben will outperform Bob. In principle, then, we should be able to bring women up to male levels of performance in cognitive tasks where men are alleged to have a biological advantage, and conversely in cases where women naturally outperform men.

Interestingly, such interventions don't necessarily involve preferential treatment. This is nicely illustrated by a recent experiment. Jonathan Roberts and Martha Bell tested men and women on a computerized version of the mental rotation task.[17] Mental rotation, you will recall, is one of those abilities that show a male advantage, which might be linked to biology. Roberts and Bell found the predicted sex difference, but then they allowed both men and women to get some practice. After that simple intervention, the sex differences disappeared. Men start out better, but women catch up. It seems that women improve from training more than men do, perhaps because men have already had more experience with spatial tasks through recreational activities like video games and construction sets.

This study drives home an important point. When we work to improve women's performance, we do not need to degrade male performance. We can train both sexes equally, give both the same amount of instruction and experience, and biological differences may level out. Eliminating differences does not require discrimination against those who have a natural advantage or even giving special treatment to those who have a disadvantage. It is possible to achieve equality by just making sure that everyone gets adequate training.

The training study presents a puzzle. If biologically based gender differences can be eliminated by practice, why do they still appear in educated adults? We have already seen that socialization may be at work, but there is also a more general ratchet effect. If a young man shows more aptitude for something than a young woman, he may get more encouragement. He is also likely to develop a stronger interest, because it feels good to excel. As a result, he will end up with more training than the young woman. Education tends to amplify differences, for this reason, rather than reducing them. To guard against this trend, it is important to encourage people who are not performing at the top of their class, since they may be able to catch up, excel and even exceed the abilities of others.

Splitting the Differences

In this section, I have accepted that there are biological factors affecting gender differences in cognition. Research on the link between hormones and cognitive skills is fairly convincing. I've simply been suggesting that culture and training can override biological differences. Biology may set default levels, but final outcome depends on experience. An adequate theory of gender differences in cognition must implicate both biology and socialization. Here's a sketch of how an adequate theory might go.

Let's start with the biological contributions. It's plausible that testosterone gives men a slight advantage over women when it comes to spatial cognition, as long as testosterone levels don't get too high. No one knows exactly why this is so. Testosterone may exert an indirect effect; perhaps testosterone makes boys more active, and that makes them more likely to explore and more likely to play with three-dimensional toys, thus giving them more experiences that are known to enhance performance on spatial tasks. Alternatively, testosterone may exert a direct affect by modulating brain systems that contribute to spatial tasks; there are neural receptors that are responsive to testosterone in the hippocampus, a brain structure implicated in spatial cognition. Testosterone may also increase aptitude for maths, and oestrogen may increase aptitude for verbal memory, and other capacities in which women tend to get higher scores. In all of these cases, the biological effects may be very modest; testosterone may boost ability only to a small degree, and it may not have that affect in all people.

Naturists would be inclined to stop with these biological speculations, but they would miss half the story. Modest biological differences in abilities tend to get augmented through socialization. Here's an analogy. Suppose a child shows a bit more interest in, or ability for, music than other children. That child might be given lessons and encouraged to pursue those natural musical talents. As a result the child will receive more training and positive feedback, and the modest differences will be magnified. In the end, the child may be a much better musician than children who showed only moderately less ability. Now, suppose that there were a disproportionate number of redheads with natural musical talent. If so, redheads might receive more musical training and encouragement. As a result, those redheads who were never especially talented might end up being more skilled musically than blondes or brunettes, because they benefit from the widespread assumption that redheads are generally more musical than others. In this scenario, we'd end up with a group that systematically showed significantly greater musical abilities than others, despite the fact that their biological advantage was small and found only in some members of the group. In this way, socialization amplifies and augments biological differences. Culture can also erase biological differences. Imagine a culture in which musical skills are considered important for everyone. In this culture, everyone would get trained in music, and those without natural talent might end up as proficient as many of the people with talent. Indeed, many of the people who had no natural advantage in music might end up outshining those with natural talent because of greater interest or enthusiasm. In this musical utopia, biological differences end up being insignificant. Blondes and brunettes can catch up to redheads.

The musical example is fanciful, but I think the story with cognitive abilities works this way. If there are biological differences between men and women, those differences probably get magnified through socialization. But it need not work that way. If we stop assuming that one sex is inherently better at certain cognitive tasks than the other, we can encourage both sexes to master the same range of important skills. Training would allow men to improve verbal skills and increase their sensitivity to contextual information, such as landmarks, and training would allow women to rival men in maths and spatial skills. If we design curricula to maximize the capacities of both boys and girls, sex differences in cognitive abilities may shrink away. Rather than blaming biology for inequality, we should blame ourselves for not taking steps to even the playing field.

Where Do Feelings Come From?

10

Fear and Loathing in Micronesia

As we have seen, culture, socialization and experience can influence how we think. The way we divide up the world and process information can depend on the language we speak, the society we belong to and where we stand in that society. This conclusion departs from the widespread view that the laws of thought are biologically fixed and universal. The mind is more flexible than your introductory psychology textbook might have led you to believe. But malleability may have its limits. One might think that certain aspects of the mind are immune to such social influence. Consider emotions. Socialization might affect how often and intensely you feel emotions – how fearless you are and whether you feel comfortable expressing strong emotions in public. But can socialization change emotions themselves, altering their character or bringing new emotions into existence? Might some emotions be socially constructed?

This sounds like an outlandish suggestion because we think of emotions as deeply rooted in biology. They are part of our animal nature, basic instincts that we share with other creatures and have little ability to control. The idea that some of our emotions might be products of culture sounds a bit like the suggestion that digestion and respiration are socially constructed. These things come to us naturally.

But the naturist perspective on emotions has been challenged. There are many ways in which culture influences how I feel, and it may even be possible for culture to instil new emotions that we would not have had if we were raised in a different time and place. Culture may also have a hand in determining what makes people happy, and in the symptoms we experience when healthy emotion regulation breaks down.

GUT FEELINGS

Before getting embroiled in the nature–nurture debate, we had better get a handle on what emotions are. This question has also been a matter of considerable controversy. There are two major approaches to the emotions: the appraisal theory and the embodiment theory. The appraisal theory has typically been associated with a nurturist take on the emotions, and the embodiment theory has been preferred by naturists. We'll see that this pairing gets things wrong. Naturists are right about what emotions are, but nurturists are right about their flexibility.

The appraisal theory of emotions says that emotions arise when and only when people appraise things. An appraisal is a judgement that something good or bad has occurred. Suppose your spouse gives you for your birthday a book on how to write a good résumé. This might be the gift you were hoping for. You think the gift was especially thoughtful. That would be a positive appraisal – a judgement that something you wanted has occurred. But suppose you were hoping for something better, like a nice wristwatch. In that case you might form the appraisal that you didn't get what you wanted. You might even judge that the gift is insulting, or you might take it as an unsubtle threat: find a job or we're through! These are all negative appraisals.

Appraisal theorists say that each emotion can be identified with a different appraisal judgement. Joy arises when you judge that a goal has been satisfied. Sadness arises when something you value has been lost. Anger arises when you find something insulting. Fear arises when danger is near.

On this approach emotions are intellectual things. They depend on how we think about an event. Appraisals can affect how we feel. You might tremble if you are afraid. But these feelings are not essential to our emotions. You can be afraid without trembling, as when you fear the outcome of an upcoming presidential election. And sometimes we tremble without fear, as when we catch a chill. So, for the appraisal theorist, emotions are not feelings at all. They are thoughts. Not neutral thoughts, but thoughts that evaluate things in a positive or negative way. They are usually accompanied by feelings, but they need not be.

I love sushi. That fact is always true of me. But the thought of sushi doesn't always fill me with positive feelings. If I have just had an ice cream sundae, I might feel ill at the thought of sushi. If I am sound asleep, I might feel nothing at all, but it's true of me on both occasions that I love sushi. That is because I have a positive view of sushi – I appraise it favourably.

The embodiment theory of emotions is diametrically opposed to the appraisal theory. Embodiment theorists deny that emotions are intellectual and insist that emotions are feelings. We can have emotions without making any judgements at all. Drinking beer, listening to classical music and skydiving can affect our emotions without the need for any judgements. If you pull the chair from under me, I will experience fear, but the event will happen too fast for me to judge that I am in danger.

Embodiment theorists say that emotions can be identified by their characteristic feelings and these feelings derive from specific changes that take place within the body. For example, when we are frightened, we get shivers, hair stands on end, our hearts race, blood rushes to our extremities, muscles tense up, eyes widen and breathing becomes constricted. Embodiment theorists say that fear is the feeling of this pattern of change. Even when we haven't made any kind of judgement. Fear can even occur when you think you are safe, as when you are watching a horror film in a safe movie theatre. But that does not mean fear has no connection to danger. The changes that take place in the body prepare us for coming into contact with predators and other threats. Changes in blood flow prepare us for flight, constrained respiration makes our breathing quieter and less detectable, and, if we were covered with fur, those goose bumps would fluff us up to make us look larger. Other emotions are associated with other bodily changes. When we're sad, we become sluggish, sulky and withdrawn; when joyful, we stand tall, spread our shoulders and relax our muscles; when disgusted, we scrunch up our faces to avoid letting anything noxious in; and when angry, fists clench, and we bare our teeth and lurch forward. Each of these body patterns feels different, and emotions are constituted by those feelings.

Defenders of the embodiment theory do not deny that appraisals can accompany emotions. You wouldn't have got mad at your spouse if you didn't find that book insulting. But such judgements are neither necessary nor sufficient for emotions in this view. You might laugh off an insult, and, in an irritable mood, you might lash out in rage at someone who hasn't done anything to insult you.

The difference between the two approaches to emotion can be summarized as follows. Embodiment theorists think that appraisal judgements often trigger emotions, but aren't essential, and appraisal theorists say that bodily feelings are often triggered by emotions, but aren't essential. One view emphasizes thought, and the other feelings.

Both of these theories have been ably defended by generations of researchers, but I think the weight of the evidence favours the embodiment theory. First, there is good reason to think that perceptions of bodily changes are necessary for having an emotion. The best argument for this conclusion was advanced by William James, the pragmatist philosopher and seminal psychologist. James was the first author to defend the embodiment theory, and his central argument is an appeal to introspection.[1] Imagine yourself in a state of terror, and then systematically subtract away each of the bodily symptoms that usually accompany that state. Imagine your facial expression and muscles are completely relaxed, your heart is beating at a comfortable rate, your breathing is calm, you have no goose bumps or knots in your belly. If you go through this mental exercise, James says, there will be nothing left that you would call the emotion. Deprived of bodily symptoms, emotions disappear. There is no terror without trembling, no anger without a disposition to clench one's fists, no grief without a lump in the throat, and no delight without an urge to smile. This intuition is confirmed by umpteen brain imaging studies. Whenever neuroscientists look at brain activity during emotional states, they see heightened responses in exactly those brain areas that are known to register and regulate bodily changes.

Against this, defenders of the appraisal theory counter that there can indeed be emotions without bodily symptoms. We saw one example already: my love of sushi. It is always true of me that I love sushi, but that emotion does not induce heart palpitations in me all day long. This is a nice counter-example, but it rests on a crucial ambiguity. When we say that someone loves sushi, we are not actually reporting an emotion. We are reporting an attitude that happens to depend on emotions. We sometimes use names of emotions to report dispositions to have those emotions. For example, I am outraged by global injustice, but it doesn't follow that I am having an emotional response right now – I am not outraged at this moment. Likewise, it can always be said of me that I am amused by *Monty Python*, saddened by Mozart's *Requiem*, disgusted by egg salad and frightened of zebras (sad, but true). But it doesn't follow that I am having any of these emotions right now. When we talk this way, we imply that, if you offer me sushi for dinner, my heart will flutter; if I catch a *Flying Circus* re-run on TV, I'll chuckle; and if Mozart's Mass plays on my radio, tears will well up. Absent these bodily reactions, you would not attribute such attitudes to me. There is an episode of *Welcome Back Kotter* in which Gabe Kaplan's character is swindled by a charlatan who poses as a talent agent and encourages Kaplan to cough up some hard cash to launch a career in stand-up comedy. At some point, Kaplan catches on and says, 'Hey, you say I am funny, but you never laugh at my jokes.' The con man taps his chest and replies, 'I'm laughing in here, where it counts.' But the point is, that doesn't count. If you really find something funny, then you will be disposed to laugh at it. Emotions always manifest themselves in the body. When appraisal theorists try to find examples of emotions without bodily manifestation, they end up undermining their case.

The foregoing suggests that the perception of bodily changes is necessary for having an emotion. Bodily perceptions are also sufficient. Research has shown that people experience emotions when they act out their bodily manifestations. If you mimic the breathing pattern of fear, you will feel mildly afraid, and, if you adopt the posture of despair, you will feel sad. Much of this research has been done using a single body part: the face. If you make facial expressions, they will affect how you feel, even if you make those expressions unwittingly. In a clever demonstration of this 'facial feedback' effect, Fritz Strack and his colleagues asked people to hold a pen in their mouths and fill out a questionnaire.[2] They claimed that they were developing training techniques for people who had no use of their arms. Some people in the experiments were instructed to grip the pen with their front teeth and others were instructed to hold it between pursed lips. The toothy method forces people's faces into a smile, though they are entirely unaware of that fact, and the pursed lips forces a subtle grimace, which also goes unnoticed. Strack found that people who were unwittingly smiling gave more positive answers on the questionnaire than people who were grimacing. In another study, Robert Zajonc and his colleagues had one group of people read a story out loud about a character named Peter, while another group read the same story, but with a protagonist named Jürgen.[3] The name Peter unconsciously forces the face into a smile-like configuration, while Jürgen induces a frown. The Peter group found the story more pleasant. This shows that facial expressions affect how we feel. It also suggests that we should use caution when naming our kids.

Defenders of the appraisal theory are not convinced. They think bodily feelings are not sufficient for emotions. They claim that bodily states are ambiguous; the same pattern of perturbation in the body can accompany entirely different emotions, so we need appraisals to tell emotions apart. The best empirical evidence for this claim comes from Canadian psychologists Donald Dutton and Arthur Aron.[4] They had a female graduate student give a questionnaire to men walking through a state park in Vancouver. She found half of her volunteers crossing a very sturdy bridge, and she found the other half crossing the harrowing Capilano suspension bridge, which wobbles unsteadily 230 feet above the river. Crossing the Capilano bridge would give anyone an adrenalin rush, and anyone feeling sweating palms and a racing heart while crossing would rightly interpret that bodily perturbation as fear. But Dutton and Aron reasoned that these physiological changes might be exactly like those that take place when people are feeling romantic attraction. Thus, when handed a questionnaire by an attractive graduate student, male passers-by might interpret the bridge-induced terror as a more amorous feeling. This is what they found. The graduate student invited all volunteers to call her later and find out the results of her study. Men on the suspension bridge were more than four times as likely to call her back than men on the benign bridge. They had mistaken vertigo for true love. Appraisal theorists take this as evidence for the view that physical arousal is not an emotion in and of itself; it qualifies as fear when construed one way and ardour when construed another. In other

words, appraisal theorists conclude that no bodily perturbation qualifies as an emotion if it is not accompanied by an appraisal.

This line of evidence does not refute the embodiment theory. First
of all, the suggestion that fear and ardour cannot be physiologically
distinguished is preposterous on the face of it. The states may overlap
considerably, but we all know that blood flows to different parts of
the body when these emotions are experienced. There are also different hormonal changes (cortisol vs. estradiol) and different behavioural
dispositions (avoid vs. approach). There are at least three compatible
explanations of the bridge effect. First of all, if there is *some* physiological overlap between fear and ardour, it may be easier for the body
to enter the latter state if it's already in the former state. Second, the
state of fear disposes the body to seek safety, and seeing another
human being during a fearful event might trigger feelings of relief and
delight, which might explain the extra phone calls. Finally, the brain tries
to maintain a state of equilibrium through what are called opponent-
processes. If you feel something negative, there will be a positive
after-effect. That's why we giggle after a fright, or feel euphoric after
skydiving. (The effect also works in reverse: ecstasy begets agony – an
effect that drug addicts know as withdrawal.) The long wobbly bridge
is a tourist attraction because it's thrilling to cross, and it's thrilling
because the body kicks into happy mode to compensate for mortal
terror. It is not surprising that men in a joyous mood are favourably
disposed to the graduate student.

There is no convincing evidence that we need appraisal judgements
to distinguish different emotions, and plenty of good evidence that
merely being in a physiological state is sufficient for feeling glad, sad
or mad. In other words, the embodiment theory is probably right.

EVOLVED EMOTIONAL UNIVERSALS?

The embodiment theory has traditionally been paired with the view
that emotions owe more to nature than to nurture. If emotions were
judgements, as appraisal theorists maintain, then it would be fairly
easy to see how culture could have an impact; cultural background

can influence our beliefs. If emotions are perceptions of bodily changes, however, there is no obvious place for culture to come in. How can culture change reactions in our bodies, especially if those reactions can arise without accompanying judgements? The point gains further support when we consider the kinds of bodily responses under consideration. Emotions are associated with physiological changes that are clearly evolved to help us survive. Consider fear again. The racing heart and tense muscles prepare us for flight, and the goose bumps make us (or, rather, our furry ancestors) look more menacing to predators. These are clearly evolved responses, not learned behaviours.

William James, who launched the embodiment theory, saw things this way. He relied heavily on Charles Darwin's observations of emotional expressions and concluded, with Darwin, that emotions are deeply rooted in our biological history. But too often such conclusions presuppose a sharp nature/nurture dichotomy. Even if we grant that evolution plays an important role in shaping our emotions, should we resist the conclusion that emotions are entirely the result of evolution? Let's have a closer look at what evolutionists say about the emotions and see why they may be pushing things too far.

Passionate Darwinism

Many of the emotions we've touched on seem to be very primitive, meaning they seem to have analogues in simpler species. Mice show a disgust-like expression when they eat bitter food, lions growl angrily when hyenas try to steal their prey, otters play happily, dogs cry when their owners leave the house, and homologues of a fear response can be found in everything from ferrets to pheasants, frogs and fish. Emotions show great continuity in nature. If you tickle a rat, it will emit a high-pitched sound that has the same acoustic profile as laughter. In honeybees, rewards and punishments release brain chemicals that are strikingly similar to what we find in human beings when we experience pleasure and pain. But some emotions are uniquely human. As far as we know, non-human animals lack moral emotions, such as guilt and shame, as well as emotions that arise in close human relationships, such as love and romantic jealousy. Such emotions seem less primitive, and perhaps less connected to behavioural instincts.

Fear looks like a quick, automatic physical response to a threatening stimulus, but love and guilt can last for years, and they are bound up with some of our most sophisticated concepts and institutions. Could these distinctively human emotions be the results of biological evolution? Ambitious Darwinists give a resounding yes. They claim that all human emotions are evolved, and they offer alluring evolutionary stories to make good on this claim.

I will focus on two representative examples, guilt and love. Many of us feel like guilt is something we would be better off living without. According to evolutionary psychologists, that would be a disaster, because guilt was evolved to serve a crucial role. The key argument was first given by Robert Trivers, a founding figure in evolutionary psychology.[5] Trivers's story begins with the observation that many human activities depend on cooperation, and cooperation presents a dilemma. If you and I set out on a joint venture, we may benefit from working together. But suppose we agree to cooperate and then I sit back while you do all the work. I will come out ahead that way, but you will come out way behind. You'd be better working for yourself than cooperating if I am not going to deliver on my end of the bargain. Of course, you know that I have a strong incentive to cheat you, because I'd come out ahead, so you are not especially motivated to deliver on your end of the bargain. And you have a strong incentive to cheat me, because, if I am dumb enough to do my share, and you do nothing, then you'll come out ahead. So we are both highly motivated to renege on our ends of the bargain, and highly sceptical that the other will deliver. Therefore, we'll probably both fail to deliver, and we'll both end up being a lot worse off as a result.

This basic insight about cooperation was first identified in the branch of behaviour science called game theory, and it was called the Prisoner's Dilemma, because it was illustrated with a story about two guys who get caught after collaborating on a crime. If the first crook rats on the other and professes innocence, he'll go free, and his partner will spend ten years behind bars as the sole perpetrator of the crime. If each crook blames the other, they'll get five years each. If both keep their lips sealed tight, they'll get a year each, but no longer because of the meagre evidence against them. That last alternative looks attractive, but not nearly as attractive as getting off scot-free.

Knowing this, they have a strong incentive to blame each other, and little incentive to stay silent because that might mean ten years in the can. So they play the blame game and get five years, missing out on that one-year sentence. Mutual cooperation leads to a better outcome, but mutual defection always looks like the safer bet.

Given this, it's a wonder that people ever cooperate. But we do. In fact, human success depends on cooperation. We enter trade agreements, build collective works, form treaties, establish enduring social bonds, farm together, hunt together and raise kids together. Almost everything we do depends on cooperation, and, without it, we'd be back in the cave, or worse. So how do we give up the very rational temptation to cheat our collaborators and freeload off the efforts of others? One answer is that, when we think hard about it, we realize we'd all be better off cooperating. But that solution depends on the assumption that we are good at postponing the short-term benefits of cheating in favour of some nebulous long-term payoff, a payoff that depends on everyone reasoning the same way as us. Not very likely. Another solution is to build social institutions that punish people for cheating. That would introduce a strong incentive (think of what happens to snitches in prison), but institutions of punishment themselves require collaboration, and there is, once again, a strong temptation to cheat. Why should I punish a cheater when I know someone else is around to do the dirty work? Trivers's solution to the problem is simple and elegant. We need a system that punishes cheaters without any need for people to do the punishment. Penalties without penalizers. We can get this if we build a form of self-punishment into human psychology: guilt.

Suppose we agree on a trade. You give me a hamburger today, and I will pay you tomorrow. We shake hands, I devour the meal and I go back home. It's a big world, and I could probably avoid seeing you again, in which case I'd have got my hamburger for free. But there is no such thing as a free lunch. The pang of guilt would be too great. Guilt is unpleasant, and we can appease it only by making amends. We feel guilty when we don't live up to our end of the social bargain, and the anticipation of guilt makes us behave much better than we otherwise would. Guilt makes us cooperate even when we can easily get away with defection. We leave tips in roadside restaurants even if we plan never to return. If we didn't feel guilt, we'd cheat whenever we could get away with it, and that is pretty often. We could steal from neighbours, lie on résumés, deceive spouses, neglect ageing parents and sleep on the job. We do some of these things, of course. Temptations are great. But when we do, there is an emotional cost, and that pang of guilt helps to keep us in line much, if not most, of the time.

The evolutionary argument is that human cooperation is nearly impossible without guilt. If guilt evolved, that would explain our ability to cooperate, and cooperation has huge advantages. We can hunt, gather and groom better when we cooperate. It's pretty hard to build houses, cast large nets and cut down large trees all alone. So once guilt comes on the scene, human beings begin to achieve great things, leading to longer, healthier, more productive lives. The payoffs are so great that the guilt-prone mutants do better than their remorseless peers. Guilt gets selected by evolution.

There are two problems with this proposal. First, a technical worry. If there were just one person born with a disposition to feel guilt, as happens when new mutations arise, then everyone would take advantage of that poor soul, exploiting his or her kindness and doing nothing to reciprocate. As a result, guilt would fail to be an advantage and get snuffed out of the genome just as quickly as it arose. There are some technical solutions to this problem in the evolutionary biology literature, but we'll see another solution in just a moment. The second problem is that Trivers's story is unnecessary. We have other more primitive emotions that can help get cooperation off the ground. It's highly plausible that, before guilt appears on the scene, we have sadness, fear, parental affection and anger. Now suppose that, during early childhood development, we do things that go against our parents' will. They will react with anger and withdraw affection. That will make us sad. Sadness is a response to loss, and there can be no greater loss than a parent's affection. We may also respond with fear because parents may threaten us physically. In effect, we'll have been conditioned to feel a blend of sadness and fear when we behave in certain ways. Perhaps guilt is just the name we give to such a blend of more primitive emotions.

Other mammals may never experience guilt because they are less psychologically dependent on parental affection. Humans have a prolonged period of dependency, which makes parent–child relationships particularly important to us. During this time, guilt is likely to emerge. But, even in human beings, guilt may not be inevitable. In anthropology, it has long been suggested that there are guilt cultures and shame cultures. Guilt differs from shame in that guilt concerns an action and shame concerns an individual. When you feel guilty that you did something, you don't necessarily feel like you are a bad person. You make amends, and the guilt subsides. With shame, you feel sullied. You don't want to make amends; rather you want to conceal yourself. The difference might emerge as a result of parental disciplinary techniques. If a child is scolded for misbehaviour, she may feel scared and sad – the raw ingredients of guilt. But suppose a child is ridiculed, made to stand in a corner or told that her actions will bring unwanted attention to the whole family. Now, the act of transgression is not the focus. Instead the child is made to feel like a freak or a monster. Embarrassment will result, coupled with an intense desire to hide. Most cultures cultivate both guilt and shame, but to different degrees. In principle, there could be cultures that conditioned just one.

If the story I've been telling is right, then guilt (and shame) initially arises in the home, and it is directed towards immediate kin. Once learned, it can be extended to others. We learn that certain forms of conduct are worthy of guilt, even when kin are not involved. This may provide us with a solution to the first problem with Trivers's proposal. The problem, recall, is that a biological mutation to feel guilt in one person would never get a chance to spread. But suppose we *all* are disposed to feel guilty towards our parents because guilt blends other more primitive emotions. Now there could be whole societies of people conditioned to feel guilt. People in these societies might, as a result, be inclined to cooperate with each other. That doesn't mean they would cooperate with complete strangers. For tens of thousands of years, we probably only cooperated with kin, neighbours, and tribe members. Subsequently, technological innovations allowed human societies to expand beyond small groups, and that may have promoted more widespread cooperation and the modern tendency to feel guilt when we mistreat strangers.

Obviously, this account is highly speculative, just like the evolutionary story, and it would need to be worked out in detail. But it illustrates two crucial points: guilt may be a blend of other emotions, and this blend may naturally arise in the familial context and then get extended. If so, we do not need to suppose that guilt evolved. The claim that guilt is a blend of other emotions is extremely plausible. Guilt arises most typically in contexts where we think our actions may lead us to lose someone we need. Such loss naturally elicits sadness. Guilt is also correlated with depression, and it can make us cry or sulk. There is also an element of anxiety in guilt. We fear getting caught. There is no distinct facial expression of guilt, and it feels a lot like these other emotions. The sadness-plus-fear story also helps explain why we make amends when we feel guilty; that's a good strategy for avoiding loss and the ire of those we care about. If all these facts about guilt can be explained by assuming it is a blend of other emotions, then there is little pressure to say it is an innate, evolved response.

A parallel conclusion can drawn about evolutionary approachs to romantic love. The economist Robert Frank has applied Trivers's theory of guilt to this, the noblest of human emotions. Frank argues that romantic relationships present us with something exactly analogous to the Prisoner's Dilemma.[6] Suppose you meet a fabulous guy or girl and, after dating a while, you decide it might be a good idea to tie the knot. If you both commit to this decision, you will be very happy. But chances are, no matter how fabulous, this person you're dating is not the most fantastic person you could ever find. There is probably someone out there who is even more intelligent, more beautiful and less, well, complicated. Suppose, after getting married, you were to meet this person. It would be more than a little tempting to have an affair. If you did that, you'd have double the joy: a spouse and a lover. But your spouse would be worse off. You'd be spending less time at home, and you might even end up with someone else's baby to care for. Before getting married, you and your partner can foresee this outcome. You know that there will be an incredible temptation to cheat when a better model comes along. Knowing this, you both realize you are doomed to a marriage of mutual infidelity, and that will make both of you miserable. So you decide not to get married, and your relationship comes to an end.

If everyone thought like this, there would be no marriage, and perhaps no collective home-building and childrearing. Society would come to a halt. But this way of thinking isn't paranoid or pessimistic; it's rational and realistic. In terms of short-term payoffs, it's rational to cheat when the opportunity arises, and it's rational to expect your partner to cheat for that reason, even if you think you can resist the temptation. So, thinking people should never exchange vows. But we're not thinking people, says Frank. We are feeling people. Romantic love evolved as a solution to this dilemma. When you fall in love, you feel as if the object of your affections is the only one for you. You feel like true love lasts for ever, and that life would be meaningless if it were dedicated to anyone else. These romantic ideas, infused with intense passion, make us choose marriage over monasticism. But all this is just a delusion foisted on us by our selfish genes. By distracting people from depressing calculus of anticipated infidelity, blind love leads us to the chapel. We are able to form romantic bonds at least for as long as it takes to have children and raise them to an age of self-sufficiency. The spell may wear off eventually, but, in the meantime, our genes have got what they wanted – duplicate copies of themselves.

Frank would have us believe that, without love, we might never propagate the species, but this whole story is very suspicious. The link between love and marriage and the monogamous ideals that he associates with romance have a distinctively Western character. In many societies marriage is arranged, so love has little to do with it, and can even pose a threat because it is frivolous and fleeting. In other societies women raise children with their brothers, so there is no need for a couple at home. In fact, in small-scale societies, the whole village may have a hand in childrearing, so a bonding emotional tie between sexual partners is hardly essential for the survival of the species. We should also be suspicious because other mammals, including the great apes, seem to get by just fine without romantic love, and without long-term pair bonding. Moreover, there are many other mechanisms that can be used to forge long-term cooperative relationships: family pressure, contracts, friendship, mutual benefits and, Trivers's favourite, guilt. If most cooperative ventures can be described as Prisoner's Dilemmas, and romantic love is rarely involved, then clearly there are other mechanisms in place to encourage cooperation.

I am not denying that love motivates some people in some cultures to get married. That's the usual formula in Western culture and many other places as well. I am suggesting that there is little reason to think that love evolved for this purpose. Frank's evolutionary story makes sense only if we assume that our ancestors lived in nuclear families where a live-in couple would be essential for raising the kids. That's just not the case.

What is romantic love, then, if not the engine behind a universal marriage instinct? Like guilt, it might be a blend of more primitive emotions. Consider the strong attachment that exists between parents and young children. This is undoubtedly facilitated by strong innate emotions. Parents have feelings of nurturance, and children have feelings of dependency. These emotions play a special role in childrearing, but nothing prevents them from arising in other contexts. In Japan there is an emotion called *amae*, which is described as a warm feeling of dependency, which adults can feel towards the corporation that employs them, their country clubs or their country. The character used to represent this emotion is derived from a Chinese pictogram depicting a suckling child, but in Japan that feeling is extended into adulthood. Likewise, nurturing feelings can be present when there are no babies to care for. We nurture pets, houseplants and art projects. Crucially, one or both of these fundamental emotions can be felt by one adult towards another. In fact, this is likely to occur if you spend a lot of time with someone and develop mutual understanding and interdependency. Couples who spend time together and enjoy the experience are likely to experience the affectionate emotions that bond parents and children. But why would a couple spend a lot of time together in the first place? The obvious answer is lust, or physical attraction. We certainly have an evolved desire for sex. If you combine attraction and affection, you get the package we know as romantic love. Lovers use baby talk between kisses; they cuddle tenderly and ache with desire.

The cocktail of attraction and affection can explain familiar characteristics of romantic love as well. When you first 'fall in love' you delight in the fact that attraction is mutual, and you experience butterflies in your stomach when courting because the outcome at that point is still uncertain. When your lover departs, you may experience longing, which is just desire for a distant object. When your lover breaks up with you, there is heartbreak because your desires are dashed. All these feelings work together to form the complex emotional landscape of love. In this story, love can certainly motivate marriage, but it didn't evolve for this purpose. Love can equally emerge when a marriage has been arranged, and it can exist between the partners in a short-term affair who have no aspiration to exchange vows. Love is just a natural outgrowth of the human capacity to combine carnality with care.

These examples illustrate a basic point. Before arguing that some emotion is innate, it's important to see whether it might actually be a combination or extension of more primitive emotions. Many of our most uniquely human emotions may be acquired during child development, rather than evolution, by using elements we already have. Pride may be a blend of joy and confidence; envy may be a blend of anger and desire; and contempt may be a blend of anger and disgust. If these blending proposals are right, then it's less likely that the emotions in question are innate. If we have the raw ingredients, we don't need evolution to make the cake. Evolutionists often just assume that these emotions are innate. But, until alternatives are explored and ruled out, that is not a safe assumption. There is little to gain from weaving evolutionary stories to explain emotions that may not even be innate.

The moral is that we should walk before we try to run. It's a good idea to start with the emotions that are most likely to be evolved – the ones that have counterparts in other species. But we'll see now that even these more primitive emotions are not entirely products of biology.

About Face

In the 1960s, it was widely accepted that culture might play a role in shaping the emotions. In those days before the dawn of sociobiology and modern genomics, people believed in the power of nurture. They were willing to believe that almost anything could be socially constructed. One of the true believers was Paul Ekman, a psychologist who set out to show that emotions vary cross-culturally. To his surprise, the evidence he collected seemed to show otherwise, and, in the decades that followed, he would become the most influential voice in the backlash against nurturism in the domain of emotions. What did Ekman find that changed his mind along with a whole generation of researchers?

The answer is smiles and frowns. Ekman travelled to a remote part of New Guinea to study an isolated group of people called the Fore, who had little contact with the Western world. Ekman wondered whether the Fore would use the same emotional expressions that we do. He decided to focus a small group of emotions – joy, sadness, anger, fear, disgust and surprise – which have become known as the Big Six. Ekman picked these because research had revealed that Americans and Europeans associate specific facial expressions with each one. In hindsight, these were ideal candidates for research on universals. They are just the kinds of emotions we might expect to find in others species – recall the angry lion and the happy otter. It is implausible that the Big Six are human inventions. Moreover, they have obvious adaptive value: joy is a reward signal that indicates when we have obtained something good for us, sadness causes us to withdraw in times of loss or defeat, anger helps us aggress against those who threaten us, fear leads us to safety, disgust helps us avoid things that are noxious, and surprise alerts us to novelty. If any emotions are the products of evolution, the Big Six are.

For each of these emotions, Ekman devised a corresponding story, which he asked an interpreter to read to the Fore. The sadness story

Figure 8. Expressions corresponding to the Big Six emotions.

involved the death of a child, for happiness it was seeing on old friend, for disgust he described a bad smell, and for fear the story described a confrontation with a wild pig. In each case, Ekman showed the Fore three photos of facial expressions and asked them to point to which one best expressed the feeling that would arise in the story. Their responses were surprisingly like ours. For the majority of these stories the face that the Fore selected most frequently was the face that Westerners would pick. This result was so unexpected at the time that it quickly became textbook knowledge. The world suddenly seemed a lot smaller.

But the results are a bit more complicated on closer examination. First, the Fore did not identify the right face for the fear story; they chose the surprise face more often when it was an option. This means that they got five out of six right, which is impressive, but falls short of perfect accuracy. The shortcoming is especially striking, since one might think fear would be so important to our survival that it would have especially high rates of recognition. Second, with the exception of happiness, which is the only positive emotion in the set, the responses were more varied than one might predict given the view that emotions are completely universal. For four out of the six emotions, there were test conditions where about 30 per cent or more gave an answer that was not predicted by Ekman. A 30 per cent error rate is pretty striking given that the Fore were given only three photos to chose from for each story, which means they had a 33. per cent chance of picking the right face just by guessing. If the Fore were able to exclude just one face of the three, such as ruling out a smile in response to a story about a dead child, then they would be left with only two options, which would mean a 50 per cent chance of making an accurate guess. Accuracy was usually over 50 per cent, which suggests the results were not mere guesswork, but given the odds of accuracy, the number of correct responses is not as impressive as the universalist might have hoped for.

The fact that the Fore were given a choice between three faces rather than a more open-ended range of options means they may have adopted a least-bad strategy, picking the face that struck them as imperfect, but better than the other options. Suppose that these emotional expressions are partially rooted in biology, but alter under cultural influence. If so, the expressions might have been close enough to the indigenous response to pick with a statistically significant degree of accuracy, but the method offers no way of testing whether any of these faces was a perfect match for how the Fore would express emotions, or for the emotions they would feel.

Ekman conducted two other studies with the Fore, which may appear to help with this challenge. If we really want to know how the

Fore interpret faces, the most direct test is to ask them. In one study that's just what Ekman attempted. Rather than having a story with a set of photos, he showed the Fore individual photos and had them select from a list of emotion words translated into their native language. When Ekman published the results in *Science*, he said the study provides proof that there are evolved universal emotions, but the data suggest a more complicated picture. The Fore tended to give the predicted answer for only four out of the six emotion categories; they interpreted the surprise faces as fear and sadness as anger. The majority of the Fore gave the predicted response on only half of the photos used in the study, and they reached a consensus of 70 per cent on only a quarter of them. And who knows what the numbers would look like if the Fore had been given more words to choose from. These outcomes are good enough to support the conclusion that the Fore interpret *some* faces *similarly* to how we interpret them, but a stronger conclusion is harder to draw.

One might worry that this study depends too heavily on language and might suffer from problems with translation. To prove that emotions have universal expressions, Ekman realized it would help to film the Fore while they made emotional faces. His first attempts to do this involved lunging at Fore children with a rubber knife, but that only induced laughter. So Ekman decided to have the Fore pose for him instead. He read them his emotional stories and asked them to produce a corresponding facial expression. He filmed their responses and took these back home to San Francisco. There he asked American college students to label the Fore expressions, picking emotion words from a short list of options. The students seemed to recognize all of the Fore's expressions except fear, and Ekman concluded that the Fore express emotions the way that we do.

These results are impressive, but they suffer from the same limitation as the other studies. Given a small set of emotion labels, Americans may be picking the least-bad choice, rather than judging that the Fore expressions are exactly like the ones they would make at home. Once again, the results might show that there are cross-cultural similarities in how emotions are expressed, but they don't show that the expressions are cross-culturally universal.

Some confirmation of this possibility comes from a follow-up study by Pamela Naab and James Russell.[7] They tried to replicate Ekman's results by showing Americans still images from the films that he had taken of the Fore. But they gave their subjects more emotion terms to choose from. For each face, they could pick one of twelve emotion labels. In this condition, performance dropped dramatically. For faces that Ekman would describe as clear expressions of one of the six emotions under investigation, American college students selected the correct label 24.2 per cent of the time, dipping as low as 4.2 per cent accuracy. Responses were above chance levels on less than half of the items in the attempted replication. This suggests that people have difficulty assigning emotional significance to the faces of people in other cultures when their options are not heavily constrained.

The ideal test for universal recognition would be a 'free response' method where people in another culture were shown photographs and asked to come up with appropriate labels. This has been attempted with very mixed results. People give widely ranging answers for the same face. In one typical free response study, Carroll Izard found that the same face might be described as pain, pity, loneliness and worry.[8]

Other research has added to this complex picture. Further studies have shown that Americans have a hard time recognizing surprise when expressed by Italians, and we are bad at recognizing three out of Ekman's Big Six emotions in Japanese faces (disgust, anger and surprise).[9] It has also been shown that we make many errors when interpreting expressions made by blind people (32 per cent accuracy on one study).[10] Furthermore, expressions in infants do not reliably correlate with the emotions on Ekman's list, and by the time babies are starting to make recognizable expressions there are already detectable cultural differences, such as differences in the degree of expressiveness, which have been correlated with parental responses earlier in life.

The overall picture that emerges suggests an interaction between nature and nurture. Nature may lead us to produce expressions that are similar to but not exactly like the expressions used in Ekman's research. Then a learning process leads us to alter these somewhat, exaggerating certain features and suppressing others. For example, evidence suggests that Americans express happiness with an exaggerated smile, while east Asians tend to emphasize the scrunched-up eyes. These patterns affect our ability to recognize happiness. We rely on eyes less than east Asians. The difference also comes out in the emoticons we use while writing emails. An American happy-face has two expressionless eyes and a big smile made with a parenthesis :) and a Japanese happy-face is made with an expressionless mouth and squinting eyes, like this ^_^.

Such differences in expression may be accompanied by differences in the emotions themselves. For one thing, if the embodiment theory is right, differences in expression directly lead to differences in how emotions are felt. Moreover, the fact that Fore responses to stories differ subtly from our own may suggest that they are having subtly different emotions. This would not be at all surprising. In the embodiment theory, emotions arise when the body prepares a response. There can surely be learned cultural differences in what to do then. What would you do if you were cornered by a snarling wild pig with menacing tusks? If you're like me you would run off screaming. The Fore might not react this way; they might stay cool, or freeze, or fight. And what would you do if you were insulted? You might start yelling aggressively. But in small-scale societies, overt anger can create dangerous rifts, so people who are insulted might be more inclined to pout. The way we react to any situation will involve a combination of instinctive responses and culturally conditioned ones, resulting in a state that could best be described as a biocultural mix. Ekman and his long-time collaborator Wallace Friesen discovered that people in Japan suppress negative emotions in public.[11] While watching a graphic film of genital surgery Americans show overt disgust, but Japanese viewers retain a neutral expression. Ekman calls this a 'display rule', implying that Japanese viewers have the same emotion as us, but resist displaying it. However, the disposition to suppress an expression is itself a bodily response, which must impact the emotion.

In addition to exaggerated smiles and suppressed nose wrinkles, there are much more overt expressive behaviours that we learn from our culture. Consider mourning practices. All people are disposed to cry when there is a loss, but crying takes different forms. Western men suppress tears, Japanese funeral-goers cry softly and Filipinos cry loudly, even hiring wailers to increase the din. In part of Papua New Guinea crying is converted into a ritualized, rhythmic hum, and mourners may also paint their bodies, cut themselves and even amputate their own fingers. In biblical times, we read of mourners falling to their knees, tearing their clothes, pulling out their hair and throwing dust over their heads. If you were raised in a culture where this practice was customary, your body would be disposed to engage in these behaviours on hearing the news that a loved one had died. We might call the resulting emotion grief, but it's important to recognize that grief differs cross-culturally, because it is embodied in different ways.

Research on expressions remains the most influential and, to many, the most compelling evidence for universal emotions. We have now seen that the data provide only weak support for universality. Recognition is too similar cross-culturally to assume that our emotions are cultural inventions – there are clearly universal biological building blocks – but differences in recognition suggest that these building blocks are reshaped by culture from the very start of life.

ENCULTURATED EMOTIONS

Back to Basics

The picture that has been emerging can be summarized as follows. The most distinctively human emotions, like guilt and love, may derive from more primitive emotions, and those more primitive emotions may have an innate base, but they are retuned by culture in subtle ways. In this view, some emotions are basic and some are combinations of those basic emotions, but even the basic emotions are culturally influenced.

This leaves us with an important question. Which emotions are basic? The most obvious answer is something like Ekman's Big Six. Ekman may be wrong to suggest that these six emotions are exactly the same across cultures, but perhaps every culture has something similar to each one on his list. In more recent work, Ekman has taken to referring to emotion families. American sadness may not be the same as Fore sadness, but they are variants of the same innate emotion, according to Ekman. But even this claim may be too strong. It is possible that the Big Six are not in fact universal or innate, even if we regard them as emotion families.

This conclusion gains support from the fact that many cultures have no word for one or another item on Ekman's list. The Chewong of Malaysia have no word for happiness, the Ilongot of the Philippines have no word for anger, the Tahitians have no word for sadness, the Utka Inuit have no word for fear, the Malay have no word for surprise, and the Polish have no word for disgust. Catherine Lutz has argued that the Ifaluk, who live on a small island in Micronesia, have no synonyms for any items on Ekman's list![12]

Of course, the absence of a word does not entail that the emotion is absent. It could be that the emotions are universal, just not universally labelled. This possibility should be taken seriously, but, as we have seen in our discussion of the Sapir–Whorf hypothesis, labelling can make a difference. By grouping certain reactions under a label, people learn what the typical reaction is supposed to be like, and, when cultures group together different examples, a new category can emerge.

Cultural variation in emotion vocabulary can also be explained by the conjecture that Ekman's Big Six are not basic emotions. Perhaps they are like guilt and love: combinations of more fundamental ingredients. These combinations may be extremely likely to arise, but not inevitable. Some cultures may have different combinations than we do.

The suggestion that Ekman's Big Six are combinations of more basic emotions has not been carefully investigated, but let's take a speculative look at how the story might go. Start with sadness. That emotion is associated with two behavioural dispositions, crying and sluggish withdrawal. But these two behaviours may come from different basic emotions. Crying arises in many contexts: we cry in mourning, pain, when listening to stories of courage and when we are overjoyed. The common denominator here is a recognition of helplessness; we cry when we can do nothing else. When listening to tales of courage, we know we cannot help the people involved. When the newly crowned Miss America gushes tears of joy, she knows that the outcome was in the hands of the judges, and she can finally stop trying to win. As a helplessness response, crying may be a supplication that beckons for help from others, but this, as these examples show, does not always involve sadness. Sluggish withdrawal accompanies tears when helplessness results from loss. In those contexts, inaction may be the only option. But sluggish withdrawal can occur without sadness, as when we give up a difficult struggle, or even in boredom, loneliness or mild contentment. Sadness blends the feeling of withdrawing effort with the feeling of helpless supplication. These are two feelings – two basic emotions – that happen to pair up naturally in times of loss.

Similar stories can be told about other items on Ekman's list. Surprise might combine a novelty-detection response (related to being startled) and interest, both of which are separate basic emotions. You can be interested in a novel, without finding it novel, and you can find my recreational interests novel, but uninteresting. We need both to have surprise.

For anger may need both frustration and aggression. The aggressive boxer is not angry, and we are not normally angry when we have trouble coming up with an answer on a crossword puzzle. But when the boxer can't get a punch in, he might get miffed, and when the crossword puzzler starts snapping pencils, mere frustration gives over to rage.

There may also be two components in disgust. That term is sometimes used to refer to physical revulsion, which is probably basic, but the term also subsumes moral disgust, which combines revulsion with hostility (perhaps the same kind of aggression we find in anger). Moral disgust integrates a disposition to aggress with a disposition to avoid contamination. The psychologist Paul Rozin showed that people would not try on a sweater that they believed belonged to Hitler.[13]

Fear is associated with two more fundamental responses in clinical psychology. One is anticipatory anxiety – which involves physiological changes that help you cope with a potential threat. The flight/fight/freezing response may be the physical underpinning of this feeling. This triad is controlled by a brain mechanism that selects the coping strategy most suitable for the situation: we flee when escape is an option, freeze when we can avoid detection and fight when cornered. But notice that we exhibit the flight/fight/freezing triad without being scared, as when we play certain sports. The response is still a form of anticipatory anxiety – we worry that the other player will catch up to us and get the ball – but it is not aversive. Fear combines anticipatory anxiety with what clinicians call panic, a physiological response that arises when a bad outcome looks inevitable. If you are fleeing in a game of tag, there is no fear, because the outcome of getting caught isn't bad. Fleeing from a stranger who is chasing you is terrifying. In the latter case, your body prepares for the worst. There is a sinking feeling in your chest, you may whimper or shriek, and you may even hyperventilate or lose control of your bladder. These symptoms of panic turn mere worry into bona fide fear.

Joy is also a hybrid. In experiments with rodents, Kent Berridge showed that liking something and wanting something can come apart. By creating small lesions in different parts of the reward systems of their brains, Berridge was able to create rats that eat compulsively but exhibit expressions of displeasure, and others that show pleasure when fed, but do not seek food. Joy may combine liking and wanting. It arises when we get what we want and like it. Consider this exchange: 'Are you happy with your lot in life?' 'Well, no, I like waiting tables, but this is not the career I wanted.'

These examples suggest that our basic emotions are more primitive than the items on Ekman's list. His Big Six may be blends, like guilt and love. If so, then there may be some cultural variation in what blends arise, and that may help explain why the facial expressions that Ekman studies are not interpreted the same way in every culture. Consider anger. For us, this is a blend of aggression and frustration. In Malay, there is no perfect translation. Instead there is one word, *marah*, which is associated with sullen brooding, rather than aggression, and another word, *amok*, which involves aggression, but of a frenzied variety more intense than the word anger implies (that's why we had to borrow the word *amok* in English). Among the Ilongot, there is also no perfect synonym for anger. Their closest word, *liget*, expresses a state that is more energetic than anger and also integrates elements of what we call sadness. The Ifaluk word *song* relates to anger, but it has a moralistic connotation that anger often lacks, and there is no implication of frustration. The Utka Inuits have no word for anger, but have one that translates roughly as childishness, which may refer to something more like a temper tantrum.

Similar observations can be made for other items on Ekman's list. Among the Gidjingali people of Australia, the word closest to fear also implies an element of shame. The word disgust has no synonym in Polish, but there is a word for physical revulsion that lacks the moral connotations. The Chewong can say I feel good (the literal translation is 'my liver is well'), which connotes a state of well-being analogous to liking, but there is no connotation of wanting, which is an essential ingredient of joy. In Malay, there is a word that translates roughly as startled (*terkejut*), and another that expresses something closer to interest, or perhaps puzzlement (*hairan*), but no single word that brings such elements together as we do with surprise.

These examples suggest that the Big Six are not biologically basic emotions or even basic emotion families, but rather compounds of states that are more fundamental. The truly basic states may be things like aggression, helplessness, startle and wanting. Notice that it's not even obvious that these are emotion terms in English. Aggression refers to a kind of behaviour, helplessness to a predicament, startle to a reflex response and wanting to an appetite or drive. Ironically, we don't really have emotion terms that correspond to the most basic emotions. Our simplest emotion terms correspond to states that are already a bit more complex and blended together under cultural influence. The Big Six are created from a universal stock of ingredients, but they may be whipped together using culturally specific recipes.

11

Gladness and Madness

How happy are you? Would you say your life is going pretty well? If you had to do it all over again, would you make any changes? These questions probe life satisfaction. The emerging field of positive psychology uses them to assess the factors that contribute to well-being with the hope that we can one day understand the formula for a good life. This research has revealed that happiness can depend on the place you call home. Culture has a hand in determining what makes us glad. Culture can also affect what drives us mad. Mental disorders are major detriments to well-being. People in psychiatric institutions consistently report that they are more miserable than others. What causes these conditions? Surprisingly, culture plays a role here too. Even when there is an organic dimension to a mental disorder, culture can affect its prevalence and symptoms. There are even disorders that are unique to particular societies. Thus, not only can culture influence how happy you are and how healthy you are; culture can even determine what counts as happiness and health.

JOY TO THE WORLD

In the last chapter, we saw that culture can foster the emergence of distinctive emotions by combining a basic stock of primitive elements. As a result, there can be an emotion that exists in one culture, but not in another. But there is also another important way that culture can have an influence. The very same emotion can be caused by different things in different cultures. For example, America is a highly individualistic and pluralistic culture, and we tend to get uncomfortable when we encounter people who we don't know. Japan, in contrast, is collectivist and homogeneous, so strangers are not a source of fear, but strangers can be a source of annoyance, because they are less likely to respect personal space. Thus, Americans are four times more likely than the Japanese to say they fear strangers, and the Japanese are three times more likely to describe strangers as a cause of anger.

Examples like this are legion. Awlad'Ali Bedouins feel shame when they are in the presence of powerful people. The Balinese are disgusted by crawling infants, and they disallow that behaviour. Edouard Manet's painting *Olympia* shocked Parisians of the nineteenth century because it depicted a nude courtesan with a confident gaze, but it is hardly shocking today.

One of the emotions upon which culture exerts this kind of influence is happiness. One might think that the same things cause happiness the world over, but this is not the case. International studies of wellbeing suggest that people in different cultures are made happy by different things. This can have profound effects. One's overall satisfaction with life can depend on what makes you happy. If culture can influence happiness, in this way, it can influence life satisfaction, and that means that people in different cultures who are equally well off in some objective sense (e.g., in their material resources) may differ significantly in subjective quality of life.

In recent years, there have been ambitious efforts to measure wellbeing across nations, and the results have been striking. The most comprehensive polling has been done by an international group of researchers called World Values Survey, directed by Ronald Inglehart at the University of Michigan. This group has found that there is a positive correlation between wealth and well-being, but many of the happiest nations are comparatively poor. When people in eighty-two societies were polled on their levels of happiness, Nigeria came in number one, and when a more encompassing measure of life satisfaction was used, Puerto Rico and Mexico got the two highest scores. The USA occupied an unremarkable fifteenth slot, with Britain lagging behind at twenty-five. The middle position was occupied by Japan, and Indonesia picked up the rear. South American countries were among the happiest despite high poverty levels, and east Asian countries were less happy even when affluent. Western countries, especially in more northern parts of Europe, and the former Soviet states were, as a group, the least happy of all.

There is no objective variable that can explain these rankings. Sometimes the same variable that promoted well-being in one place is irrelevant elsewhere. For example, the degree of autonomy a person had was more likely to increase well-being in individualist nations than in collectivist ones. As this example shows, what makes people happy is subjective. It depends on preferences that are culturally instilled.

The psychologists Ed Diener, Shigehiro Oishi and Eunkook Suh and their collaborators have been trying to identify how happiness varies across cultures. Much of their work looks at the contrast between Western countries and the Far East, and the findings have been striking. In the West, happiness depends more on pleasure than in the East, and one's current condition is more important than working towards future goals. Westerners are also more preoccupied than Easterners with self-satisfaction and with being consistent. Easterners gain more happiness than Westerners from helping others and being accepted.

One consequence of these striking results may be that people in the Far East are less happy than people in the West because some of their goals are harder to achieve. Ironically, this implication, which may strike us as a minor tragedy, is probably regarded in less dire terms by people in the Far East. There, happiness is often less important than it is here. In fact, happiness can even be seen as a bad thing. In China, 16 per cent of those polled say that positive emotions are undesirable. If happiness is getting what you want and liking it, then in China, happiness does not make people happy.

Naturists find such findings perplexing. They tend to assume that happiness has universal correlates. In fact, one popular claim among those who favour nature over nurture is that each person has a biologically predetermined level of happiness. This is called set-point theory, and it got off the ground in 1978, when Philip Brickman and two collaborators published a study of lottery winners and victims of accidents who ended up paralysed.[1] As might be expected, winning the lottery leads to euphoria and paraplegia leads to misery, but Brickman tried to show that these effects are short-lived. After a while, lottery winners and paraplegics seem to settle back to moderate happiness. This was interpreted as showing that each person has a biologically set happiness level, which can be shifted by major events, but only temporarily. This finding is interesting and important. It shows that the human spirit is resilient; we can bounce back from tragedy. It also shows that money doesn't buy happiness. Hitting the jackpot does not eliminate all of life's woes and may even introduce some new ones. But the study also has limitations. A closer look at Brickman's data suggests that lottery winners are a bit happier than the rest of us, and paraplegics are considerably less happy. Other studies have shown that there can be an enduring reduction in happiness for widows, prostitutes, homeless people and the unemployed. Also, there are things that may explain Brickman's results without assuming a biologically fixed set-point. Lottery winners may find that wealth brings new challenges (distinguishing true friends from the greedy horde), and accident victims may find a new appreciation of life and pride in their ability to cope with adversity. It would be a leap to infer from Brickman's results that happiness levels are biologically determined. The alternative view that I've been exploring here says that happiness depends on values that are established, in part, by one's culture. It is possible that victors and victims come up with similar assessments when they weigh their status against prevailing norms.

The fact that we find significant cross-cultural differences in well-being confirms that happiness is not set by genes. If you compare people within the very same culture, it may be that biology makes a contribution. If you fix the environment, heredity can account for some variance. But environment has a huge effect when we look cross-culturally. In Switzerland, when people are polled about their level of well-being, the average is 8.36 on a ten-point scale, and it's 5.03 in Bulgaria. That is the difference between feeling richly fulfilled and feeling as if life is teetering between tolerable and intolerable.

The moral is that culture has a big hand in determining what happiness is and how happy you are. How good you feel depends on where you were raised.

MAKING MADNESS

Just as culture can play a role in determining how happy we are, culture can influence when we feel bad. There are obvious ways in which this is so. Cultural conditions can be highly aversive. War, poverty and oppression can make life miserable for a population. But there is also a less obvious way that culture can contribute to human misery. The societies we live in play a role in creating psychiatric disorders. There is cultural variation in the prevalence of particular disorders and the symptoms they present. Societies can also have an impact on determining what counts as a disorder, and there are disorders that are found in some cultures, but not others.

The Medical Model

The term 'madness' is no longer used in psychiatry. It is considered outmoded and even offensive. It is certainly a stigmatizing term, because it identifies people with their disorders. A madman is not a person who happens to have a condition, but a person defined by that condition. But the term did not disappear because of stigmatization. The change came when unusual psychological traits began to be characterized as medical conditions. Since the nineteenth century, there has been an effort to understand psychiatric disorders as illnesses. We describe people as mentally ill. On the medical model, psychiatric disorders are compared to organic diseases, such as small pox or cancer. They are treated with drugs, surgery, shock therapy and other physical interventions. The mind is construed as a machine, and psychiatric disorders are malfunctions in the machine that can be diagnosed and repaired. There is a lot of good science behind this model, and there have been major advances in classification and treatment. But the medical model can be taken too far. It underestimates the role of culture.

The medical model faces an immediate difficulty when shifting from organic diseases of the body to the conditions that are treated by psychiatrists. With organic diseases, there is usually a pathogen, a lesion, or an anatomical abnormality. We can find a foreign agent that invades the body or else something structurally anomalous. With psychiatric diseases, this has proven much more difficult. Most are not virally transmitted, and there are no physical features of the brain that are regularly and reliably used for diagnosis. No scan can confirm the nature of the disorder. In fact, when psychological symptoms are found to have physical causes, they are usually not regarded as psy- chiatric disorders any more. Alzheimer's disease, aphasias and petit mal seizures are not listed in the current edition of the *Diagnostic and Statistical Manual of Mental Disorders* (DSM IV), precisely because they have known physical causes.

How can mental disorders be regarded as diseases if there is no anatomical abnormality that one can point to? Let's consider two of the main ways that defenders of the medical model have tried to address this problem. The first is one you've certainly encountered before. There is a widespread belief, promoted by the pharmaceutical industry, that mental disorders result from chemical imbalances in the brain. A chemical imbalance is not a lesion, so it would be easy to miss using the standard ways of finding an organic disease, but it is nevertheless a physical condition. If mental disorders are chemical imbalances, the medical model would be vindicated.

There is reason to be very sceptical of this approach. The concept of chemical imbalance is poorly defined. Those who use the term don't bother to tell us what qualifies as a balanced brain. What is sup- posed to be balanced with what? There is no formula that specifies the ideal ratios of neurotransmitters, and no test to tell whether optimal chemical levels have been achieved. Chemical imbalance plays no meaningful role in diagnosis or in accounting for psychiatric symp- toms. While it's true that some symptoms involve brain systems that metabolize certain chemicals, we have no idea what would count as a deficiency in those chemicals and what would count as an excess.

The emptiness of the chemical imbalance construct is easiest to see by looking at a specific case. Consider depression, the ailment that is most associated with this terminology. You have probably heard that depression involves the neurotransmitter serotonin. It has frequently been described as an imbalance in serotonin levels, though we are never told what those levels are supposed to be in balance with. In a recent review of the literature, Jeffrey Lacasse and Jonathan Leo find that efforts to link depression to serotonin have come up empty.[2] Studies have failed to show correlations between depression and serotonin levels, and brain lesions that affect serotonin production do not cause depression. The main reason for thinking that depression involves serotonin is that symptoms of depression can be alleviated by taking a class of drugs called selective serotonin reuptake inhibitors, such as Prozac, Lexipro and Zoloft. But it is a mistake to infer the cause of a disease from an effective treatment. Headaches are not caused by aspirin deficiencies, infections are not caused by aberrant antibiotic levels, and pimples are not caused by an imbalance in benzoyl peroxide. We also know that many other interventions are effective for treating depression, including dopamine reuptake inhibitors like Wellbutrin, electroconvulsive therapy and exercise. Looking at published efficacy studies, Irving Kirsch and his colleagues found that antidepressant drugs are only marginally more effective than placebos for moderate to mild depression, and these efficacy rates drop below clinical significance when unpublished trials are taken into consideration.[3]

On reflection, the idea that something as complex as depression could be reduced to something as simple as imbalance in serotonin levels is not even plausible to begin with. In the DSM IV, which is the main guide for diagnosing mental disorders, major depression is defined by a list of symptoms. There are nine in all, but a person need only have five to be classified as depressed, provided those five include one of the first two symptoms on the list, which are depressed mood and lack of happiness. If you do the maths, it turns out that there are 104 different ways to satisfy the criteria for major depression. Some of the nine symptoms actually include multiple variations, so this number is conservative. For example, one symptom is insomnia *or* hypersomnia (sleeping too little or too much). Consequently, two people can be clinically depressed, but share no symptoms in common. One might have a depressed mood, suicidal thoughts, excessive guilt, hypersomnia and significant weight loss, while the other suffers from lack of pleasure, indecisiveness, observed restlessness, insomnia, fatigue. Accounting for these diverse and highly specific symptoms by appeal to one chemical would be impossible.

The claim is not that serotonin has no relationship with depression.

For the severely depressed especially, these drugs have some impact. But the relationship may be very indirect. For example, if depression had an environmental cause, like unemployment or losing a loved one, manipulation of serotonin could blunt the effects by having a global impact on the brain's ability to process the emotional significance of life events. Other drugs that alter brain function, such as hallucinogens, narcotics and nicotine, can have an impact as well. This underscores the fact that we cannot infer that chemistry is the root of the problem from the efficacy of chemical cures, and it also draws attention to another important point. The focus on chemistry often serves to distract away from external causes of depression, such as life events and the way that a person thinks about those events. There is nothing wrong with treating symptoms with an effective drug, but it is a potentially harmful mistake to infer that psychiatric disorders are merely chemical.

These problems have led some defenders of the medical model to abandon the idea of chemical imbalance in favour of some other approach. One leading alternative is the idea that mental disorders are dysfunctions. Dysfunction is a technical term in biology. It implies that there is some function that an evolved mechanism was naturally selected to serve. The function of the heart is to pump blood, and the function of the lungs is to pump oxygen. Heart disease is a dysfunction because it prevents the heart from doing its job efficiently, and likewise for lung cancer. It is tempting therefore to equate illnesses with dysfunctions. To extend this idea to psychiatry, we would need to establish that psychological mechanisms evolved for specific purposes, and that psychiatric disorders disrupt their ability to do so.

This has some initial plausibility. Consider anxiety disorders. Anxiety is fear about future outcomes, and fear is clearly adaptive. Fear warns us about dangers and helps us prepare for them. Anxiety is plausibly an evolved response, and abnormalities in normal functioning can be regarded as dysfunctions in the technical sense. This is even more obvious in the case of phobias. Many of the classic phobias, such as acrophobia, arachnophobia and social phobias, concern things that could have been sources of serious danger in our ancestral past: heights, spiders and other people. But these evolved mechanisms may be dysfunctional for some people in the modern world. That can occur because of an internal malfunction in the mechanism, but it can also occur for external reasons. Dysfunctions arise when there is a mismatch between an intact mechanism and the environment. Fear of heights is dysfunctional when you are staring out of the window in your office building, because there is no risk of falling. Fear of spiders is dysfunctional when your hometown has few venomous species and several top-notch hospitals. Fear of people is dysfunctional when you are not a hunter-gatherer competing with violent neighbours for precious resources.

This approach, labelled evolutionary psychiatry, has some very attractive features. For one thing, it offers an account of psychiatric disorders that is consistent with physical disorders: both are dysfunctions. Also, it does not presuppose that psychiatric disorders involve anything like a lesion in the brain, or a chemical imbalance. Often, the dysfunction arises because of a mismatch between ancient mechanism and the modern world. When you put a cave man in a glass skyscraper, something may go awry. This implies that the root cause of a disorder may be external. Unlike the chemical imbalance approach, which tries to treat the brain, evolutionary psychiatrists argue that psychiatric symptoms can arise in response to unhealthy environments – external conditions in which our minds were not evolved to thrive. The cure can involve changing one's circumstances.

The contrast between the chemical imbalance approach and evolutionary psychology can be dramatically illustrated by considering the case of depression. One of the most puzzling facts about depression is that incidence seems to be rising. Depression has been nearly doubling every ten years in industrialized countries. Some of this may be an artefact of increased diagnosis, but chances are there has been a significant and steady increase in the number of cases as well. Why? Is fast food leading to more chemical imbalance? Evolutionary psychiatrists say we should not look for an inner cause. Rather, we should think about what function depressive symptoms might be evolved to play. Why would we ever evolve a mechanism that makes us feel hopeless and lethargic? Evolutionary psychiatrists offer an answer. Randolph Nesse and John Price have argued that depression evolved as an adaptation to cope with losing battles.[4] Suppose you are living in a hierarchical group and you decide to take on the alpha male or alpha female, and you lose. After nursing your wounds, you might try again, and again, and again. After all, alpha status brings huge perks. If you keep trying unsuccessfully, that suggests your chances are pretty much nil. Worse still, each attempt consumes considerable energy and results in injury. These injuries could even be fatal. So evolution furnished us with a mechanism that makes us recognize when we are pursuing a lost cause. It makes us feel hopeless, and we withdraw, so we can recover and downscale our ambitions.

In our ancestral past, depression would have been adaptive, not dysfunctional, but something has changed. In the modern world, we are surrounded by alphas. Everywhere we look, we see people who are richer, prettier, smarter and more powerful. The mechanism that once helped us avoid the occasional aspiration to gain rank is overactive in the media age, because we are inundated by reminders that we are inferior. Every time you turn on your television set, you see dozens of people who outrank you outrageously, and that is enough to make anyone feel like a loser. This explains why depression is on the rise. People spend more time than ever watching TV, seeing movies, looking at advertisements and visiting websites. The rapid increase in media consumption means that we all have dozens of daily encounters with our social superiors, and that can be very depressing indeed. There may be other factors as well, such as crime, divorce and pollution, that remind us we are fighting a losing battle. Even when these things do not threaten our lives, they trigger evolved mechanisms that make us want to curl up into the foetal position and weep.

Charming as it may be, this account of depression is somewhat implausible. For one thing, if depression were a universal response to lost bouts for increased status, we might expect everyone with a TV to be depressed. That is simply not the case. Nor does there seem to be any evidence that people are more likely to feel depressed after an evening of TV or night at the movies. In any case, it's not clear why merely seeing fabulous people should make us depressed. In small-scale societies and species that live in small groups, the alphas are frequently present and visible. Moreover, people in contemporary industrialized societies actually live more comfortably than just about anyone in human history, so modern life could hardly be described as a losing battle. The theory does correctly predict that depression will be higher among the poor than the wealthy (incidence is twice as high for the poor), but this only confirms the platitude that hardship makes us blue. Postulating an evolved mechanism adds little to what we already knew. Indeed, even if evolution did have a mechanism for dealing with lost-cause battles, social withdrawal and suicidal tendencies would hardly seem helpful, because both diminish reproductive success. A better mechanism would make low-status individuals gleefully accept their lot after suffering a defeat. Evolutionary logic predicts that the poor should be revelling.

Evolutionary psychiatry does no better with phobias and anxiety. The claim that phobias involve dysfunctions in mechanisms that helped us avoid dangers in the past is plausible when we consider cases like fear of heights, but many phobias are more exotic, such as trichophobia (fear of loose hairs), phagophobia (fear of swallowing) and ataxophobia (fear of untidiness). Evolutionary psychiatrists could get into deep water if they postulated an evolved mechanism corresponding to each phobia. They would also get in trouble if they pushed the idea of environmental mismatch too far. Consider anxiety, which might plausibly be regarded as an evolved mechanism for coping with threats. In the modern world, anxiety is often triggered by things that are either benign or totally unlike the physical threats for which the anxiety machinery in our brains was evolved. We experience anxiety at horror movies, when taking exams, and while bidding for items on eBay. In each case, the perspiration and racing heart suggest that the flight/fight/freezing response has kicked in, but that clearly won't help us cope with these modern challenges. The theory of psychiatric disorders defended by evolutionary psychiatrists seems to entail that these are cases of dysfunction, since there is a mismatch between an ancient mechanism and a contemporary concern. But it is absurd to suggest that eBay-anxiety is a mental illness.

This last observation suggests that dysfunction (defined as a departure from the role selected by evolution) cannot be sufficient for disease. Cases are easy to multiply. Masturbation uses sexual organs in a way that departs from their evolved function, and the desire to masturbate was once regarded as a medical disorder. John Harvey Kellogg, who co-invented cornflakes (with his brother) and ran a sanatorium, was also famous for his campaign against masturbators;

he advocated surgery and painful acid applications as a treatment. The dysfunction theory of mental disorders entails that Kellogg's view was right, even if his methods were a bit excessive. The theory also entails that homosexuality is a mental disorder, as it used to be described in medical manuals and older editions of the DSM. In fact, even sex using contraception is dysfunctional from an evolutionary point of view. Clearly departing from evolutionary design is not tantamount to illness.

The most embarrassing objection to the theory that dysfunctions are disorders comes from psychiatric practice itself. Suppose anxiety is an evolved response to deal with threatening situations. As a result, a police officer, who faces constant danger, might experience anxiety and take Xanax to cope with it on the job. Doing so would interfere with the evolved function of anxiety, and that means evolutionary psychiatry is committed to the outlandish view that taking Xanax is a mental disorder. The cure becomes a sickness in this view!

Similar concerns have been raised by Jerome Wakefield, one of the leading defenders of the evolutionary approach to mental illness.[5] Wakefield recognizes that many things that qualify as dysfunctions, on the technical definition of that term, are not mental disorders. We should not send masturbators to the madhouse. He advocates a hybrid theory that combines the evolutionary approach with something more subjective. For Wakefield, disorders are *harmful* dysfunctions. Taking Xanax, masturbating and stressing out during a bidding war on eBay are not mental disorders because they are not harmful. Wakefield admits that harmfulness is a value judgement. Value judgements can be influenced by culture, and what is harmful here might be regarded as benign elsewhere. Thus, Wakefield's theory attempts to save the evolutionary approach by bringing in a cultural dimension.

Is Mental Illness a Social Construction?

Wakefield admits that values are important to the concept of mental illness, but he might not go far enough. Once we admit that the difference between mental health and mental disorder depends on value judgements, why not drop the evolutionary component entirely? Wakefield thinks that being dysfunctional, in the technical sense, is necessary but not sufficient for calling something a disorder, but this claim can be challenged. Consider psychopaths. By all accounts, psychopathy is a mental disorder. Psychopaths are people who lie, cheat, steal and sometimes kill without the slightest bit of remorse. They are resilient to rehabilitation, and they don't seem to recognize that their behaviour is wrong. Clearly, these individuals are both harmful and abnormal, but is it clear that psychopathy is a dysfunction? From an evolutionary perspective, the answer may be no. Some evolutionists have argued that psychopathy belongs to the normal range of human variation. Psychopaths often have more children than the rest of us (though they don't care for them very well), and they can live long and successful lives. We dislike them, but they seem to be perfectly functional. Does it follow that they are not ill? Or consider post-traumatic stress disorder, which can have symptoms that are profound and devastating. But these symptoms may actually be adaptations. A soldier who shuts down or acquires psychologically generated paralysis may be experiencing the effects of an evolved mechanism that puts people out of commission in the period following a trauma. Would the discovery that this mechanism is evolved entail that post-traumatic stress disorder is not a disorder? It seems very counter-intuitive to make the clinical status of such conditions hinge on the outcome of evolutionary hypothesis that may be impossible to confirm. We don't need to know the biological history of these conditions to know that the people who have them are not well.

For this reason I think we might be better off dropping the dysfunction criterion, and the entire medical model. When we describe some syndrome as a mental disorder we are not committing to the existence of an organic disease or a departure from evolutionary design. Rather we are saying that the syndrome violates our norms and expectations in a particular way. Mental disorder is not a concept that belongs to natural science. Rather, it is a social category that reflects our attitudes towards certain traits. According to this view, there is no mental illness in nature. There is only variation. And some variations we regard as normal, whereas others are pathological.

This purely evaluative approach to mental disorder is sometimes regarded as radical. It is associated with the anti-psychiatry movement. Authors like Thomas Szasz and Michel Foucault argued that mental illness is a myth propagated by the ruling elite to ostracize certain members of society and deprive them of their rights. I am saying no such thing. Mental disorders are certainly real, as the people who have to cope with them on a daily basis will surely attest. But the fact that a condition is real does not imply that it is an organic disease. Mental disorders are not medical in this sense. They are medical only in the much more important sense that medical science offers some of the most promising treatment options. We do not try to treat mental disorders because they are deviations from evolved biological regularities. We treat them because they are bad. They are things that we regard as harmful impediments to human flourishing.

Wakefield objects to such a purely evaluative approach. He notes that there are many things that we regard as bad, but don't consider to be disorders, such as poverty, ignorance and the pain babies experience when teething. He also argues that our values are sometimes mistaken. In the nineteenth century, the term drapetomania was applied to slaves who had a tendency to run away. This was devalued and regarded as pathological by slave owners, but we now think they were gravely mistaken to regard the desire for freedom as a mental disorder. If being a disorder was simply a matter of being devalued, all of these things should qualify, but clearly they don't.

Wakefield's counter-examples to the evaluative approach are compelling, but they work only against the most flat-footed versions. Clearly it's not enough to say a disorder is anything we dislike, or even anything that happens to be regarded as abnormal. I think the key concept to understanding what disorders are is intelligibility. Mental disorders are, in an important sense, unintelligible. We cannot make sense of why people with disorders act, think and feel the ways that they do. Their behaviour is not just unusual; it is unintelligible. We cannot make sense of the phobic who fears harmless things. The fear seems irrational and insensitive to reason. Likewise for the paranoid person whose suspicions have no basis in reality, or the depressive who says that life is unbearable even though things seem to be going no worse than they are for the rest of us. In contrast, we can make sense of the person who is depressed about a divorce, or who has religious beliefs, or even someone who fears flying. It's not necessarily that these cases are more rational in any objective sense than the pathological cases. Maybe it's deeply irrational to believe in God or fear flying. But these cases fall within the range of variation that we, the evaluators, can understand. I can project myself into the mind of a theist, the mind of someone with opposing political values, or even the mind of someone who hates sushi. Those attitudes are intelligible to me. But my ability to project has limits. I might not be able to get into the mind of a serial rapist, members of a religious cult or someone who washes her hands incessantly because they can never be clean enough. In these cases, we're inclined to see the behaviour as pathological.

On this analysis, mental disorders are relative to our values. What is pathological in this culture may be healthy in another. If a person you know claims that her flu is a spell that has been cast by a witch, you might think she needs serious psychiatric treatment. But the same belief expressed by someone in an isolated Nigerian village might seem normal. We have expectations about what people in our culture should believe that don't necessarily travel across oceans and continents.

Notice that this account has no difficulty with the counter-examples that Wakefield uses to critique evaluative accounts. Poverty is not a disorder in my view, because there is nothing about a poor person's thoughts, feelings and actions that is unintelligible. Likewise for the pain of teething – it strikes us as an appropriate response to the growth of new teeth. Slave owners might say that slaves who try to escape are foolish, but they would be lying if they said they cannot make sense of a slave's desire for freedom. They are wrong to call drapetomania a disorder.

The intelligibility account has two implications that might seem undesirable. First, there is the aforementioned relativity. What I find opaque may be perfectly intelligible to others. The cult looks pathological to me, but, from their perspective, I am the one who is unintelligible. I think this consequence is consistent with how we use the concept of disorder. Different groups regard each other as crazy, and, from their perspective, they are right. But this relativism does not mean anything goes. The drapetomania case illustrates this point. You can't call something a disorder just because you want it to stop. It really has to be unintelligible. Kellogg could surely understand the desire to masturbate. He just thought it was bad. But that value judgement does not warrant calling something pathological. Indeed, some pathologies are not bad at all. The mad artist who compulsively builds a cathedral from bottle caps is pathological, but pleasantly so. We value some behaviours that we can't understand.

The second implication that some people might want to resist is a kind of fuzziness. Intelligibility comes in degrees. Some behaviours are on the borderline. To someone with conventional sexual tastes, it's fairly difficult to understand the sex life of someone interested in S&M. Is it a pathology to desire spankings? Hard to say. When we look at specific cases, an unusual practice can become more intelligible or more opaque depending on the details. But, in some cases, we may remain on the fence. The category of mental disorders can be irresolvably vague as a result. But that may be okay. Perhaps the link between normal and pathological is graded. This fits with our intuitions and allows us to retain a category of extreme eccentricity, which we cannot firmly locate on either side of the madhouse walls.

To summarize, I've suggested that mental disorders are real conditions that include patterns of thought, feeling, or behaviour that are regarded as unintelligible. Unintelligibility is relative to the judge. The judges may be clinicians, family members, employers, teachers, lovers or juries. There may be considerable convergence across these judges within a particular culture, but there is no guarantee that judgements are cross-culturally consistent. Consequently, there is a sense in which having a mental disorder depends on the attitudes of those in a position to judge you. Some conditions, like schizophrenia, may be regarded as pathological almost everywhere, but others, like speaking in tongues, might not be.

Cultural Conditions

We've been looking at ways in which culturally informed values can determine whether something is normal or pathological. Mental disorders might be a social category, rather than a biological one. Culture can also contribute in other ways. Culture can impact the incidence of a disorder and its symptoms. In some cases, a particular constellation of symptoms will be unique to a culture. Thus, culture does not only help us decide when something is pathological; it can also influence the form that our pathologies take.

Consider depression again. As noted earlier, depression is on the rise. For an American twenty-five-year-old in 1955, there was a 2 per cent chance of having had an episode of major depression; a twenty-five-year-old in 1965 had chances closer to 5 per cent; by 1975, a twenty-five-year-old had about 10 per cent chance; in the 1980s rates inched up to 15 per cent, and now the number is about 25 per cent. One in every four American twenty-five-year-olds has had a bout of major depression. Numbers are also increasing in industrialized countries around the world. Evolutionary psychologists say this results from an increasing exposure to media images that remind us of our lowly social status, or convince us that life is a losing battle. I've expressed doubts about that. If TV were so depressing, people might not watch so much of it. Others have implicated increased work hours, high divorce rates or, in the case of women (who have much higher depression rates than men), the challenge of balancing career and childcare. But this is unlikely to be the complete story since depression is rising in younger age groups, who are yet to face the challenge of marriage, motherhood and money-making.

Consider an alternative explanation. Perhaps we are learning how to be depressed. Each symptom of major depression can occur naturally, and each can be viewed as a coping response to the challenges of life. When things are going badly, we often feel down, sleep badly and lack energy. These symptoms can co-occur in rough times. It may be, however, that culture taps into our natural capacity to experience such symptoms and increases the odds. For everyone, life is full of stress and struggle. From minor setbacks to major hardships, there is much to be depressed about. But that has been true throughout human history. The difference now might have something to do with the fact that we all know about depression. That psychiatric category is familiar. We know about its symptoms, and we have seen depressed people in our lives or in mass media. Everyone has learned how to be depressed. We have learned the depression script. We have learned that a certain set of symptoms co-occur with an intensity that makes life painful and difficult. Explicit knowledge of a culturally disseminated script may increase prevalence of the disorder.

The depression script is not like a movie script that people voluntarily act out. Rather, it's an involuntary coping mechanism that can be triggered in anyone. But what triggers it? Life events and genetic predispositions could play some role, but that can't explain why depression is on the rise. For many of us who have experienced depression the sad truth is that culture may be teaching us how to be sick. Culture can do this in three ways. First, it can make the cluster of symptoms salient. Without culture, we might experience one or two of the nine diagnostic criteria for depression, but the likelihood of experiencing five or six may rise as this pattern becomes associated in our minds. Explicit knowledge of what depression is like begets unconscious knowledge that this complex coping strategy is available if life throws a curve ball. Second, culture helps define the relevant curve balls. Familiar examples of depression teach us contexts where these symptoms are likely to arise, and that may increase the chance, through associative learning, that symptoms will strike us when we enter one of the precipitating contexts. Just as culture can teach us to feel unconscious and automatic embarrassment under certain circumstances (Am I too fat? Did I pass gas? Is my fly open?), it can teach us to unconsciously enact the depression script. Finally, culture can train us to construe things in a way that increases the likelihood of getting depressed. If you form the belief that you have no value, that life is pointless or that the future is grim, you are more likely to become depressed. Many circumstances in life can be construed as evidence for these beliefs. In fact, they may all be true. But not everyone draws these conclusions. Culture plays a role in helping us interpret things, and culturally inculcated patterns of pessimistic thinking could make depression more likely.

To be clear, the suggestion is not that people who are depressed are faking it. The symptoms are real and agonizing. They can make it difficult to function, and they greatly increase chances of suicide. The claim is that these symptoms belong to a larger stock of coping mechanisms available for responding to life's challenges. These mechanisms depend on a history of associative learning that can make the depression response more likely than it would otherwise be. The patient isn't responsible for being depressed, but her culture may be.

If you find this hard to swallow, you can lubricate your throat with a bit of cross-cultural psychiatry. Diagnoses around the world differ considerably. Some of these diagnoses are called culture-bound syndromes, because they are found in only one population, but there is often overlap across the globe, suggesting that each culture cobbles together different clusters of coping mechanisms that are potentially available to all of us.

Even depression is not experienced the same way everywhere. In east Asia, especially among the least Westernized individuals, physical aches and pains are much more commonly associated with this disorder. This is called somatization, and Western doctors sometimes imply that people in Asia focus on bodily symptoms because of the stigma associated with mental illness. It's also possible, however, that a focus on the body reflects a more holistic conception of health, in which mind and body are less dualistically distinguished. In some east Asian countries depression-like symptoms are more frequently found in a different disorder called neurasthenia. The symptoms include depressed mood, but also anxiety, impotence, body pains and headaches. This condition was once so common in the United States that William James called it Americanitus. It was gradually displaced by depression here, from which James himself suffered, and neurasthenia was exported. Known as *shenjing shuairuo* in China and *shinkeisuijaku* in Japan, neurasthenia remains quite common in Asia, with incidence rates of 10 per cent in some communities. Now depression is on the rise in Asia, and neurasthenia may be on the way out. These look like distinct but overlapping coping mechanisms that may wax and wane with changes in cultural values.

Some cultural syndromes are more unusual. For example, in the Indian subcontinent, there is a condition in men called *dhat*, which is characterized as a semen-loss anxiety. In ancient Indian texts, semen was said to be produced in the body through a very complicated process; for every forty drops of food, a man produces one drop of blood; and for every forty drops of blood, there is one drop of flesh; forty drops of flesh then make one drop of marrow; and with forty drops of marrow one drop of semen is produced. Each stage takes forty days. In this tradition, semen is also believed to be a vital source of energy. If it gets wasted through masturbation, nocturnal emissions or urine, weakness is said to ensue. Men with *dhat* suffer from anxiety, weakness, fatigue, loss of appetite, guilt and sexual dysfunction. Without the culturally specific beliefs, anxiety about semen loss and this particular constellation of symptoms probably wouldn't arise. There are disorders like *dhat* in other places where people believe that semen is a precious energy source. Similar anxieties may also have plagued Americans in the late nineteenth century. The minister Sylvester Graham believed that non-procreative semen loss was harmful and could be cured by eating a bland diet, which would reduce inflammations in the stomach, the organ from which lust emanates. With this medical objective, he invented a snack food, known now as graham crackers. Americans no longer associate semen-loss with weakness, and we don't consume pies with graham cracker crust to quell lascivious urges. As a result, clinicians here don't have too many patients reporting anything like *dhat*.

A related culturally specific anxiety is *koro*, which is found in east and south-east Asia. *Koro* is a manic fear that the genitals will retract into the body, up into the abdomen, eventually causing death. The term derives from the Malay word for tortoise and alludes to the way a tortoise's head retracts into the shell. There was a major outbreak of *koro* in Singapore in 1967, when panic broke out due to rumours that genital retraction could be caused by consuming pigs that had been inoculated for swine-fever. In ten days, 469 people sought treatment, and men tied weights or chopsticks to their private parts to prevent them from shrinking inwards. Anxieties about genital retraction can be found in other parts of the world as well. In 2003, men in Sudan became concerned that shaking hands with foreigners would cause their penises to melt, as part of a Zionist plot to curb Sudanese procreation.

There are also culture-bound syndromes that predominantly affect women. One example is *latah*, a condition affecting Malay women who are middle-aged, poor and disenfranchised. *Latah* is described as a hyperstartle disorder because it is triggered by a startle response. The startle causes a trance, and in this state sufferers exhibit echolalia, echoproxia and coprolalia – that is, they echo words said to them, mimic body gestures and shout obscenities. It is theorized that these symptoms may help relieve the stress of conforming to societal norms.

Many other conditions have been well described in the archives of cultural psychiatry. There is *windigo,* a condition that was found among Algonquin Indians, in which a person is stricken by a desire to eat human flesh. Sufferers would often beg others to kill them so as to prevent them from acting out their cannibalistic compulsions, and these requests were frequently obliged. One of the last autonomously ruled Algonquin clans in western Ontario was taken over by white control when Canadian authorities executed their chief, Jack Fiddler, for killing a man with *windigo*. Among the Gururumba of New Guinea there is a condition that translates as being a wild pig, because the young men who suffer from it believe they have been bitten by a pig spirit. During an episode, these men go into a trance and burglarize their neighbours' homes, then they are captured and held over a smoking fire, which causes them to come out of the trance with no memory of what they have done. Interestingly, the group allows these men to keep what they have stolen, and the condition might have emerged as a mechanism for helping men who are struggling with the pressures of independence that come at this stage of life. One further possible example is zombification in Haitian voodoo. Zombies speak very little, have stilted movements and sometimes follow others' commands. They are believed to be dead people who have been brought back to life. Sometimes the condition is attributed to the mildly poisonous chemicals used in some zombification rituals, but it is more likely that the symptoms owe to the power of suggestion backed up by culturally entrenched beliefs about the reality of zombies and their behaviour.

It is tempting to conclude that the syndromes found in other cultures are exotic. One might concede that these syndromes are culturally constructed while insisting that the diagnostic categories used in the West are purely biological. After all, Western medicine is more scientific than medicine in many other cultures, and we have given up many of the supernatural beliefs that are prevalent elsewhere. This assessment would be mistaken. Many clinical conditions that we find in our culture might look equally bizarre if viewed from afar.

Consider hysteria, now known as conversion disorder. Though no longer widely diagnosed, this was once one of the most common syndromes affecting Western women. Symptoms include fits, pains, numbness and paralysed limbs, and have no physical causes. In the 1880s, an American doctor developed an instrument for treating this disorder: the vibrator. In the 1890s, an Austrian neurologist came up with another cure: talk therapy. In other words, hysteria is a syndrome that involves paralysis without biological causes, and can be effectively treated by masturbation and dream analysis. Both the sickness and its cure would surely look bizarre to a member of another culture.

Another example is anorexia. Eating disorders of this kind have cropped up in different times and places, but it is particularly prevalent and destructive in contemporary Western societies. Victims starve themselves and acquire extreme body dysmorphia, seeing emaciated limbs as grotesquely overweight. Extreme social pressures, including the idealization of skinny physiques and the devaluation of young women, along with well-publicized cases, contribute to making this a widespread cultural phenomenon.

Even more striking is dissociative identity disorder (or multiple personality disorder, as it used to be known). Individuals who have endured sexual abuse or other traumas sometimes develop multiple selves, which have entirely different personalities, each having limited knowledge of the others. A person with this condition can suddenly switch from one self to another, changing temperament, voice, age and even gender. Incidence has risen steadily in North America since some well-publicized cases in the twentieth century. There are concerns that the condition can be induced by suggestion in psychotherapy and might also arise as a coping response when people who have been victims of abuse learn about the condition in movies and television talk shows. No one chooses to have multiple selves, but witnessing the disorder makes this option available unconsciously. Like depression, a culture may help people learn how to divide the self, and that can result in the spread of the syndrome, which is hardly known elsewhere.

This exploration of cross-cultural psychiatry should make it clear that many mental disorders are not purely biological. Their contours are culturally delineated. Human beings have many physical and psychological resources for coping with life's challenges, but the way these resources align and the circumstances under which they arise can vary with time and place. Some psychiatric syndromes may turn out to be more consistent across cultures. Schizophrenia and bipolar disorder are possible examples. But even in these cases, incidence, duration and symptoms are culturally variable. Perhaps we can improve our own prospects by studying the sources of variation.

Gladness and madness are not in the genes. Or not entirely, anyway. We may have genetic predispositions that affect temperament and our ability to cope with pressure, but that is only part of the story. How happy we are also depends on what we want in life, and those preferences can be shaped by socialization. And how we break down depends on what coping mechanisms are salient and socially sanctioned. The same genetic predisposition could lead to major depression here, neurasthenia in Japan and being a wild pig in New Guinea. And no specific genetic predisposition is necessary for any of these outcomes. All human beings are malleable enough to fall victim to the most extraordinary ailments, and, when we do, genes may have less to do with it than the talk shows we've been watching.

Where Do Values Come From?

12

Coping with Cannibalism

You probably wouldn't eat your neighbours. Why? It just seems wrong. We have an overwhelming and immediate intuition that cannibalism is a bad thing to do. Such intuitions suggest that we have a moral sense. We can see and hear and smell, but we can also sense when things are right or wrong. Where does this capacity come from? One possibility is that we are born that way. This view has a long history in philosophy, and it is gaining popularity in the cognitive sciences. Psychologists, ethnologists, criminologists, economists and neuroscientists are actively exploring the possibility that morality is innate. Recent research has led to important discoveries about how people make moral judgements, but the case for innateness has been underwhelming. There is simply too much moral variation in the world. We find cannibalism horrifying, but others find it appetizing. Therefore, even if human beings are capable of moralizing in virtue of our innate psychological endowment, there seem to be few constraints on the content of our moral values. To understand why we recoil at the thought of eating our neighbours, we must move beyond biological universals and explore cultural particulars.

MORAL INTUITIONS

Before we can figure out where our moral attitudes come from, we need to get clear on what exactly they are. What goes on in the head when we judge that an action is good or bad? Here are some cases. For each one, just answer yes or no. Is it okay to: decapitate a stranger? Pull a bird's wings off while it's alive? Give a donation to Oxfam? Ask a prepubescent child for oral sex? Volunteer at a homeless shelter? Run a marathon for cancer research? Spit into your friend's beer? Spend the afternoon with a relative in a nursing home? Punch someone in the face for insulting you? For most of these, your answers probably came pretty quickly. Some of the actions are clearly good, and others bad, though you might have hesitated here and there, wondering if extenuating circumstances might make a vicious act acceptable. I chose these acts, in part, because even the ones that seem abhorrent to us are considered morally acceptable in some cultures. But the main point of the present exercise is to experience what it's like to make moral judgements. What are you doing when you deliver your response?

Philosophers have thought about this question for a very long time, and there are two competing answers. According to some, we arrive at a moral judgement through a purely rational process. We bring cool principles to mind and deliver a verdict. According to others, we consult our emotions. The cases elicit strong feelings, and we deliver a verdict based on whether those feelings are positive or negative.

The emotional approach has been especially attractive to Empiricist philosophers like David Hume. Hume, you will recall, wanted to reduce all human concepts to sensory states, or mental images. Abstract concepts, like good and bad, resist such reduction because there is no way to paint a picture of goodness, and there is no distinctive aroma of the bad. Good deeds do not look different. Bad deeds have no telltale scent. But Hume realized that there may be a common denominator if we consult our emotions. Good things feel good and bad things feel bad. Emotions, we have seen, can be identified with sensory states. They are perceptions of changes in the body. If we can tell whether something is good or bad by feeling how our bodies react, then these very abstract ideas can be reduced to something viscerally concrete.

But Hume had detractors. For example, Immanuel Kant argued that emotions interfere with our ability to make moral judgements, and that we should try to assess whether things are good or bad using rational principles. Kant claimed that we can and should try to make moral judgements without paying attention to our emotions, and when those emotions arise, they are not the source of our moral attitudes, but rather distracting accompaniments that sometimes happen to stir up our insides.

It is hard for anyone to deny that emotions are aroused when we imagine acts of charity or violence, like the ones mentioned above. So the real debate concerns the role of these feelings. Are they the essence of our moral attitudes or do they just come along for the ride? Philosophers have traded intuitions on that question for centuries, but neither side seems to convince the other. Now we are in a position to settle the question scientifically, and the evidence suggests that Hume was right.

Feeling Bad About Badness

Morality used to be a subject matter for philosophers, some theologians and the very occasional policy-maker. In recent years, it's become a topic for cognitive scientists. Morality is an important part of human life. It pervades our psychology. Moral values influence what we wear, what we eat, how we earn our bread, who we sleep with, how we treat our friends and who we vote for on election day. Anyone interested in studying human behaviour must ultimately take a look at morality. That is finally happening.

One breakthrough came when Joshua Greene, a Ph.D. student in philosophy, teamed up with a group of neuroscientists to study what goes on in the brain when people make moral judgements.[1] Greene became a neuroscientist himself when he finished his degree, but from this first study on he has called heavily on his philosophical training. Philosophers often probe moral intuitions by constructing thought experiments. In moral philosophy, some of the most famous thought experiments are dilemmas involving runaway trolleys. Suppose a trolley car is speeding down the tracks and the driver has lost control of the brakes. A short distance away, five workers happen to be standing on the tracks unaware of the impending danger. If you do nothing, they will be killed. But you happen to be standing next to a lever that would divert the trolley to an alternative track. There is one worker on that track who will die if you pull the lever, but five others would be saved. Is it okay to pull the lever? Call this the Lever Case. Now contrast it with another. Once again a runaway trolley is speeding towards five workers, but this time there is no lever. However, you happen to be on a bridge above the track, and there is a very hefty stranger standing next to you. You calculate, with perfect certainty, that if you push him off the bridge, the trolley will be derailed by his size. He'll die, but five will be saved. Is it okay to push him? Call this the Pushing Case. If you are like the vast majority of people, you have the opposite intuitions in these two cases; pulling the lever seems good and pushing the stranger seems bad, even though in both cases, one person dies and five are saved. For philosophers, that reveals something about moral values. For Greene, it was a chance to see the moral brain at work. He and his collaborators gave people dilemmas like these while measuring neural activity in a functional magnetic resonance imaging (fMRI) scanner.

For contrast, Greene also included a bunch of dilemmas that had nothing to do with morality. For example, people were asked if it's okay to buy a generic brand headache medicine if you can't find the name brand. They found that, as compared to non-moral dilemmas, the moral cases caused activation in brain areas associated with emotion,

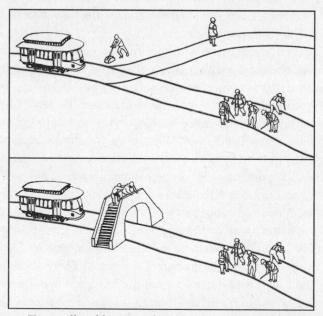

Figure 9. Two trolley dilemmas: the Lever Case and the Pushing Case.

such as the temporal pole at the front of the temporal cortex, which is known to play a role in associating emotions with mental images of events. This was the first neuroscientific proof that emotions play a role in moral judgements.

Greene's conclusion was a bit more qualified. He noticed that the level of emotion activation in the Pushing Case was much higher than in the Lever Case, and he concluded that we may use emotions to settle moral dilemmas of the first kind, when we have to consider directly causing a person's death, and no emotions in the second case, when the resulting deaths will be caused by pulling a lever or some other indirect means. But there is another interpretation, which strikes me as more likely. In all these moral dilemmas, there are two options: you can save five people (that's good!), or you can do something that results in one person dying (that's bad!). When you think about doing something good, you may have a positive emotional response; it feels good to help people. When you something bad, your response is negative. These two emotional forces battle it out in the brain and the stronger one wins. In the Lever Case, positive feelings outweigh the negative. Pulling a lever is such an innocuous act that it hardly feels like killing. But when you imagine pushing a rotund stranger off a bridge – wham! – that's horrifying. You get a thunderbolt of negative feeling that overshadows the joy of helping. In this interpretation, moral dilemmas always recruit emotions.

Support for this interpretation can be found in Greene's own data. The Lever Case did cause emotional activation in the brain, after all. And there is further corroboration from other studies. Two different labs have measured brain activity when people imagine giving to charity, and they both found emotional activation. That suggests that helping people triggers an emotional response. Further evidence comes from Michael Koenigs, Liane Young and their collaborators.[2] They gave trolley dilemmas to people with brain injuries in the ventromedial prefrontal cortex, a brain area that plays a role in weighing one emotion against another. It is known that these patients have a hard time adjusting goal-directed behaviour in light of negative feedback. If they are gambling, for example, they will keep pursuing the allure of a big payoff, even after sustaining even bigger losses. That's exactly what Koenigs and Young found with the moral dilemmas.

In both the Lever Case and the Pushing Case, these patients said it was good to save the five people. The reward of saving people could not be overshadowed by the thought of killing in people who cannot adjust positive goals in light of negative costs.

The claim that emotions are involved in all moral judgements has now been corroborated in dozens of neuroimaging studies. Neuroscientists have given people a wide range of moral judgement tasks while scanning their brains. Jorge Moll asked people to distinguish sentences that say something right from those that say something wrong.[3] Hauke Heekeren asked people to distinguish sentences that were morally incorrect from sentences that were semantically incorrect.[4] Carla Harenski showed people morally charged photos, such as a driver holding a beer bottle.[5] Sylvie Berthoz had people read stories about individuals who intentionally violate social norms.[6] Alan Sanfey had people play economic games with other players who occasionally acted unfairly.[7] John King had people play a video game in which they performed actions that were good or inadvertently bad.[8] In all these studies, and others like them, moral reflection is correlated with activity in emotional areas of the brain.

This research shows that emotions are active when we make moral judgements, but Immanuel Kant might not be moved. According to his view, moral judgements can be made without consulting the emotions. That doesn't preclude the possibility that, once we make a moral judgement, emotions follow. Nor does it rule out the possibility that emotions co-occur with moral judgements. The fact that emotions are active when we make moral judgements does not show that they actively contribute to those judgements. We need further evidence if we want to determine whether Hume's position is right.

The needed evidence comes from psychology. Recent studies have shown that emotions influence our moral judgements. They are not merely along for the ride. Simone Schnall and her collaborators showed that you can influence a person's moral evaluation by making them feel disgusted.[9] One study seated people at a desk and had them rate how bad it would be to do certain things. For example, suppose you accidentally run over your pet dog as you are pulling your car out of the driveway. Is it wrong to chop Fido into bits and cook him for dinner? Most people say this is wrong, but Schnall found that people gave this scenario a higher rating of wrongness if they happened to be sitting at a filthy desk. People at a clean desk said the behaviour was moderately bad, but people found the behaviour significantly worse if their desk had a chewed pencil, a used tissue and a sticky soft-drink cup sitting on it. Schnall also induced harsher moral judgements when she had people recall physically revolting events, watch a film clip with a seriously gross toilet scene (taken from *Trainspotting*), or when she sprayed trace amounts of fart spray near by. In a similar vein, Kendall Eskine was able to harshen moral judgements by giving people bitter beverages, which suggests that we should all be careful when deciding what to serve our dinner guests.[10] When people are disgusted, their moral judgements become less forgiving. The effect works for other emotions too. Jennifer Lerner used film clips to induce anger (a scene with a bully from *My Bodyguard*) and found that the subjects were more eager to punish people for minor transgressions. Pushing in the other direction, Piercarlo Valdesolo and David DeSteno showed people a hilarious Chris Farley clip from *Saturday Night Live* and then asked them about trolley dilemmas.[11] People who watched the clip were three times more likely to say it's permissible to push the fat guy in front of the train.

These studies show that when you push people's emotions around, their moral evaluations change. That means people must be consulting their emotions when they decide what's right and wrong. If an action makes you feel disgusted or angry, you judge that it's wrong. If your feelings of outrage are suppressed by a good joke, evil acts can look more innocent.

Kant would have to concede the point, but he might try to lobby for a hybrid theory. He might say emotions are playing a role here, but sound principles may be at work as well. A defender of Hume must show that the mere presence of emotions can be sufficient for making a moral judgement, even when no rational principles are available. Such evidence is available. Thalia Wheatley and Jonathan Haidt induced disgust in a group of subjects through hypnosis and then asked them to make moral evaluations of a group of characters whose behaviour they described.[12] One character was Dan, a student council representative who invited speakers to his school that would appeal to faculty and fellow students alike. People who were feeling disgusted said there was something bad about Dan, but they couldn't say what it was since the description was so positive. Some just invented rationalizations for their harsh judgement after the fact: Dan seems to be up to something, or he is just kissing up to his professors.

In another study, Haidt and collaborators asked people to consider the case of an adult brother and sister who decide to have sex with each other.[13] Almost everyone said this was wrong, but when they were asked to explain, Haidt defused every justification by elaborating on the case. They will have deformed babies, many feared. Impossible, Haidt rebuffed; they use contraception. They will be traumatized, some argued. Nope, Haidt replied; they do it just one time and it brings them closer together. The community will be corrupted, some countered. Won't happen, said Haidt; it's their little secret. But the Bible condemns incest, a few desperately urged. Where, Haidt scoffed sceptically; can you recite the chapter and verse? Is it in the story about Abraham marrying his half-sister or Lot sleeping with his three daughters? At this point, most people's lines of justification gave out. A handful conceded that consensual incest between siblings might be allowable, but 83 per cent stuck to their guns. Consensual incest is wrong, they insisted. Why? Because it's disgusting! People made direct appeals to emotion or showed their attitude with looks of outrage and revulsion. In the absence of any justifying principles people are still willing to make strong moral judgements provided their emotions tell them something is wrong.

At this point, Kant might acknowledge that people sometimes make judgements on the basis of emotions without any rational principles, but he would insist that, on other occasions, people can use rational principles in the absence of emotions. In other words, Kant might concede that emotions are sufficient for making moral judgements, but deny that they are necessary. Here too, however, science is on Hume's side.

Some of the best evidence comes from the study of criminal psychopaths. One of the central diagnostic symptoms of psychopathy is 'flattened affect': they rarely have strong emotional responses. They show deficient responses when looking at disturbing photographs, diminished capacity for fear-conditioning, reduced response to pain and even impairments in the ability to recognize emotions in other people. This emotional deficit results in a deficit in moral emotions, such as empathy, guilt and remorse. As a result, psychopaths tend to engage in a wide range of criminal activities when they can get away with it, and there is no known method of rehabilitating them, because they lack the emotional machinery. The problem is not simply that psychopaths give in to their temptation to misbehave. Chillingly, they can't comprehend why people think their actions are wrong. Tests have shown that psychopaths draw no distinction between moral rules and arbitrary social conventions. When asked why it is wrong to harm someone, they point to the risk of getting caught, not the suffering of the victim. In a series of studies, James Blair has shown that psychopaths think there is no difference between the prohibition against hitting people and the prohibition against speaking in a classroom without raising your hand.[14]

Such findings strongly suggest that psychopaths don't understand what morality is all about. Hervey Cleckley, who wrote a classic study of psychopathy back in the 1940s, said that the psychopath has no sense of good and evil:

It is as though he were colourblind, despite his sharp intelligence, to this aspect of human existence. It cannot be explained to him because there is nothing in his orbit of awareness that can bridge the gap with comparison. He can repeat the words and say glibly that he understands, and there is no way for him to realize that he does not understand.[15]

When psychopaths say that killing is wrong, they don't really know what this means. They understand that people in their communities regard killing as wrong, but they don't grasp why. Cleckley attributes this moral blindness to the deficit in emotions. Without a capacity for remorse, psychopaths fail to comprehend why their actions are wrong. When a psychopath who was imprisoned for theft was asked whether he regretted stealing jewellery from his mother, he replied, 'I'm the one who spent time in prison, not my mother ... Anyway insurance covers all that crap.'

It seems, then, that emotions are necessary for understanding morality. If Kant were right, psychopaths would be in a good position to become ideal moral agents, undistracted by misleading emotions. Instead they become self-serving criminals. We have also seen that people make moral judgements when they have no rational principles to back them up: emotions are enough. Thus emotions are necessary and sufficient for moral attitudes. This conclusion echoes the one that Hume came to back in the eighteenth century. The judgement that something is morally good or bad consists in an emotional response. To judge that killing is wrong consists in the horror or outrage we feel when we consider acts of murder. You can't sincerely judge that something is wrong without feeling that it is wrong. That would be like proclaiming that fast food tastes bad while gleefully eating it. Evaluations that don't express feelings are empty, dishonest or confused.

This account is called Sentimentalism in philosophy, because it says that moral evaluations are sentiments. Judging that something is bad is feeling bad about it. This is the only account that makes sense of all the data we have been considering. But many philosophers still find it objectionable. Before moving on, let's address a few nagging concerns.

Three Objections

Under the weight of the evidence, Kant might be forced to admit that moral judgements are based on emotions, but he could still object that they *shouldn't be*. He could say that people should replace their emotional way of making moral judgements with a cool rational procedure. Perhaps Kant is right about this. Our goal here is to figure out how the human mind works, not to devise a better kind of mind. But Kant may also be wrong. Without emotions, we might be indifferent to morality, just like the psychopath. And it's not clear that reason ever tells us what to do. As Hume argues, we cannot derive an ought from an is. Reason can only tell us how to act if we already know what we want to achieve. Rational principles cannot tell you to pick crème brûlée over tarte tatin without first establishing whether you prefer vanilla or fruit. Finding a rational foundation for morality might be as hopeless as finding a rational foundation for ordering dessert.

A related objection concerns the role of reasoning in morality. When Haidt found that people couldn't justify their condemnation of consensual incest, he concluded that we don't base moral conclusions on reasoning. But that's hard to reconcile with the fact that people reason about morality all the time. We have big public debates about moral issues such as abortion and capital punishment, and some moral choices require lengthy deliberation: Should I eat meat? Should I support pre-emptive strikes against potential terrorists? How can the Sentimentalist view make sense of this?

First, it should be noted that some reasoning is really rationalization. We present arguments as mere rhetoric to convince ourselves we are more right than our opponents or to rally others who are on our side. This is even true in cases like abortion. Would you really change your mind if science convinced you that all this time you've been wrong about whether or not a foetus is a person? Would deterrence statistics sway you on capital punishment if they went one way rather than the other? Perhaps. But we are prone to self-deception in this domain. Sometimes deeply held feelings exist independent of the ethical reasoning we employ. I think all of us should have the humility to admit that our favourite knee-jerk arguments for any given position could probably be ripped to shreds by people who are more informed about the issues.

That said, there is a crucial role for reasoning even on a Sentimentalist picture. Some of our values are basic: don't murder, don't be cruel, be fair and so on. Most people would find it difficult or impossible to find good rational arguments for these. Philosophers have been trying, with little success, for millennia. But other values are derived. We don't have a basic value about eating meat. Those who think meat eating is wrong make the case by arguing that it is a form of murder or cruelty. Extensive reasoning is often required to figure out which if any of our basic values apply in a given case. Most public policy decisions are like this, and they require extensive reasoning as a result. But once we establish which basic values are applicable, reasoning stops and passion settles the issue.

Some people object to Sentimentalism by trying to come up with counter-examples. Suppose you were raised in a homophobic community and now you think homosexuality is morally acceptable. Still, you might feel a shudder of disgust when you see public displays of affection between members of the same sex. You judge that the behaviour is fine, but you can't suppress this bad feeling. Cases like this have been studied by psychologists. For example, Haidt and his collaborators told people a story about a man who buys a chicken carcass at a supermarket and then masturbates into it.[16] People say this is disgusting, but not wrong. Doesn't that refute Sentimentalism?

In both of these cases, I think there is another explanation. Maybe we should say that the emotional reactions to homosexuality and deviant masturbation are, in fact, moral judgements. People who have these reactions really do, at some level, find the behaviours wrong. But these reactions are incompatible with a more deeply held value, which says that we should not condemn harmless sex acts. This is also an emotionally grounded value. You might feel squeamish around gay people, but you would be outraged at anyone who endorsed homophobic public policies and you are ashamed of your own residual homophobia. You have conflicting values, but your deep convictions about tolerance trump the traces of your intolerant past. When the values come into conflict, your convictions about tolerance are stronger, and they dictate your overt behaviour. By comparison, a person who shows implicit racism on a psychological test is a bit racist, even if it's also true that this person thinks racism is terribly wrong and would fight hard to stop racist tendencies in herself and others. We often have conflicting values and, when one clearly outweighs the other, we sometimes deny that the conflict exists, but that's not being completely honest with ourselves.

These examples suggest that emotions are sufficient for moral values. We can say that recovering homophobes still harbour moral misgivings about homosexuality, which they should work to overcome. But it is important to add that Sentimentalism is not committed to the view that every negative emotion is a moral attitude. You might hate doing the laundry, but you don't think laundry is evil. You may be repelled by maggots, but you don't blame them for what they do. You might enjoy surfing, but that does not mean you think it is morally praiseworthy. Genuine moral judgements require moral emotions. But which emotions are moral?

Three Ways of Being Naughty

David Hume sometimes talks as if there is a single moral emotion, what he calls disapprobation or disapproval. But there is no single feeling of disapproval. Psychological research has shown that we have different moral emotions for different kinds of transgressions. Disapproval comes in many flavours. To see this, first imagine how you would feel if I cut in front of you in a queue to buy movie tickets. Now, instead, imagine that I confess to you that I like to perform sex acts with my pet cat. You will disapprove in both cases, but the feelings involved are different. In one case you feel angry, and in the other you feel disgusted.

Paul Rozin is the world's expert on disgust. He has shown in his lab that people will not drink from a sterilized cup if they believe it once contained urine or a cockroach. Disgust protects us against physical contamination. But, in many human societies, disgust also plays a role in morality. We find consensual incest disgusting. We also have a disgust response to bestiality, sex slavery, cannibalism and axe murders. So disgust can serve as a form of disapproval. Rozin became very interested in this phenomenon, and he and his colleagues made an important discovery: moral disgust arises only in response to transgressions of a particular kind. The list just offered may look like a hodgepodge, but there is a common denominator. We are disgusted by acts that violate the body. All these acts do something deviant with human bodies, and we feel that both victim and perpetrator are defiled by them, or rendered impure. In this respect, moral disgust is just like physical disgust. It extends our mechanism for protecting the body against contamination to cases of moral contamination, and these are cases where an immoral act does something to the body that we consider unnatural. We sometimes call these unnatural acts or crimes against nature.

In contrast, queue cutting causes anger. Your body is not violated when I cut in front of you. The act is wrong because it is unfair. On a grander scale, we get angry about injustice. If a government denies people the right to free speech, we express angry indignation. We also get angry when people steal. If I pinch your wallet, you will be irate, not disgusted. And we get angry about physical violence as well, providing bodies don't get too horribly mauled. If I go on a shooting spree, you will be outraged, but if I go on a chainsaw massacre, rage turns into revulsion. Unfairness, injustice, stealing and killing share something in common. In each case they take something away from a person who is entitled to it. We are entitled to our position in line, free speech, ownership of our possessions and, above all, our lives. Rozin refers to transgressions that trespass against such rights as crimes against persons.

In addition to crimes against nature and crimes against persons, there is a third category: crimes against community. Consider a vandal who destroys a public park, an inconsiderate neighbour who keeps an untidy lawn or a politician who violates public trust. In each case, the crime hurts the community in some way. Community norm violations also arise when people are insufficiently sensitive to the status of others. Examples include a teen who eats dinner before her parents get to the table, or a young man who takes up two seats on the train and refuses to make room for an elderly passenger. Parents and older people deserve respect. We frown on those who take their sacrifices for granted. In these cases, the disrespect directly affects an individual person, but the integrity of the community is also at stake. Failure to respect certain groups of people represents an assault on the way a society is organized.

In Rozin's research, the emotion that people direct towards those who commit crimes against community is contempt. We sneer and look down our noses at those who don't show adequate concern for public resources or for people who have earned a position of respect. Arguably, contempt is a blend of other emotions. In fact, it seems to combine both anger and disgust. The snotty teenager who is rude to her parents is both irksome and repellent. This explains the appearance of contempt. Anger makes us want to confront her, but disgust makes us withdraw. The result is a stern confrontational posture with the head cocked back, which is why we look down our noses. Anger makes us lower our brow and glare, but disgust makes us curl up our lips and wrinkle our nostrils, hence the sneer. The blending story also explains why contempt is so related to social hierarchy norms. Social hierarchies are seen as the natural way to organize people in a culture – the natural order of persons. In a hierarchy violation, someone steps out of place, which is both unnatural (hence, disgust) and offensive to others (hence, anger). When vandals defile public property and politicians betray their proper role as defenders of public interests, we have reason to feel both angry and disgusted, so we experience the blend known as contempt. And we have contempt for the two-faced hypocrite whose duplicity seems both unnatural and unjust.

Putting this together, there are two fundamental kinds of transgression: crimes against nature and crimes against persons. These elicit disgust and anger. But there is also a hybrid category, crimes against community, which elicits the blend of disgust and anger that we call contempt.[17] Cross-cultural research by Richard Shweder suggests that these may be the only kinds of crimes that people identify.[18] Nature, persons and community may exhaust the things we can transgress against. Some communities talk about crimes against God, but for theists, God determines the natural order, so crimes against God can be subsumed under crimes against nature, and, predictably, the emotion that enforces such divinity norms is disgust. We must remain pure before God. So this taxonomy may be complete.

Still, one piece of the story is missing. Disgust, anger and contempt are things we feel towards other people. But what happens when we ourselves do something wrong? There are two well-known emotions of self-blame: guilt and shame. Now ask yourself, which of these would you feel for the following situations: you cheat on your time-sheet at work; your roommate catches you masturbating while stroking your pet cat; you step on a friend's foot while being careless; you have recurrent sexual fantasies about someone much, much older or younger. As you can probably confirm for yourself, the two sexual cases instil shame, and the others, guilt. Shame seems to be the counterpart of disgust, in that it arises when we feel like we have sullied our bodies in some way, and guilt is the counterpart of anger, in that it arises when we do something to harm a person. These emotions are usually distinguished by noting that in shame we feel like a bad person, and with guilt we feel like we have done a bad thing. This fits the model. Committing crimes against nature makes us feel like there is something wrong with us, but harming a person makes us focus on making amends for the act.

Intuitions are less clear about community norms. What would you feel if you painted graffiti on a public building and later realized this was wrong? You might feel both guilt and shame. You feel guilt about having damaged something that others use and value and ashamed because the public nature of your transgression draws negative attention from the whole community and that makes you feel like a bad person. In English, there is no good word for the blend of shame and guilt, but it is likely that such a blend arises when we commit crimes against community. If contempt is a blend, so is its self-directed counterpart.

These meditations on moral emotions provide some meat for the Sentimentalist skeleton. I think the science of moral judgement leads to the following picture. Basic moral values are moral sentiments directed towards various acts. To believe that stealing is bad is to have a negative moral sentiment towards stealing. To have a negative moral sentiment is to be disposed to feel one of the other-directed moral emotions when someone else commits the act, and a self-directed moral emotion when you commit the act. The specific emotion you feel will depend on what kind of act it is. Since stealing is a crime against persons, the belief that stealing is wrong consists in a disposition to feel anger or outrage (intense anger) at thieves, and guilt if you filch anything yourself. From this basic value, you can derive non-basic values. If you are a libertarian, you might conclude that taxes are wrong on the grounds that taking money from taxpayers is tantamount to stealing. A lot of reasoning would be required to defend that inference.

There is just one final detail. With all this talk about negative moral emotions, you might wonder about the good side. What goes on in our heads when we applaud heroes and saints? The answer is, we don't really know. Far less research has been done on this question. We know that hearing stories of moral goodness can instil a feeling of elevation; we feel moved in an uplifting way. But this feeling may be reserved for cases of unusual altruism. For more mundane cases, moral praise may take the form of admiration, or, perhaps, gratitude if a good deed comes your way. When you yourself do something commendable, it feels good. Reward centres of the brain sparkle. We don't know the precise identity of this feeling, but it may be a kind of pride, or perhaps it's just plain happiness. Helping others is something we like to do.

BORN TO BE GOOD?

The Sentimentalist account of moral judgement that I've just outlined closely resembles a view that David Hume defended in the eighteenth century. Hume was an Empiricist, and, as such, he was highly sceptical of innate knowledge. But there is nothing about Sentimentalism that precludes innateness. In fact, Hume's mentor Francis Hutcheson was also a Sentimentalist, but he believed in an innate moral code. God has equipped us with a moral sense, he conjectured, that permits us to recognize good and evil by attending to our emotional reactions.

In contemporary cognitive science, moral nativism has come back into vogue, and it is often coupled with a Sentimentalist theory of moral judgement. In the modern approach, God is replaced by natural selection. Evolutionary ethicists propose that morality is an evolved faculty built on innate principles and shared by normally developing members of our species. Let's consider four arguments that have been put forward for this position.

Universal Rules

One could try to argue for the innateness of morality by showing that some specific moral principles are found in all human cultures. The mere existence of universals would not prove that something is innate, of course. Fire, clothing, shelter, religion and art are found in most cultures, but they may not be innate. Still, universal moral rules would provide some evidence for innateness. After all, humans live in very different conditions, including conditions that are very different from the ones we evolved in. If moral rules were invented and learned, we'd expect to find considerable variation.

Are there universal rules? Consider this one. It might be supposed that there is a universal prohibition against harming people. Something like: don't hurt innocent people. It's hard to imagine anyone would disagree with a rule like that. It's obviously bad to harm the innocent. Or so we think. It turns out this rule is far from universal. Many societies are extremely violent. In pre-colonial New Guinea, for example, Richard Wrangham tells us, male homicide rates ranged between 20 and 35 per cent, which is more than double the mortality rates in the worst-affected countries during the Second World War.[19] Similar numbers have been reported among the Yanomamo of the Amazon basin, where 50 per cent of men participate in killing and 25 per cent die that way. Many small-scale societies live in constant warfare. They orchestrate raids against their neighbours, or mass together in unclaimed territories to hurl poison arrows or spears. Some of these groups do not think they are killing innocent people when they engage in acts of war. They blame their neighbours for everything from past raids to crop-killing witchcraft. But some of the violence is clearly perceived as directed against innocents. The Yanomamo men will kidnap women as wives, and spouse beating is reported to be common. The Gahuku-Gama of New Guinea claim there is nothing wrong with killing someone in another group, and there are numerous head-hunting cultures that kill people without any moral, political or spiritual motivation. For example, the Ilongot of Luzon in the Philippines will take a head to relieve stress or cope with depression when a loved one dies. Or consider the practice of *tsujigiri* in medieval Japan (outlawed in 1603), in which a samurai would test a new sword by killing an innocent passer-by.

If there were an innate prohibition against harm, it would be much more psychologically difficult to witness human cruelty than it evidently is. In ancient Rome, people would crowd into arenas to watch innocent slaves, dressed in gladiatorial garb, brutally kill each other. Tens of thousands would attend a single event, and the practice continued for 600 years. An innate harm-avoidance mechanism would make such spectacles unbearable for viewers. Or consider public torture and execution, which continued in western Europe into the eighteenth century and is still practised in some parts of the world today. Indeed, the very fact that we can watch graphically violent movies as entertainment suggests that we don't have any innate resistance to seeing innocent people suffer.

Another candidate for an innate rule is share and share alike. In human beings, it is common for people to divide resources. For example, a small group of hunters might share their quarry with the whole tribe. Is there an innate moral injunction to share? Probably not. Parents are innately disposed to share with their children, but this innate disposition is underwritten by feelings of parental affection and concern, not a sense of moral obligation. We don't need morality to be kind to our kin. Moral rules come into play when we share with people who are non-relatives. Such sharing is commonplace in human societies, but probably results from social pressures rather than innate rules. The impact of socialization can be inferred from the fact that there are sizable cultural differences in what counts as a fair split. In small-scale societies, resources are often shared pretty equally, but, with the rise of social stratification, gross discrepancies arise. In the United States, 10 per cent of the population owns 70 per cent of the wealth, but humans clearly don't have an instinct to fight for the downfall of capitalism. What have you done for the revolution lately?

Recently economists have been intensively investigating how different societies think about resource sharing.[20] To measure this, they use something called the Ultimatum Game: one person in a society is told that he or she can divide a resource with a stranger in the same society; if the stranger accepts the split, they both get their share, and, if not, they both walk away empty-handed. Presumably, the person making the offer will aim for a fair division, because unfair offers can be rejected out of spite. In Western nations, people tend to offer between 40 per cent and 50 per cent of the resource. Across the globe, numbers vary dramatically. Among the Machiguenga of Peru, a small society of independent farmers, the average offer is 26 per cent – an offer half the Americans polled would reject. One interesting finding in this research is that, in some cultures, it's common to reject *generous offers*. Suppose someone gives you $100, and you offer $60 to a stranger. If that stranger is Russian, it's not unlikely that he'll reject your kind split, and you will both walk away with nothing![21]

Cultures also vary in their preferences for who deserves to get a bigger piece of the pie. Suppose you are the CEO of a company and you have to distribute bonuses at the end of the year. You could distribute equitably (giving more to people who produced more earnings for the company), equally (giving the same to everyone), or on the basis of need (giving more to the employees who are least well off). When asked to consider these options, Americans prefer equity, Chinese prefer equality, and Indians prefer need. Cultural variables such as capitalism, communism and widespread poverty seem to play a role in shaping intuitions about how to share. If there were an innate sharing instinct, we might expect it to decide how to distribute resources. After all, an instinct that commands us to share, without saying how, wouldn't be especially useful. If Dad tells Sally to give some of her cake to her little brother, how much should she give? That's a question we probably can't answer instinctively. Sally might give Billy a bite, but it will probably take a lot of conditioning before she splits on the basis of anything like equity, equality or need. And the choice between these three options will depend on whether she's being raised in Boston, Beijing or West Bengal.

Even if sharing is not innate, there may be innate rules that compel us to reciprocate, or so evolutionary ethicists would have us believe. Many cultures have something like the golden rule: do not do to others what you would not like to be done to you. Interpreted in terms of reciprocity, this might be better formulated as, if you scratch my back, I'll scratch yours. Reciprocity is the key to cooperation. Without it, one party will take advantage of another, and collaborative efforts will break down. Human beings seem to have a reciprocity instinct. This can be seen by studying how people respond in iterated Prisoner's Dilemma games. A Prisoner's Dilemma, recall, is a situation in which the payoffs for cheating your partner are higher than the payoffs for cooperating, but mutual cooperation has a higher payoff than mutual defection. Both players have an incentive to cheat, but, if they both do so, they miss out on the benefits of teamwork. In an iterated Prisoner's Dilemma, people engage in several consecutive games with this payoff structure. One crucial finding is that people tend to play a tit-for-tat strategy. If you cheat me, I'll cheat you the next time around; but if you cooperate, I will too. Tit-for-tat can lead to sustained cooperation.

Some researchers believe that reciprocity comes to us naturally. We believe innately that it's good to pay back favours. We even exhibit what economists call deferred reciprocity. We do good things even for people who cannot pay us back, because they lack resources, or because we will never see them again. We operate under the principle that what comes around goes around. If I help people who can't reciprocate, someone may help me when I can't reciprocate. This is a remarkable bit of behaviour since human beings, like most animals, have a hard time with delayed gratification. Why should I do something for someone if I don't know when or whether I'll get rewarded? Mathematical models suggest that this strategy pays off for us in the end, but there is a question about how we ever come to it psychologically. Nativists answer that there is an innate moral rule that tells us to treat others well even if we incur costs doing so and reciprocate when others are nice. Such evolved altruism would make sense of what researchers have observed in the lab.

There are several problems with this proposal. For one thing, reciprocity may not be a moral rule. It is possible to engage in a reciprocal exchange without viewing it as a moral obligation. There could be an innate programme that makes us reciprocate without any attitude, or we might even reciprocate out of rational self-interest. Anyone who can calculate the benefits of cooperation can see that it is an optimal strategy (better than defection) provided your trading partner will reciprocate too. In fact, evidence shows that psychopaths, who are blind to morality, will play the tit-for-tat strategy on iterated Prisoner's Dilemma games. If rational self-interest suffices, then reciprocity may not be an innate rule, much less a moral one.

In response, the nativist might reply that most healthy human beings do, in fact, have moral attitudes towards reciprocity. Failures to reciprocate elicit anger and, in the first-person case, guilt. You feel guilty if you forget to leave a tip at a restaurant. This shows that, for us, unlike psychopaths, reciprocity is a moral rule. But it doesn't follow that this rule is innate. As we saw in chapter 10, it would be difficult for biological evolution to endow us with a guilt response when we fail to cooperate. The first person who felt guilty would be exploited by those who don't. Therefore, it is more likely that we learn to feel guilty through a process of cultural conditioning. There are mathematical models that show how reciprocity could emerge through a process of cultural evolution, rather than biological evolution. Cultures that encourage cooperation may do better than those that don't, and such cultures may be more likely to endure.

Finally, the claim that reciprocity norms are innate predicts that performance on Prisoner's Dilemmas will be pretty consistent across the species. This is not the case. For example, in iterated games, Americans are considerably less cooperative than people from China, and people from Belgium are more competitive than Americans. There is also evidence that women are more cooperative than men, under some conditions, which may reflect differences in how men and women are socialized. More generally, the strategies people pursue on iterated Prisoner's Dilemmas are extremely flexible. Social psychologists Aaron Kay, John Bargh and Lee Ross ran a study in which students played an iterated Prisoner's Dilemma that was introduced as 'The Co-operation Game' for half of then, and as 'The Wall Street Game' for the other half.[22] The name change brought a dramatic reduction in cooperation rates, suggesting that we don't have an innate programme driving reciprocity, but rather adjust strategies in dramatic ways based on culturally meaningful contexts. Such flexibility smacks of learning, not innateness.

These examples show that it is very hard to find universal moral norms. The most obvious candidates are difficult to defend, given massive cultural variation. In fact, I would venture that there is no specific moral rule that is universal. For every society that prohibits some act, there has been another that either tolerates it or encourages it. I say 'specific moral rules' because we might find some universals if we formulate rules in a sufficiently general way. Every society may have rules against harming some innocent people. But there is massive variation in who counts as innocent. Usually, the people protected under a rule like this are members of the in-group. But, even so, many societies tolerate gross mistreatment of certain in-group members, including women and the poor. Moreover, we don't need biological evolution to explain why societies prohibit harming some of their members. A society with no such prohibitions wouldn't last very long! Likewise, most societies may have rules about resource distribution, reciprocation and marriage, but that's because such rules are a precondition for social stability. The variation we find in such rules suggests that there are few innate constraints on what form they take.

Cheater Detection

So far, we've been seeing that moral rules vary from place to place. Thus, there is no way to prove that morality is innate by appeal to universal rules. But let's consider another argument. Linda Cosmides and John Tooby have argued that, even if no specific moral rule is innate, we have an innate capacity to enforce moral rules whatever their content happens to be.[23] According to their view, society may come up with the rules, but the capacity to understand what rules demand of us and of others is innate.

More specifically, Cosmides and Tooby claim we have an innate mechanism for catching cheaters. They argue for this mechanism by showing that people are much better at reasoning about social obligations than structurally similar rules that don't involve obligations. Suppose a sociologist observing a classroom proposes a correlation between watching TV and waking up early. Her study concludes with this hypothesis:

> If a child watches a lot of TV, that child will get up early.

You decide to test this hypothesis on kids in your own family. You know about the TV habits of some children in your family, and the sleeping hours of others, but you often don't know both bits of information. Which of the following individuals would you need to test the hypothesis?

    A. Your niece Annie, who watches tons of TV;
    B. your nephew Ben, whose wakes up at 6 a.m.;
    C. your niece Carmen, who hates TV;
    D. your nephew Dennis, who sleeps as late as possible.

If you are like most of us, you quickly recognize that you need to test Annie. If she is a late riser, the hypothesis is false. But you might not have noticed that you need to test Dennis. If Dennis watches a lot of TV, he disproves the hypothesis as well. Ben and Carmen can provide no evidence one way or the other. Cosmides and Tooby have shown that most people overlook Dennis in examples like this. But they don't make this mistake with social rules. Suppose you were entrusted to enforce the following rule in your family:

> If a child watches a lot of TV, that child must get up early.

Cosmides and Tooby find that people immediately recognize that Dennis must be investigated if they are enforcing this rule. They conclude that we are innately wired to look for cheaters.

Cosmides and Tooby rest their case for innateness on the following argument. Human beings are good at reasoning about obligations, but bad at reasoning about other kinds of rules, even when these rules are structurally very similar (they are both expressed: 'If P, then Q'). So it looks as if we have a specialized mechanism for thinking about obligations that functions independently of our more general cognitive abilities. The problem with this argument is that obligations and other kinds of rules are very different, despite superficial similarities in how they are expressed. To enforce a rule that says, 'You must do X' you need to find people who are not doing X; anyone who is already doing X won't need to be disciplined. But rules of the kind above merely express correlations or causal connections, and we don't learn about these by observing counter-instances. Consider the rule 'If something contains alcohol, it will get you drunk'. We don't confirm this by testing every non-intoxicating beverage for alcohol content. The rule is learned by the positive cases, and it is refuted if one of those positive cases fails to have the usual effects (weak drinks may not get you drunk).

If rules about social obligations and rules about correlations between features are totally different, then they may be acquired in different ways. Rules about obligations may be learned by parental discipline. Suppose a parent says, 'If you want dessert, you'd better eat your veggies'. The child learns what this means by disobedience. No veggies results in no dessert. It's pretty easy to see how episodes like this would instil comprehension of how obligations work. And how do we learn about rules that express correlations? The likely answer is that we don't. If any capacity is innate, it's our ability to track conditionals like 'If A occurs, B does too'. The innate mechanisms for associative learning yield knowledge in that form, but, as noted, we don't learn that As co-occur with Bs by tracking cases where there are no Bs. That would be like learning cookies are sweet by determining that spinach is not a cookie.

If this analysis is right, then Cosmides and Tooby have things exactly backwards. Our capacity to think about correlations is probably innate, and our capacity to think about obligations is learned. It's easy to see how we get so good at catching cheaters without supposing that this capacity is innate.

Selective Deficits

If morality is innate that means there are specialized psychological mechanisms dedicated to thinking about morality. And if there are specialized mechanisms, then there should be cases where these break down, but leave other psychological capacities in place. Some people argue for moral nativism by claiming that such selective deficits exist.

Valerie Stone and her collaborators have made this claim, building on the work of Cosmides and Tooby. They claim that cheater detection can be selectively impaired, leaving our capacity to think about other rules intact. Their support comes from a study of a patient who sustained a brain injury that left him unable to successfully reason about obligations.[24] This looks like strong evidence. If a brain injury can lead to an impairment in some aspect of moral reasoning and leave everything else intact, that would be grounds for concluding that the aspect of moral reasoning does not derive from more general (non-moral) capacities, and thus might be innate. But the study in question establishes no such thing, because the patient examined there has widespread injuries, affecting the orbital frontal cortex, the temporal pole and the amygdala. The orbital frontal cortex is known to play important roles in emotion regulation, the temporal pole allows us to assign emotional significance to memories and images of hypothetical scenarios, and the amygdala is implicated in fear responses and other emotions. Cosmides and Tooby do not test for such emotional impairments, but they do report that the patient has difficulty recognizing faux pas. It is overwhelmingly likely that the patient suffers from broad emotional deficits as a result of his brain injuries. Given the link between morality and emotion, it is hardly surprising that he performs poorly on cheater detection and faux pas detection. These errors may stem from a general inability to assign emotional significance to social events.

Consider another case, which has been suggested by the psychologist James Blair, one of the world's experts on psychopathy.[25] Blair claims that psychopaths have a selective deficit in morality. They have intact cognitive capacities, but they are blind to morality. Moreover, this moral deficit is highly heritable, suggesting that there might even be a gene for morality that is missing in this population.

This argument is based on a faulty assumption. Psychopathy is not a selective deficit in morality. Psychopaths do have other impairments. As we saw already, they have profound emotional impairments. Hervey Cleckley finds that psychopaths have difficulty appreciating art, music and literature.[26] They don't seem to form many strong interests or preferences at all. He also suggests that they have problems with decision-making. Their lives seem disorganized, and their decisions can be erratic. Psychopaths have also been found to have cognitive deficits. For example, they make many errors when doing mazes, following blind alleys and crossing over walls. This suggests a problem with impulse control. The pattern of anti-social behaviour and moral incomprehension in psychopaths almost certainly arises from these more general problems.

It seems that every case of moral impairment co-occurs with other impairments. Thus, the argument from selective deficits is actually self-defeating. Failure to find such cases may be evidence against the claim that there is an innate moral sense.

Saintly Apes

Is morality uniquely human? Some people think not. There are remarkable tales of altruism in the animal world. Rats will press a lever to help another rat in distress, vampire bats will regurgitate blood for unrelated bats, and monkeys will starve themselves to prevent another monkey from being shocked. Some of the most impressive acts of decency are found in our closest relatives, the chimpanzees. Franz de Waal has shown that chimps share food, reciprocate grooming, console each other after sustaining injuries and kiss and make up after fighting.[27] There is remarkable behavioural continuity between chimps and humans when it comes to pro-social behaviour. Felix Warneken and Michael Tomasello showed that both chimps and young children are very helpful.[28] If you drop something on the ground and reach for it, both chimp and child will pick it up and hand it to you. This suggests that human morality may not be learned after all. It has homologues in other species and may have evolved in some ancestor that we share with chimps.

There are two problems with this argument. First, the continuities may be overshadowed by discontinuities. There is no good evidence that chimps have key moral emotions such as guilt, shame and moral disgust. These are crucial for human morality. There is scant evidence that chimps engage in pro-social behaviour with animals who are not related or in the same social group. Humans help unrelated strangers in distant places. There are only a few anecdotal cases of chimps intervening when they are not directly affected. Human morality extends to third parties; it's not about how you treat me, but how you treat others. These are qualitative differences. Morality concerns the formulations of general rules that govern a whole society, or people everywhere. Chimp behaviour most often arises in dyadic relationships: helping a friend and punishing a foe. The underlying psychology seems to be quite different, and there are sometimes remarkable differences in behaviour. Joan Silk found that chimps who had lived together for decades would pass up on opportunities to give each other food, even if doing so cost them nothing.[29] Suppose you and your oldest friend are at a bar, and the bartender says, 'It's Friends-Drink-For-Free Day. Would you like a free beer for your friend?' If you were a chimp, you'd shrug your shoulders with callous indifference.

The second problem with the ape argument is even more serious. A crucial distinction must be drawn between doing something good and doing something because it is good. There are all kinds of reasons for doing good things. You might do them for reciprocity, you might fear reprisal if you don't, or you might be biologically programmed to do good things without reflecting on it. Bees, for example, will sacrifice their lives to save the hive if it is threatened, but they are not driven by a deep moral conviction that their sacrifice will serve the common good. Similarly, there is no reason to infer an understanding of morality from the fact that chimps do things we might applaud. Some chimp kindness may be motivated by fear. For example, there is a 'tolerated theft' model of chimp food sharing according to which they allow others to take food because they might be attacked if they didn't. Chimp consolation and helping behaviour might be driven by genuine empathy, but even so, that wouldn't entail that it is driven by a moral sense. There is a difference between doing something because you hate to see others suffer and doing it because it's right. If I help you paint a mural on your wall, it's not because I feel a moral obligation to do so, and if I give you aspirin when you're sick, it's not because I'd be racked by guilt if I didn't. Much human decency has little to do with morality, so we shouldn't assume chimp decency is morally motivated. In fact, given their apparent lack of guilt and concern for enforcing general rules, we should assume that chimp decency is not mediated by moral convictions, even if we think their behaviour is admirable.

The Ubiquity of Morality

Even if there are no clear precursors of morality in apes, there is plenty of evidence that morality is widespread in humans. Perhaps it is unique to our species, but it is also ubiquitous. Cultures that are completely isolated from each other all end up with systems of moral rules. Doesn't this suggest that morality is innate?

The answer is no. To see why, it's helpful to consider an example mentioned in passing earlier: religion. All societies seem to have religion. This is taken by some as evidence for the conclusion that religion is innate. But that is a minority opinion. After all, it's not clear why we would evolve to have religious beliefs. Why, then, is religion universal? The prevailing answer has been most influentially advanced by the anthropologist and psychologist Pascal Boyer. He argues that religion is a nearly inevitable byproduct of other human capacities.[30] When we hear a sound in the woods, we normally believe it is caused by some living thing, but when there is no perceived cause, we believe there is a hidden spirit. When misfortune strikes, there is often someone to blame, but, when no villain has been caught, we blame witchcraft. We form close social bonds with family members, and when they die, those feelings remain, and we believe their ghosts are still with us. We sometimes consume mind-altering substances or undergo physical stressors that cause bizarre experiences, so we conclude that external agencies are entering our heads. We entertain ourselves with stories that defy physical laws, like a person who can walk through walls, and these are so fascinating that we recall and repeat them for generations until we come to think they are true. In all these ways, and many others, we are prone to posit the existence of supernatural agencies, which form the core of most religions. The cognitive mechanisms that allow us to think about real people are over-sensitive and lead us to posit people and forces that are invisible and magical.

The example of religion illustrates how a human universal can emerge without being innate. Religion is an especially remarkable case because it's not clear that it has any payoff. Our supernatural beliefs don't lead to greater reproductive success or help us hunt and forage better. Morality, in contrast, is pretty useful. Moral rules help create stable, cooperative, productive societies. So it's somewhat less surprising that every culture has ended up with morality. But the mere fact that morality is useful doesn't fully explain why it's universal. Tractors are useful too, but most human societies in history haven't had them.

The ubiquity of morality is best explained by the confluence of two non-moral traits. First, we have a rich emotional life. According to the Sentimentalist theory of morality, moral judgements are based on emotions, such as anger, disgust, guilt and shame. It's possible that animals have analogues of anger and disgust, but guilt and shame are probably uniquely human. These two emotions are likely to emerge out of others that are not initially moral emotions. Shame probably emerges out of embarrassment, and in many languages one word is used for both. Apes lack embarrassment; they do not blush. Shame may be acquired when embarrassment becomes extremely unpleasant, as it would in contexts where we are punished. Guilt, I have suggested, is a blend of fear and sadness. Apes probably have both of these emotions, but, when apes are punished, they mainly experience fear. Human children realize that, when they are punished, they may lose their caregivers' affections. That may lead to a blending of fear and sadness. Thus humans acquire moral emotions.

The other factor may be imitation. Human beings are great imitative learners. Apes can do some imitative learning, but far less effectively and studiously. This may be one of the main reasons why humans have advanced so far. In the moral domain, imitation has an important implication. We may imitate the attitudes of our parents. If you punish a dog, it will learn to obey, but it won't punish other dogs. The same is probably true of apes. Even in captivity, if you teach an ape a skill, it will rarely transmit it to others. But suppose you punish your daughter, Sally. Before you know it, she may be punishing her little brother, Billy, for the same misdeed. Disciplining dogs results in obedience, but disciplining children leads to the creation and spread of a general rule.

Human emotions and the human capacity to imitate may be the key ingredients for morality. That would explain why morality is ubiquitous.

In every culture, parents get irritated with their children. That elicits emotions of self-blame in children and, over several repetitions, the disposition to get irritated by the very thing that irritated their parents. Rules that spread in a family can be calibrated and coordinated across a whole village, leading to conformity in the group and a genuine system of morality.

Nativists will resist this story and say that morality can emerge without instruction. But there is one final reason for thinking that the innateness hypothesis is wrong. With innate domains, there isn't much need for instruction. Innate traits emerge on their own. In the moral domain, instruction is extensive. Children are not very well behaved. They do some sweet things, but they also grossly violate norms all day every day. They are loud, destructive, inconsiderate and violent. If an adult behaved like a toddler, we'd send that adult to prison. But kids learn. And they learn through incessant correction. Between the ages of two and ten, parents correct their children's behaviour every eight minutes or so of waking life. Much of that correction is moral in nature, and much of it has an impact. In due course, our little monsters become lovely little angels. More or less. This, more than anything, gives us reason to think morality is learned.

THE HISTORY OF MORALS

We've just had an explanation of why morality is likely to emerge in every culture, even though it's not innate. Human emotions plus human social learning co-conspire to make morality practically inevitable. But what explains why we end up with the particular values we have? Why do cultures end up valuing different things? To answer this question, we must shift our discussion away from general features of the human mind and get into particulars. Historical particulars, that is. Each moral value can be viewed as an artefact, and each artefact has a history.

Consider cannibalism. Most of us are horrified at the idea of eating human flesh, and the taboo is so strong that tales of cannibalism have often been concocted by one group to demonize another. Still, evidence for cannibalism as a practice is incontrovertible. In South America, the

Wari' Indians are known to have practised funerary cannibalism, in which relatives' remains were consumed. Funerary cannibalism also existed among the Fore of Papua New Guinea (the same people that Paul Ekman studied in his work on universal emotion expressions), and this may have resulted in an illness called *kuru*, whose symptoms include headaches, joint pain, trembling, pathological bursts of laughter and ultimately death. *Kuru* is related to mad cow disease, which arises in humans who have eaten cows who were fed infected cow brains. Like these cows, Fore had a custom of eating the brains of dead kin, which is believed to have emerged during a period of food shortages.

There are also credible reports of warfare cannibalism in other parts of Papua New Guinea, in which enemies would be killed after violent conflicts. Living tribe members still recall these practices, and there is a first-hand report from an American anthropologist among a group called the Akamaras. There is also converging indigenous testimony confirming warfare cannibalism in nineteenth-century Fiji. In contemporary Congo, there have been reports that the Mbuti Pygmies have been hunted by their enemies and consumed, and they have called on the UN to define cannibalism as a crime against humanity.

Cannibalism sometimes exists because of supernatural beliefs about the powers attained by consuming enemies and kin, but given its pervasiveness in the anthropological record it is hard to pin the practice on a specific belief system. The anthropologist Marvin Harris has speculated that cannibalism initially appears in small-scale societies for nutritional reasons.[31] Suppose two hunter-gatherer groups have a violent skirmish over resources and one group is victorious. If anyone has been killed, the bodies could be left to rot, but that would be a waste. Hunter-gatherers know the value of meat, and they try to use all available resources. Therefore, they are likely to consume the bodies of their enemies. If they have taken any captives, they are likely to eat them too. If they let their captives go, they will seek revenge, and if they enslave their captives, they will have to feed them and police them, which is costly and risky. So captives must be killed, and, again, dead bodies should never be wasted. It's not surprising, then, that many small-scale societies are reported to eat dead bodies.

According to Harris's theory, cannibalism makes perfect sense on a cost-benefit analysis. People are nutritious and perhaps even delicious. So the hard question is, why aren't we all cannibals? Harris's answer is taxation. As societies become larger and more complex with the advent of agriculture, they develop trade relations with neighbouring groups as well as a professional warrior class, who can coerce neighbours to pay tribute. The material benefits of regular tax collection far outweigh a one-time feast of flesh. Moreover, when intensive agriculture comes on the scene, another vicious institution emerges: slavery. For hunter-gatherers, slavery is too costly, but farming societies can put slaves on their fields, where they can produce more calories than they are able to consume. When slavery becomes cost-effective, killing your neighbours becomes outmoded. Apparently, high taxes safeguard us against cannibalism.

Harris's analysis fits well with the pattern observed by anthropologists, but there is one resounding exception: the Aztecs. They had intensive agriculture and slavery, but they also had the most extreme form of institutionalized cannibalism in human history, killing and consuming up to 50,000 people annually, according to one estimate. The anthropologist Michael Harner gained notoriety for suggesting that this practice may have existed to compensate for a protein deficiency in the Aztec diet: they had no domesticated animals.[32] But given the fact that many cultures have thrived in Mexico without eating people, this seems unlikely. A more plausible suggestion is that the Aztecs lacked the military force to police their massive empire, and used cannibalism to terrify people from their many subordinate territories into paying tribute.

The explanation here is an example of what Harris calls 'cultural materialism': it explains a cultural practice by appeal to the material conditions that made the practice viable rather than the beliefs that were used to sustain the practice. Harris's view is that we can explain the history of morals without describing the myths and beliefs that justify moral practices in the minds of their practitioners. I think that pushes things a bit too far. The Aztecs were able to sustain their cannibal empire by convincing people that cannibalism appeases the gods and wards off the encroaching apocalypse. Myths may even serve to sustain a practice after its material benefits have waned. But Harris is right that moral systems can usually be linked to some payoff, and, even if practitioners are unaware of that fact, the material rewards allow the practice to continue.

Consider the gladiatorial games in ancient Rome. Why would the Roman government promote such a murderous form of recreation? One answer is that the games helped spread and preserve the empire. Gladiators were often slaves who had been captured on the empire's outskirts. They were made to wear outfits that represented quasi-mythical kingdoms that had been conquered in the past. The main virtue emphasized in the battle was valour, and more specifically courage in the face of death. Because of their courage, the gladiators became respected celebrities. Collectively, the games promoted a number of values crucial to Rome's success: it's good to conquer foreign lands, so the games can continue; it's good to be courageous, so joining the front lines is desirable; once captured, foreigners become Roman, so we should get over our differences. In sum, the games promoted militaristic nationalism in an otherwise pluralistic and bourgeois society, and that helped keep the empire in place.

Some early Christian leaders tried to ban the games because their newly adopted religion preached that obedience to God, rather than military prowess, was the cardinal virtue. But the bans were not entirely successful, and games continued sporadically, often funded by the Imperial purse. The ultimate reason for their disappearance may be bound up with the fall of Rome itself. Over-extended and ravaged by a crippling trade deficit, the empire lost its interest in continual expansion and became vulnerable to foreign invasion. With a weakened military, the Christian leadership may have realized that the preservation of political influence depended not on military valour, but on proselytizing their humanistic creed.

We should not infer that Christians were wholeheartedly committed to non-violence. Dying for the Church was seen as a virtue, as was the punishment of heretics and other wrong-doers. This brings us to another major change in moral values. Well into the eighteenth century, European nations practised public execution and torture. Torture was used both to punish and to extract confessions, and methods included everything from stretching bodies on the rack to breaking limbs and weaving people into large, suspended wheels. In these days, when water-boarding is seen by many as extremely cruel, it's sobering to recall that judicial pratice used to include the most extreme methods of mutilation that human beings have ever devised. Then, suddenly, things changed. Torture that had been practised for centuries, sometimes with public spectators, disappeared in the space of a few decades. This was partially the result of influential critics, including the Italian legal philosopher Cesare Beccaria, who condemned torture on rational and humanistic grounds. But, we have to ask, why were people receptive to such critics at this particular time and place in history.

Various factors may have been in play. A big one is the Thirty Years War, a century earlier, in which almost 10 per cent of the European population had been killed. The war effectively ended the control of the Holy Roman Empire and led to an increase in secularism. Citizens resented the heavy taxes that were levied after the war to pay for damages and military upkeep. Faith in Europe's monarchies started to wane, and there was a major move toward democratization. Within this context, the violence of torture began to lose its appeal, but more importantly, it made less sense. A monarchy can preserve power by grand displays of force, but when governments are taken over by the people, force is no longer needed to establish authority. But the new concept of government by the people did not eliminate the need for punishment and control. This presented a challenge. Ruling powers needed to distance themselves from the brutal excesses of the enfeebled monarchies, but they couldn't abandon the penal system. There was a shift then from punishment as a visible display of strength to punishment as a concealed process, hidden behind prison walls. And in this setting, torture, which had been discredited as a means of obtaining reliable confessions, no longer served any function. Instead, the approach that seemed most concordant with the emerging ideology was to deprive convicts of what was increasingly toted as the most valuable of all commodities: liberty.

The Enlightenment preoccupation with liberty was not strong enough to undermine another institution, which had been around since the dawn of state-scale human societies: slavery. In the eighteenth century, slavery was still widespread in Europe, and widely accepted. Some Enlightenment thinkers began to wonder whether it was consistent with the principle that 'all men [sic] are created equal', but many saw no tension. By this time, slaves were mostly coming to

Europe and the American colonies from Africa, and racist doctrines allowed wealthy white people to see this population as inferior and in need of white protection in the form of enslavement. The factor that initially tipped the scale in favour of anti-slavery sentiments was the industrial revolution. Technological breakthroughs in Britain begat a new economy based on wage labour. The destitute could seek security by working in factories and mines. Critics were quick to notice, however, that labourers were enduring horrific conditions – worse, in some respects, than slaves. Industrialists had a quick rebuttal, however: at least they are free. With this defensive posture and a new model for how to grow an economy through wage factory production, slavery came under attack. That attack led to the fall of slavery in Britain, and it fuelled anti-slavery arguments in America, on both economic and moral grounds. But, by that time, 30 per cent of the United States economy was based on slave-grown cotton. Industrialization appealed to Northern manufacturers, but Southern planters could smell disaster, and they were willing to fight a bloody war to keep their system in place.

What's striking about this period is that we see a remarkable revolution in attitudes towards slavery. First, the practice is widely accepted, then, by the early nineteenth century, deep doubts emerge, followed by anti-slavery legislation. There is also a rise of pseudo-scientific arguments for racial differences as part of an effort to justify the practice, and finally slavery collapses in Europe and the Americas. How could such a radical shift in moral values take place? It couldn't have been the refutation of scientific racism, since whites began enslaving blacks long before pseudo-scientific ideas about racial inferiority came into vogue. It seems that a major economic upheaval was the catalyst. Owning another person simply stops making sense with the rise of industrial capitalism.

The Civil War ushered in an economic and moral revolution in the United States. The South, which had been the backbone of the economy, was badly beaten and declined into poverty. Well before that war, there were already two cultures in the USA, Northern and Southern, and that division continues to this day. One remarkable fact is that Northerners and Southerners still subscribe to different moral values. One difference is that Southerners are more violent. In the South, homicide rates among whites in small cities are three times higher than in New England. These homicides also differ in that they are often the result of relatively minor provocations, such as insults, whereas, in the North, homicide usually takes place during robberies or other crimes. Southerners are also much more likely to oppose gun control, to favour corporal punishments and to believe that people have the right to kill in defence of property. Where do these values come from?

The social psychologists Richard Nisbett and Dov Cohen took up this question and came up with a surprising answer.[33] Southern white violence is not a consequence of slavery; it correlates negatively with regions that had sizeable slave plantations. Agriculture seems to make people less violent, even when human bondage is involved. Nisbett and Cohen ultimately traced Southern violence back across the Atlantic. Whites in the South were Scots-Irish. They had immigrated from places like Northern Ireland, where they had lived by herding animals. Northern Ireland had little government support, so people had to police themselves. They developed a strict honour code, and it became advantageous for each man to cultivate a tough-guy image and a hair trigger, lest his precious animals be stolen by rustlers. Similar cultures of honour are found among herders throughout the world, from Bedouin nomads to west African tribesman.[34] The Scots-Irish imported this culture to the American South, and, remarkably, it took hold even though social and economic conditions changed radically. It thrived especially in towns in the hills, cut off from farming, and it was spread by cowboys across the frontier.

These examples show that there is dramatic variation and fluidity in morality. But that also gives insight into how moral values emerge and change. Morals help people organize their societies, and they often conform to the prevailing economic conditions.

CONCLUSION: COPING WITH RELATIVISM

The claim that morality is not innate might strike some people as depressing. It makes morality seem so contingent, so optional. This anxiety is exacerbated by the observation that morality varies across time and space. Our values would have been completely different if we had been raised elsewhere. Perhaps we can radically change our values, or dispense with morality entirely.

I don't think this bleak view is warranted. The fact that morality is contingent doesn't make it useless. Morality is an extremely valuable tool. It helps us organize our societies, and it may be essential for social stability. But I do think the lessons of this chapter have some important implications.

We should recognize that our own values are just some of the many values that have existed. Perhaps our values are better than others. Perhaps they are worse. Perhaps they could be improved on. We should guard against the belief that we are in possession of Moral Truth. People whose values differ from ours are not dumb or evil. They were just reared in different communities. Like everyone, we learn morality through cultural inculcation, long before we engage in careful rational reflection, and there is a risk that our most treasured arguments are rationalizations, not justifications. Perhaps better arguments and open-minded reflection could lead us to revise the convictions with which we so passionately identify.

The recognition that morality is a tool should provoke us to wonder whether the tool we've inherited from our cultures and subcultures could be improved. Does the morality we treasure serve us well? Does it promote social stability? Does it allow us to flourish? Does it reflect our current position in the world or retain baggage from a bygone time? Does it have our interests in mind or was it imposed on us by a ruling elite who benefit from our subjugation?

These are hard questions to ask, because our moral values are emotional values. We have internalized them by learning to feel outraged when they are violated, and ashamed when we question their authority. But the history of culture is a history of moral transformations and should remind us that we are not stuck with the values we learned on Mother's knee. Together with our communities, we can explore the possibility of moral reform. The flexibility of morality does not condemn us to an anything-goes moral nihilism. It frees us from intolerance and moral stagnation and allows us to improve on what we have.

13

In Bed with Darwin

If anything comes naturally to us, it's sex. We wouldn't be here without it. Evolution has clearly ensured that we pursue sexual partners and procreate. It would not be surprising, then, to find that this aspect of human behaviour is heavily influenced by our biology. What is surprising – though quite evident – is that sexuality is also heavily influenced by culture. Culture affects everything, from whom we desire to how we find partners and what we do with them once we've found them. No aspect of human life better illustrates the way in which something natural can be at the same time thoroughly cultural.

WHOM DO YOU LOVE?

The first step in the mating game is finding a partner. We need to figure out who we like and how to pursue them. Forming preferences might seem like a trivial matter. The birds and the bees don't need to be socialized to figure out what they want. Nor, presumably, do we. But socialization has a profound impact.

Beauty

What's hot and what's not? All species have natural aesthetic preferences. Depending on your species, you might go in for bright plumage, big horns, a pungent musk or swollen behinds. Some of these preferences may have emerged because they are indicators of health. The most famous case is the male peacock, whose fabulous tail feathers look preposterously maladaptive. But carrying that heavy load correlates with physical fitness, and females are biologically programmed to pursue the fittest males. The role of female preference in mate choice is a source of evolutionary pressure – called sexual selection – that can turn a modest rump into a spectacular fan of glistening colour. In addition to these lavish displays, we have more basic ways of revealing the most basic fact necessary for a reproductive encounter: our gender. Males and females look different from each other through the animal kingdom, and equally so in us. From breasts to beards, gender is easily perceived. The traits that make it visible are linked to sex hormones, and they reveal facts about our level of fertility.

Evolutionary psychologists have studied what humans find attractive, and they argue that our preferences track indicators of reproductive fitness. Fertile women seem to like broad shoulders and prominent chins, which correlate with testosterone levels.[1] Women also like darker complexions, since men are naturally darker than women, and this preference increases during peak fertility periods in the menstrual cycle.[2] Men are drawn towards youthful features,[3] such as big eyes and full lips, and a relatively low body-mass index, which is the ratio weight to height.[4] By seeking youth, evolutionary psychologists say, men can maximize the number of reproductive cycles in their partners.

Some details of the conjectures have been challenged. For example, the female preference for prominent chins and other testosterone markers has not held up in every study. Some women may unconsciously perceive high levels of testosterone as a threat. Also, male preferences for youthful faces may turn out to be an artefact of averaging. The visual system recognizes objects by averaging together what it sees. When you encounter an object, say a tree or a turtle, that is more like the average, it is easier to recognize, and that facility of processing boosts ratings of attractiveness. The faces men like turn out to be, well, just average. Morph together the photos of fifty random women, and the result will look like the most stunning supermodel you have ever seen.[5] The looks that we like may have more to do with this averaging process than anything to do with fertility. Experiments have shown that averaging even drives our aesthetic preferences for lab-created dot patterns, but average dot patterns are not more fertile![6]

There are also some striking cross-cultural variations, which cast doubt on any simplistic evolutionary approach. Consider body-mass index. For white, middle-class men in Manhattan, thin is in. But men in many other societies prefer a more corpulent physique. A preference for curves has been documented among the Shiwiar of Ecuador, the Hadza of Tanzania and the Sámi of Scandinavia. Among the most isolated members of the Matsigenka people in Peru, bigger is better, but this preference disappears after exposure to Western media.[7] Similarly, Zulus who move from South Africa to London develop a taste for the scrawny Kate Moss look.[8] Fijians liked full figures until the arrival of Western television. After just three years of TV, 74 per cent of Fijian women felt they were too fat, and many began showing symptoms of eating disorders. Body mass preferences have also changed in the West, from Rubens's voluptuous vixens to Twiggy's hunger-strike chic. Studies trace the changing waistlines in beauty pageant winners and *Playboy* centrefolds, and show gradual decrease in body-mass since the 1950s.[9]

Weight preferences often fluctuate with societal variables, such as wealth. Opulent times often opt for skinnier bodies, saying you can never be too rich or too thin. In tougher times, or times of economic recovery, a well-fed figure can signal access to resources. Evolutionary models account for this by saying we are programmed to like bodies that are commensurate with current resource availability, but this can't be a complete explanation. Women who are too thin are often less fertile, and societal wealth is a factor that should promote an increase in reproduction, so the preference pattern is actually the opposite of what evolutionary models predict.

Moreover, evolutionary models are hopeless when it comes to predicting many more obvious variations in standards of beauty. Some societies use scarification to attract; others (like ours) use tattoos. Kayan women wear neck rings, which gradually weigh down the shoulders and dramatically lengthen the neck. Mursi women wear enormous lip plates, which leave huge holes in the face and make it harder to eat. In China, women bound their feet for 1,000 years, and men became sexually attracted to the deformed, 3-inch ideal, as well as the foul smell caused by chronic gangrene.

When it comes to evolutionary theory, there is no accounting for taste. These variations are difficult to explain in evolutionary terms.

Darwinians point out that they help distinguish the sexes, advertise a person's ability to endure pain and mark group identity. Such explanations are difficult to test and somewhat problematic. Do we really need cosmetic enhancements to distinguish women from men? Is pain endurance always sexy? Does the preoccupation with group identity reflect evolutionary pressures, or did it emerge with population growth after we had fully evolved? More importantly, evolutionary theory cannot predict the specific forms that beauty takes. For that, a cultural story tracing the history of our preferences is required.

For example, foot binding may have emerged as a result of the fact that the Chinese kinship system made daughters more valuable as brides than as labourers, and the men who were wealthy enough to pay for brides also wanted to control them. Mursi lip plates may have initially emerged as a way to turn off slave traders. American tattoos were first imported by soldiers and sailors who had sailed the South Seas. Then they became part of anti-establishment subcultures when some Second World War vets started forming biker clubs. Bikers became fashion icons when 1960s political turmoil fuelled anti-war sentiments and teenage rebellion. Musicians began to adopt elements of the biker look, and tattoos gradually became a staple in urban counter-culture and finally trickled up into mainstream fashion. This last stage in the process is fascinating in itself, because fashion primarily used to trickle down from the wealthy elite to the folks on the street. Changing conceptions of status in modern democracies are influencing the direction of diffusion in the domain of taste. None of this shows that evolutionary models are wrong, only that we may learn more about human conceptions of beauty by looking at culture and history.

Which Sex?

In addition to influencing what looks we like, culture can influence what sex we like. Most people in most places at most times in history have been attracted to the opposite sex. Biology certainly has something to do with that. But, from a biological perspective, it may not matter if some of us prefer the same sex, or if all of us swing both ways. Homosexual acts are widely observed in other species, and, for the bonobo chimps, just about anything goes. In human beings, people clearly have variable preferences, and that raises the question: where does this variation come from?

One trendy answer is that sexual preference is in the genes. Nobody has bothered to look for a heterosexual gene because it's presumed that heterosexual desires are statistically dominant in all species, and there is probably no single gene involved in orchestrating pursuit of the opposite sex. But there has been a concerted effort in recent years to find the genetic basis of homosexuality. Most of this research has been done on gay men, perhaps because men still dominate in the sciences. There have been numerous twin studies concluding that male homosexuality is highly heritable, and, in 1993, Dean Hamer published a study in *Science* that tried to link homosexuality to a particular gene.[10] Hamer claimed that families with two or more gay brothers were somewhat more likely to have a particular allele at the position labeled Xq28 on the X-chromosome. Another study, published in 1999, disputed these results, and it has been back and forth ever since. At present the weight of the evidence suggests that there are modest correlations between male homosexuality and a few gene sites including Xq28. This has led the press to conclude that there is a 'gay gene', or perhaps several gay genes.

Some members of the gay community welcomed this finding since, to them, homosexuality never felt like a choice. Christian groups denounced the results, saying that homosexuality is not genetic destiny. In one sense, the Christians are surely right. The idea of a 'gay gene' is clearly a media simplification. No complex human psychological trait has been pinned on any single genetic marker, and, in any case, the correlations that have been found are small. Many straight men have the alleged gay genes, and, even if your identical twin is gay, there's a pretty high chance that you won't be. Something other than genes is playing a role. This does not mean that homosexuality is a choice. The non-genetic factors may include other biological variables (such as hormone exposure *in utero*) as well as cultural and biographical influences that are beyond anyone's control. Moreover, choice is a red herring in the ethical debates about homosexuality. We don't need to find a 'race-mixing gene' to argue that it's morally acceptable to date someone from a different ethnic group. If you have that preference, wherever it comes from, there's nothing wrong with it. Likewise, the dispute between Christians who oppose homosexuality and defenders of gay rights really concerns the question of whether homosexuality is a bad thing, and the arguments on that question do not hinge on whether homosexuality is genetic. If you believe the relevant passages in the Bible, it shouldn't matter if homosexuality is genetic; it's still wrong to indulge in it. And if you don't believe those passages, then you shouldn't think homosexuality is worse than any other sexual preference, be it for blondes, brunettes or redheads.

Another problem with the concept of a 'gay gene' is that the category 'gay' is very difficult to apply. We often think about homosexuality in terms of sexual desire. A person is gay (or bisexual) if attracted to members of the same sex. So defined, the category seems to have little relationship to biology. Talk of a 'gay gene' gives the impression that we need some unusual biological feature to gain sexual gratification from members of the same sex. That is patently absurd. Anyone who has ever masturbated has had a sexual encounter with the same sex. There are also cultures in which same-sex relationships are widespread.

In classical Greece, Athenian men often had young male partners. They did not engage in sodomy, but they did have sex using the younger male's thighs as a sexual organ. Similar practices of older men taking young males as partners were reported among the Azande of Sudan before Christian missionaries converted them. Or consider the Sambia of Papua New Guinea.[11] In that culture, sexually mature teenagers are fellated by younger boys in order to get semen, which is believed to have important powers. A similar practice is found among the Etoro, also of New Guinea, who permit heterosexual sex for only 100 days of the year and never in the home.[12]

Even within Christendom, homosexuality has been tolerated at various times. Evidence suggests that strong anti-gay attitudes became more central to Church doctrine in the eleventh century, when there was a concern that the Church was losing power and had to ensure the moral integrity of priests.[13] Before that there may have been whole monasteries where male monks regularly enjoyed each other's sexual favours. In the tenth century, there was a heretical group called the Bogomils, doctrinally similar to the Manichaeists, who are believed to have encouraged sodomy (both male and female). In this way, they hoped to avoid procreation in preparation for the apocalypse which was forecast for the first millennium. The Bogomils were centred in Bulgaria, which meant they were Bulgars, from which the English slang term 'buggery' is derived.[14]

In most of these cases, the men who engage in homosexual practices also have female partners. Consequently, it seems wrong to call them 'gay'. But these men clearly enjoy homosexual acts and have no difficulty discerning male beauty. Men clearly don't need a special gene for that. It's equally absurd to suppose that there is a gene that causes a preference for gay traits of a non-sexual nature, such as a gay persona or musical taste, since these affectations vary across individuals and cultures.

Perhaps genes can increase the likelihood that a man will *dislike* women sexually or *prefer* men. Or perhaps genes make some men more effeminate, which can increase the probability that they will be culturally expected to prefer men. It may be premature to rule out these possibilities, but even the prevalence of men who only like men can vary with culture. One factor is tolerance. There are more gay men in urban areas than in suburbs, and there are more gay men in countries that tolerate homosexuality than in those that don't. In a growing number of western European countries, Canada and South Africa, gay marriage is legal, and in some Islamic countries, like Saudi Arabia and Nigeria, it's punishable by death. There may be fewer gay men in these countries, not just fewer who admit they are gay. If identical twins do not always share the same sexual preference, then environmental variables may have an impact, and intolerant countries may push the odds in favour of heterosexual outcomes. One factor believed to have an impact is pro-natalism: societies that promote procreation. Societies that are more pro-natalist are more opposed to homosexuality, and societies, such as those in the contemporary West, that have cut back on procreation tend to be much more tolerant. Having a preference for same-sex partners may increase if there is social pressure against reproduction.

Even if genes contribute to exclusively homosexual preferences, they seem to have little to do with the prevalence of homosexual acts and relationships. There is massive variance in sexual behaviour, and the best predictor of whether an arbitrarily chosen man will have sexual relations with other men is not his DNA, but his culture. This suggests that human sexual desire is not under strict genetic control. I have focused here on male homosexuality, because most of the science has been done by men on men, but the lesson clearly extends to women. If anything, the science of lesbianism suggests that women are more sexually ambidextrous than men are, so there may be even less involvement of genes in determining the contours of female desire. For both men and women, there is presumably an appetite for the opposite sex in most individuals, but where we wind up on the bisexuality scale may be profoundly affected by geography.

THE BATTLE OF THE SEXES

We've been looking at sexual attraction, and finding that culture has an influence on what turns us on. Ideals of beauty can be culturally informed, and, more surprisingly, so can our degree of interest in having sexual relations with someone whose intimate anatomy resembles our own. It was conceded, however, that most people across the world's cultures are attracted to the opposite sex, and this is clearly the result of biology. But what attracts us in the opposite sex? What do we look for when playing the dating game?

What We Want in a Partner

It is often suggested that men and women want different things when it comes to relationships. Guys want as many partners as they can get, and they like women who are young and attractive. Women like guys who are older and affluent. They care more about love than sex, and they are very choosy about their intimate partners. This is the stuff of trashy soap operas and sitcoms. You've heard the clichés a thousand times. But are they really true? Evolutionary psychologists answer with an unabashed yes. They say cross-national data confirm every stereotype, and evolutionary theory can explain why.

The evolutionary story begins with an obvious fact: women gestate and men do not. If women are programmed to replicate their genes, they need to be very discriminating when it comes to mate selection. To get through a pregnancy and early childrearing, it's good to have a reliable partner with plenty of resources. It's hard to go at it alone. Partners must be chosen carefully, and sex must be limited so as not to get pregnant with a partner who won't help out. Women can only get pregnant a few times in life, so they need to be extra sure these times work out. When picking a partner, age doesn't matter much, because men remain fertile for a long time, but resources do matter. This makes older men a better bet, because they have had more time to accrue wealth.

If men are programmed to replicate their genes, there is a simpler strategy: have lots of sex with lots of partners. By spreading their seed, they could potentially have many offspring, and, since they don't get pregnant, they have very little invested in their children. As a result, they don't need to be very choosy. But, when they do choose, youth is a plus because it's a sign of fertility. Youth becomes especially important if a man decides to settle down, because young women have more reproductive cycles ahead of them.

It's a seductive story because it provides an explanation of the stereotypical sex differences. And it turns out the stereotypes are true. In an ambitious series of studies, David Buss and his collaborators have shown that men and women have very different preferences.[15] Consider the five most touted findings:

Male Seed Spreading. When asked how many sexual partners they would like over the lifespan, the average female response was five, whereas the average male response was close to twenty.

Female Choosiness. When asked whether they would sleep with an attractive stranger, 75 per cent of men said yes, but every one of the women said no.

Male Preoccupation with Looks. Men across the globe rate physical attractiveness as more important than women do.

Female Preoccupation with Wealth. Women value wealth more than men do.

Age Differences. Across the globe, men prefer younger women, and women prefer older men.

All these results confirm existing stereotypes and fit nicely with the predictions of the evolutionary approach to mate selection. Men want as many young, fertile women as possible, and women want good providers.

There is, however, another explanation. All these results can be explained on the assumption that men enjoy economic and political dominance over women in the societies that have been investigated. The male dominance model predicts that men and women will have different priorities in finding partners, and it also predicts that women's responses on questionnaires will tend to conform to male ideals. Let's have a closer look.

Why do women express less interest in multiple sexual partners than men do? In male-dominant societies, women are often treated as male property, and men control female sexuality. Male ideals of chastity are almost universal in such societies, and the slut/stud double standard is pervasive. If this explanation is right, then the reported differences in the number of partners desired may not reflect actual preferences. Women may desire as many partners as men do, but they may be worried about stigmatization in expressing that preference, and men may feel that boasting about the desired number of partners gives a positive impression of dominance. Research supports this alternative explanation. Michele Alexander and Terri Fisher tested the impact of social stereotypes by asking male and female college students to report the number of sexual partners they have had under two conditions: either while another student could hear them or while hooked up to what they believed to be a lie detector.[16] When others could hear them, men reported more sex partners than women, but the difference disappeared with the lie detector, and in fact women reported having slightly more partners than men. In another study, Gillian Brown reviewed reproduction records in eighteen societies and found strong evidence that men and women have an equal number of partners.[17] Others have claimed that Buss's data may be more consistent with this social explanation. Buss showed that when average responses are calculated men want more partners than women. But average scores can be affected by outliers: if a few people exaggerate preferences in conformity with social ideas, the average scores for men and women will differ. To overcome this concern, William Pedersen and his collaborators looked at the *medians* instead of the *means*; he calculated the percentage of men and the percentage of women who picked each possible number of desired sex partners.[18] When calculated this way, the gender differences disappear.

Similar conclusions can be drawn about the finding that men express more willingness to sleep with strangers. Buss found a massive gender gap here, with the vast majority of men saying they would sleep with someone they just met and no women admitting to this at all. But this is clearly a case where the stigma of being viewed as promiscuous could influence women's answers. Moreover women can reason, with no need for an evolved preference, that sex with strangers carries risks. Getting impregnated by someone you don't know is more challenging financially and socially than having an accident with a long-term boyfriend. So female reluctance may reflect prudential reasoning on their part. In any case, such self-report measures are not very reliable. The combination of stigma and prudence could make women overestimate their degree of sexual restraint. There are hard data to prove that women are much more willing to sleep with strangers than they let on. In a study of college students, Catherine Grello and her colleagues found that one-third of casual sex encounters were between men and women who had just met. These numbers are hard to square with the evolutionary predictions, and they may reflect the fact that birth control now allows women to have sex without the risks.

What about the male preference for beauty? Why do good looks matter more to men? On the male dominance model, this is partially a function of the fact that men don't depend on women financially to the same degree that women depend on men. The data show that looks are higher up on the male priority list, but that may just mean that they don't need to be as concerned about earning potential. For women, their ultimate income bracket often depends on securing an affluent mate, and a massive cosmetics industry has emerged to help women make themselves more attractive. In small-scale societies, where men and women both contribute to subsistence, the gender difference disappears. This has been confirmed by Elizabeth Pillsworth among the Shuar of Ecuador,[19] and in a study by Frank Marlowe with the Hadza.[20]

The fact that women value wealth more than men do is explained in a similar way. Men are the primary wage earners in male dominant societies, and they do pretty well financially without any female assistance. In fact, men improve their earnings after getting divorced. Women's earnings drop substantially and, even in our society, which enjoys high degrees of female employment, it is hard for single women to make ends meet. In the USA, women are much more likely to live in poverty than men, especially if they have children. Financial dependence on men was even greater in past decades, and women's preferences may gradually catch up with improving economic prospects. One way to show this is to look at how the female preoccupation with earning potential relates to financial gender equality. Alice Eagly and Wendy Wood reanalysed Buss's data in relation to indices of women's empowerment and found a strong negative correlation between women wanting wealthy guys and having more job opportunities.[21]

Against this hypothesis, Buss has argued that affluent women in the USA are even more preoccupied with getting rich partners than middle-class women are. But there are two good explanations for this. Wealthy women are more likely to be materialistic (that's how they got wealthy), and they are more likely to face one of sexism's uglier stigmas: the unthinkable prospect of a woman making more than her husband.

Turn, finally, to age differences. The evolutionary and cultural approaches offer a similar explanation of female preference; women want older men because they are more likely to have needed financial resources. But the explanations of male preferences diverge. Evolutionary psychologists say that men want younger women in order to maximize fertility. According to the male dominance theory, the preferred explanation is that older men find it easier to dominate younger women. This account explains three findings that are harder for evolutionists to explain. Ironically, all three findings emerge from data obtained by evolutionary psychologists.

First, Douglas Kenrick and Richard Keefe found that the male preference for younger women increases over the lifespan, with older men wanting to have even more seniority over their partners than younger men.[22] Evolutionists explain this by supposing that young women are the most fertile, so old men should retain their preference for youth. The trouble is, men who have reached an advanced age report a preference for women who are considerably younger, but still too old to reliably reproduce. Men do not just spend the lifespan fixated on twenty-year-olds. In a male dominance model, this is explained by the fact that age gaps are a way of dominating spouses. A sixty-five-year-old can dominate a fifty-year-old, even if he can't have children with her. On the other hand, a sixty-five-year-old would have a harder time dominating a sixty-year-old to the same extent that a twenty-five-year-old could dominate a twenty-year old, which may be why the preferred age gap broadens over the lifetime.

Second, Buss's theory predicts that the male preference for younger women should correlate with the male preference to have many children, since he interprets the preference for youth as a preference for fertility. In an effort to confirm this prediction, Buss looked at male preferences for youth and for offspring in thirty-seven societies and found the predicted correlation. But the correlation was statistically driven by just two countries, Nigeria and Zambia, which happened to be the only polygamous cultures in his sample. Polygamy often correlates with extreme male dominance, as it does in these cultures, and women's roles are often relegated to breeding children. The crucial finding is that there is generally no correlation between preference for a younger bride and a preference for many kids, which is strong evidence against the evolutionary explanation of age preference.

Finally, in their re-analysis of Buss's data, Eagly and Wood found that preference for big age gaps correlated negatively with female empowerment. This is predicted by the male dominance model and hard to reconcile with the view that preferred age differences reflect innate strategies for maximizing reproductive success. Unsurprisingly, Eagly and Wood also found that the societies in which men wanted to have greatest seniority were also ones in which they wanted women to be good at cooking and housekeeping. Most of the Buss gender differences correlate with this measure of female servitude. Unless they suggest that the male desire to have a housewife is innate, these statistics suggest that gender differences vary with a culturally created preference that is strongly associated with male dominance. The fact that gender differences are variable already suggests that they are culturally driven. The fact that they co-vary with male dominance basically settles the case.

Poisonous Passions

We've been looking at the preferences that go into selecting a romantic partner. Once you find a person you like and convince that person to enter into a relationship with you, there is nothing left to romance but a lifetime of bliss. Or maybe not. Relationships are complicated, and many factors prevent them from being the glorious equitable partnerships that we might have envisioned.

One challenge that lovers face is jealousy. John Dryden said that jealousy poisons passion, and Shakespeare said it is more deadly than a mad dog's tooth. Most agree that jealousy is unpleasant and noxious. It can destroy a loving relationship and even lead to violence. This raises the question, why do we have such an untoward emotion? The obvious answer is that jealousy exists to protect us against infidelity. It may destroy some relationships, but it preserves others, guarding us against partners who stray.

Evolutionary psychologists say that jealousy is an evolved response. Jealous lovers are highly possessive and work actively to prevent infidelity. This makes sense from an evolutionary perspective because our genes are designed to replicate. To do that, they must make sure that we take good care of our own children and don't waste precious resources investing in other people's children. Jealousy helps us ensure that outcome. If we didn't guard against cuckoldry, we might not produce viable offspring, and genes that tolerate that don't survive.

One piece of evidence for this claim is an alleged gender difference. As evolutionary psychologists like to remind us, women get pregnant and men do not. When the baby is born, the mother knows that it's hers. She was there, after all, when the kid popped out. Her genes know it too, and they want to make sure their duplicate copies make it to a viable age, so they make mom keep a close lookout on her bread-winning husband. From their perspective, the most important thing is that dad sticks around to support childcare. Dad's genes face a different problem. They can't be sure the baby is his. After all, mom could have got a little action on the side. This is what evolutionary psychologists call uncertain paternity. Dad's genes need to make sure his wife is faithful, because they don't want to invest in caring for another man's baby.

This difference between the predicaments faced by men and women leads to a prediction. There are two kinds of infidelity: romantic and sexual. A partner can have casual sex on the side, but remain committed to the relationship, or a partner can fall in love with someone else, which would jeopardize the relationship even if no sex were involved. From the genes' point of view, men need to be more concerned about sexual infidelity than romantic infidelity. If a man's wife is sleeping with others, he might inadvertently end up taking care of someone else's genes. Women need to be more concerned about romantic infidelity. If a woman's husband falls for someone else, he might stop supporting her financially, and the baby will be in danger. Thus, evolutionary psychologists predict that men will be more bothered by sexual infidelity in their partners than romantic infidelity, and women should show the opposite pattern.

Buss and his collaborators asked college students what form of infidelity would be worse.[23] They found that men were more likely to say that sexual infidelity is worse, which seems to confirm the predictions of evolutionary psychology. They also measured emotional reactions using a device that records galvanic skin responses. Galvanic skin responses are increases in electrical conductivity due to a trace increase in perspiration. The same technology is used in lie detectors. Buss found that men show a big spike in galvanic response when they are asked to consider a scenario where their partner has sex with someone else. Women don't show that effect. Buss interprets this as a physiological confirmation of the claim that men are more bothered by sexual infidelity.

There are numerous problems with these conclusions. Evolutionary logic does not only predict a gender difference. It predicts that men will find sexual infidelity worse than romantic infidelity. That is not what the data showed. Men seem to be bothered by both forms of infidelity more or less equally. In fact, when the study was replicated in China, Japan and Germany, men were much more bothered by emotional infidelity than sexual infidelity. In the Netherlands, men were asked whether it was worse to imagine their lovers' falling in love with someone else or trying out different sexual positions with someone else. 72 per cent said love infidelity would be worse. This is striking, because the risk of sexual infidelity may be high in a country like the Netherlands, where people have a very relaxed attitude about sex, so selfish genes should go into overdrive and make men very sexually jealous. The very fact that there are cultural differences is hard to explain in the view that the response is innate.

In response, evolutionary psychologists will be quick to point out that everywhere this study has been done, a gender difference appeared. Men are more bothered by sexual infidelity than women. This needs to be explained. One possible explanation is that women are reluctant to admit that they would be more bothered by sexual infidelity because there are cultural taboos against confessing an interest in sex. In a male-dominant society, women are supposed to be highly emotional (hence irrational) and also willing to look past their male partners' indiscretions. So women are under a lot of cultural pressure to say they care more about love than sex. Moreover, women who reflect on the situation may realize that they are better off tolerating a fling here and there than emotional betrayal. If women depend on male partners financially, sexual infidelity is hurtful, but emotional infidelity means it will be hard to pay the rent.

If women are trying to conform to cultural expectations or calculating damages, then their responses may not be an accurate reflection of their gut reaction to the two forms of infidelity. To test this, David DeSteno reran the Buss study, but this time he had everyone retain a list of numbers in their head while thinking about the dilemmas.[24] Holding numbers in your head is known to interfere with conscious reasoning because it consumes cognitive resources, so responses under these conditions should reflect pure attitudes unfiltered by concerns about social stigmatization or economic calculations. Sure enough, in this version of the study, women's responses resembled the male pattern, and male responses remained unchanged. If the gender differences were innate, they should not be affected by a task that interferes with conscious reasoning.

Evolutionary psychologists will object that the gender differences must reflect deeply rooted instincts because men show strong galvanic skin responses and women don't. But there is another possibility. Galvanic skin responses show emotional reactions, but they don't reveal whether the reactions are positive or negative. It's even possible that men show a galvanic skin response when they hear about sexual infidelity because men get excited whenever they think about sex. To test this, Christine Harris measured galvanic skin responses while asking men to imagine their partners having sex with someone else – a condition from Buss's study – and to imagine having sex with their partners themselves.[25] Harris found a comparable galvanic response in both conditions. Maybe the word sex alone would do the trick. This shows that male bodies respond somewhat differently to sexual imagery, but it casts doubt on the claim that men are more bothered than women by sexual infidelity.

Finally, it must be noted that there is something suspicious about the logic that evolutionary psychologists are deploying here. They are assuming that women need men to raise babies and that men will fare badly if they invest in babies that aren't theirs. But this may not be true from an evolutionary perspective. The nuclear family is rare in nature, and, in species that do form long-term bonds, 'extra-pair matings' are commonplace. In some ostensibly monogamous birds, 40–60 per cent of the offspring come from such secret trysts. Among chimps, monogamy is unheard of, and childrearing is often a group activity. Likewise, in small-scale human societies, it takes a village to raise a child. When mom is busy, others help. This means that women don't depend on their husbands. In this system, no one needs to worry about investing in other people's babies because they can be confident that their own babies will be well taken care of by the group. Parents who invest in the whole community may actually be more likely to raise viable children than parents who selfishly care only for their own. If early human societies followed this model, then there would have been little evolutionary pressure for gender-specific jealousy mechanisms. Indeed, there would be little need for jealousy!

It's very possible that jealousy is not an innate emotion. Like so many emotions, it could be a blend of more primitive responses. Suppose your partner cheats on you. You might get mad at the betrayal, worried about the future of your relationship, sad about the prospect of loss and disgusted by the thought that another person has been intimate with the one you love. This toxic cocktail may be the emotion we call jealousy. Notice that, in the right cultural setting, all these other emotions are likely to arise in response to infidelity, so there is no pressure for evolution to create a whole new emotion. And, in cultures where fidelity is less valued, these negative responses may not even arise.

Why, then, does jealousy arise here and in all the other cultures examined by Buss? One answer is that our systems of childrearing and marriage make spousal relationships extremely important. Losing a partner can have devastating implications having to do with child custody, livelihood and preservation of an intact estate. Men and women should develop strong negative reactions to anything that threatens the bonds that structure our society. For both men and women, loss of love is more serious, since it carries the greater risk of divorce. But the preference order could potentially switch in societies with extreme forms of male dominance, where women are treated like property. In that arrangement, men may think of their wives as physical objects rather than loving partners, so sexual control may be more meaningful than emotional loyalty. Women in male-dominant societies depend on men, so they are forced to tolerate sexual infidelity even if it bothers them intensely. Male dominant societies also control women's sexuality prior to marriage, because a daughter who has children out of wedlock will be very hard to sell to suitors, who will resist the extra financial burden. This promotes a powerful tendency to suppress women's interest in sex. Contemporary Western cultures have made great progress on women's liberation, but full equality has not been achieved, and a dark past colours contemporary values. If there are gender differences in jealousy they may be a sinister cultural residue rather than divergent defence mechanisms of selfish genes.

Technologies of Dominance

Throughout this treatment of gender differences we have been looking at the contrast between evolutionary explanations and cultural explanations. The culture explanations keep coming back to the same theme: male dominance. At this point, the evolutionary psychologist might retort that male dominance is itself a consequence of evolution. To counter this disturbing possibility, the nurturist must show that there is a cultural explanation for why men dominate women in such a large proportion of recorded societies. Doesn't this pervasive pattern suggest nature rather than nurture?

There are many explanations of why male dominance is so widespread. I will consider just one, which was expounded by the anthropologist Marvin Harris.[26] You might recall from the discussion of cannibalism that Harris is a proponent of cultural materialism, the idea that environment, technology and other material resources drive cultural change. Harris offers a materialist explanation of how human groups transformed from small bands of hunter-gatherers into large agricultural states.[27] Along the way, there were a number of important changes in economic organization and the distribution of power, including changes that result in male dominance over women.

The story begins with a basic biological fact: human males are physically stronger than human females. On average, men are 7 per cent taller than women and, depending on the muscle group, about 20 or 30 per cent stronger. For this reason, in most hunter-gatherer societies, men become the primary hunters. Women in such societies do the majority of the gathering, and that is a major contribution, since foraged food constitutes most of the diet. Therefore, women are highly valued. Because of that, some hunter-gatherer societies enjoy considerable equality between men and women. Spousal abuse is comparatively rare, though not unknown, and men and women make many decisions together. Examples include the Mbuti of Congo, the Cheyenne Indians, the Montagna-Naskapi of Labrador and the !Kung of the Kalihari (the exclamation point signifies a clicking sound, which has no letter in the Roman alphabet).

Unfortunately, gender equality is hard to sustain. If men are the hunters, they are also likely to be the weapons manufacturers. They create and control the instruments of death. There are societies in which weapons are rarely used against human beings; they are reserved for hunting. But this is usually not the case in societies that compete for resources. If one band of hunter-gatherers is competing with another band of hunter-gatherers, and both have weapons, it is likely that they will use those weapons against each other. Societies that engage in armed combat sometimes do so deliberately; they organize groups of warriors to attack neighbouring villages to kill competitors or steal stored resources. If there are several societies in one geographical area, and one of them wages war, the others will have to begin training warriors as well. They too will go on raids to obtain resources and exact revenge. In warring societies, men are trained to be aggressive and fearless, and the men take on a new role as protectors. These changes transform gender relations. Men are more violent and they attain a new sense of importance. Without their protection, the society would collapse. Female dependency on men increases, and men exploit this fact. Men in warring societies often capture women from neighbouring villages during raids and often take multiple wives. They also tend to be patrilocal, which means that the household is organized around the extended family of a group of brothers and their male offspring, the brides of these men are separated from the own blood relatives, who might have an interest in protecting them. In this way, men can dominate women with impunity. Examples include the Aborigines of Queensland in Australia, the Sambia of Papua New Guinea and the Yanomamo of the Amazon basin.

Warfare often cultivates belligerent men who are domineering and abusive, but this state of affairs can change if warfare is taken to a more ambitious level. Sometimes, groups of neighbouring villages that wage war will join forces. Suppose one village becomes very powerful and manages to force neighbours to surrender. When this happens, the dominant village can force its neighbours to pay tribute, thus ending the cycle of raiding that persists when neighbouring villages are equally powerful. When this happens, the village-sized society expands into a larger centrally organized collection of villages, which anthropologists call a state. States often continue to wage war, but not internally. The villages within a state do not fight each other. Rather, their men group together to raid other societies that are farther away. These raids often require that men leave home for long periods of time, and that leads to an improvement in women's lives. If a husband is leaving home for a month-long raiding expedition, he may be apprehensive about leaving his wife with his brothers, so he will encourage her to stay with her natural family, who will protect her. This strategy effects a shift from patrilocal organization to matrilocal organization, in which women live in the same household as their sisters and daughters, and husbands of these women join them from other households. In a matrilocal household, it is harder for a husband to get away with spousal abuse. Women attain domestic authority. This was the case, for example, among the Iroquois.

Harris's story does not end there. The status of woman can remain reasonably high in states where women continue to make major contributions to the economy through the cultivation and collection of food. But that all changes when societies develop certain technologies for farming. Some farming is done with small, hand-held tools. A hoe and a spade can be used effectively by both men and women. If a farming society has only these tools, men and women will retain equal importance in food production, and women can sustain high status. The Yoruba of west Africa are an example. Suppose, however, that the society devises the plough and domesticates large animals for pulling it. These technologies were used in many parts of the world. When animals are used for farming, heavy yokes and ploughs are required, and farmers must be able to control animals that push and pull with tremendous force. Women are at a serious disadvantage with plough farming, because they are far less efficient than men. When societies acquire the yoke, women are almost invariably relegated to less significant roles in the production of food. Men become the primary providers. In controlling the technologies of yoking animals, men also control animal-drawn vehicles, and therefore they become the primary traders. Trade promotes the development of writing, which is essential for good bookkeeping, and the men, who are doing the trading, are the ones to become literate. In societies like this, women and men usually have very different status. Women may continue to do minor farming, but they usually spend more time doing domestic work and rearing children. There is no need for them to read and write and there are no jobs outside the home in which they could compete as equals with men.

According to Harris's account, there are two major factors that contribute to male dominance. With the rise of warfare, which is an outgrowth of hunting technology, men in small-scale societies come to dominate women physically. With the invention of the plough, men in larger societies come to dominate women economically. In small societies without war and large societies without ploughs, male dominance is less prevalent.

Other factors may exacerbate or ameliorate gender inequality. Every culture has its own history, and cultural change is a complex process. Specific events in history, religious beliefs, technologies for producing wealth domestically and gender ratios can all have some impact. Harris's account certainly isn't complete, but his two sources of male dominance are especially prevalent in the anthropological record. The nearly universal dominance of men may be a function of the fact that war and the plough have been so widespread throughout world history. Male dominance is nearly universal, but that does not mean it is biological. Harris's story explains how facts about gender relations can nevertheless have a cultural basis. The biological differences between men and women play an essential role in this story, but physical strength is not a sufficient condition for male dominance. Technological advances also play an essential role. Harris's story is both biological and cultural.

In our own society, technological changes have allowed women to compete with men economically for the first time since the invention of the plough. That has been a major boon for women's rights. Unsurprisingly, with the industrial revolution, we also see the emergence of the women's suffrage movement. Harris's theory predicts this kind of correlation. It predicts that cultural and economic variables will have a dramatic effect on relations between the sexes. But we are still in a transitional time. In contemporary societies, misogyny is exacerbated by the fact that women, who are still regarded as inferior to men, are attaining greater economic equality. We have invented derogatory categories for women who are ambitious ('the bitch') or critical of male dominance (Rush Limbaugh refers to 'the femi-Nazis'). But the radical changes of the past century should give us hope that technological advances will bring sexism to an overdue end. Such changes also tend to confirm that male dominance is cultural rather than biological. Like evolutionary psychologists, Harris has given us a just-so story, which should be carefully tested against the historical record. But, whatever the details, it's clear that relations between the sexes have been transformed historically, and that shows that nurture has a dramatic impact on this relationship.

SETTLING DOWN

We've looked at how culture can influence sexual attraction as well as the features that men and women look for in their romantic partners.

We have also seen how relationships can have power dynamics that are affected by culture, including the over-arching fact that men have been able to dominate women. There is one more profound way that culture can shape our relationships. Culture determines the rules of marriage. When we decide to settle down, culture tells us who can enter into a matrimonial union and what form such unions can take.

Most cultures have rules for forming long-term romantic bonds between people who have sexual relations and produce offspring. For this reason, social scientists sometimes say that all cultures have marriage. However, this may be a misleading conclusion. The famous anthropologist Clifford Geertz called marriage a 'pseudo-universal'. Marriage arrangements vary so greatly across cultures that we can hardly say it's the same institution. In some cultures, marriage involves co-habitation, and in others not; in some, it involves exclusive sexual access, and in others not; in some it's initiated by a ceremony, and in others not; in some it's a bond for life, and in others not; in some it's chosen by the partners, and in others it's arranged; and so on. Such differences show that marriage means very different things in different places. But one common denominator is that in all these cases there are restrictions on who can marry whom. It's those restrictions that I want to consider now.

Who is Off Limits?

Every culture has marriages that are allowed and others that are disallowed. Until recently most countries disallowed gay marriage, and, not too long ago, interracial marriage was taboo. Many religions continue to demand marriage within the faith, though this restriction is rarely state-sanctioned. The two factors that may be most heavily regulated are age and blood relations. But, in both, there is considerable variation.

One factor that varies across cultures is marriageable age. Now, most nations have laws that require a minimum age for marriage partners of between eighteen and twenty-one, unless they get special permission. But many societies traditionally allowed younger children to marry, or, in male-dominant societies, young girls were married off, for a price, to adult men. This practice is still relatively common in some parts of the world. In some African communities as many as half of the girls get married off before they turn fifteen. In India, nearly a quarter of the women may get married before they are eighteen. In Saudi Arabia, clerics allow marriages between prepubescent children and between young girls and adults. In Yemen, a ten-year-old girl named Nujood Ali recently became an international celebrity when she filed for divorce. *Glamour* magazine named her woman of the year. In ancient Rome, the legal age for marriage was twelve. Child marriage is correlated with male dominance and, given the variation in dominance, it is not surprising that there are sizeable differences in marriageable age.

Somewhat more surprising is that there is massive cultural variation in rules pertaining to unions between blood relatives. It is often supposed that there is an innate taboo against incest. Many species avoid incest, and most human societies do as well. Incest avoidance within the immediate family is good for the genome, because inbreeding can lead to the proliferation of harmful recessive traits. For this reason, we seem to have evolved a disposition to avoid incest, but it is important to distinguish incest avoidance from incest taboos. There are lots of things we avoid innately, such as contact with rotting food or noxious smells, but no one suspects that there are innate moral rules against these things. We don't need innate rules for things we naturally avoid (there is no rule against chopping off your own limbs, and no need for one). In fact, if you survey world cultures, only 44 per cent have explicit rules against incestuous unions.[28]

There is also massive variation in what counts as incest. In the Christian world cousin marriage is strictly verboten. In the eleventh century, the Church banned cousin marriage up to the seventh degree of relatedness, so descendants of your great-great-great-great-great-grandparents were off limits. There wasn't even a method for tracking that distant degree of relatedness. In contrast, cousin marriage is viewed favourably in many places, including most Muslim societies. An estimated 20 per cent of the world's population lives in places where cousin marriage is encouraged.

Attitudes even vary on sibling incest. Sibling marriages have been common among royalty, but rare among ordinary folk. One striking exception is Ptolemaic Egypt, where census records reveal that many of the Graeco-Roman citizens were married to their siblings. The reason for this is not known, but there is one good theory. In the Ptolemaic period, the indigenous people of Egypt came under Graeco-Roman occupation. The occupiers turned Egypt into an apartheid state. They were morbidly afraid that Greek and Roman immigrants would marry native Egyptians and lose their loyalty to the foreign regime. As a result, they outlawed intermarriage and made it illegal, on pain of death, for an Egyptian to adopt a Greek or Roman name. But this had a drastic effect on the marriage pool. Graeco-Roman immigrants were a small minority and didn't have many partners to choose from. It seems, then, that the Ptolemaic rulers began to encourage sibling marriage as a simple solution. It wasn't hard for them to convince citizens that they had a positive attitude towards sibling marriage, since eleven out of fifteen Ptolemaic pharaohs were married to their full sisters. The combination of royal pressure and royal example seems to have worked, as sibling marriage became commonplace. In some urban areas, up to 30 per cent of the married Graeco-Romans citizens could claim a brother or sister as a spouse.[29]

The Egyptian case is a rare exception, but an interesting custom also existed in Taiwan during the first half of the twentieth century. There, poor families could save on exorbitant marriage costs by adopting daughters to marry their sons. These daughters would be raised as adopted children in the household and then married when both children came of age. These have been termed 'minor marriages', because they don't follow the more traditional form of marriage, but they are culturally accepted. The couple are not related by blood, which means this is not a case of incest from a biological point of view, but in terms of familial roles it certainly qualifies.

Arthur Wolf, an anthropologist, has shown that Taiwanese minor marriages have higher divorce rates and lower fertility rates than ordinary, or 'major', marriages.[30] He interprets this finding as evidence that there is an innate taboo against sibling incest. In the nineteenth century, Edward Westermarck argued that the innate taboo is triggered by co-habitation. Wolf resuscitated the Westermarck hypothesis, and, more recently, the evolutionary psychologist Deborah Lieberman has tried to obtain experimental support.[31] Lieberman's studies have shown that the idea of incest is especially disgusting for people who were raised with opposite-sex siblings, suggesting that an innate incest taboo was triggered by living under the same roof. But the evidence is not convincing. Since we live in a society where sibling incest is taboo, it's not surprising that this norm is more deeply entrenched for people with opposite-sex siblings. For them, the taboo matters for their behaviour so they are under more pressure to internalize it. They may not receive more explicit instruction about incest than same-sex siblings, but it is likely that inadvertent inappropriate contact between opposite-sex siblings elicits more adamant interventions from caregivers.

As for Taiwanese minor marriages, some non-biological factors may contribute to their lower reproductive rates. These marriages do not result from love affairs or financial arrangements between two families, and that means the desire and pressure for success may be lower than in more conventional marriages. The couples in minor marriages also tend to be poor, since they come from poor families. Poverty is both a source of stress in a relationship and a reason to resist having children. Plus, more traditional marriages begin with an exchange of money that can be used to start up a successful family. The whole point of these marriages is to bypass that monetary exchange. In addition, there is the more prosaic fact that children who grow up together have to compete for resources at home. Years of sibling rivalry can instil a degree of contempt. This is especially likely when the children are close in age and competing in the first few years of life, when the sting of having a competitor can be greatest. Statistics show that these variables correlate with low fertility in minor marriages.

I am not suggesting that there is no biological tendency to avoid incest. There probably is, and it may even contribute to the fragility of minor marriages. In fact, the presence of incest avoidance mechanisms makes minor marriages even more remarkable. The very fact that such marriages have existed shows that cultural pressures can overcome biological predispositions to a surprising degree. Incest avoidance may be universal, but taboos vary, and, when it comes to marriage, the degree of cultural variation is remarkable.

The most extreme case of cultural permissiveness when it comes to incest may be the ancient Zoroastrians.[32] They evidently had no incest restrictions whatsoever. Brothers could marry sisters, fathers daughters, and mothers sons. Everything was allowed and every arrangement occurred. This is well documented over centuries, and Zoroastrianism was once one of the largest religions in the world. Indeed, there is some textual evidence that Magi, the class of Zoroastrian priests immortalized in the Gospel According to Matthew, were supposed to be born of unions between mothers and their sons. Freud notwithstanding, such unions have been rare in human history, and the Zoroastrian case is clearly an exception. But it testifies to the extraordinary variation that can exist in incest laws.

Clearly, what one culture finds deeply repellant and unnatural can be allowed or even encouraged by another. We giggle with faint disgust when we learn that Charles Darwin was married to his first cousin, but this would be unsurprising if he had been a Muslim rather than an Englishman of Christian descent. Had Darwin been an Englishman of Pakistani descent in contemporary Britain, the probability that he would have been married to a first cousin would be 55 per cent. Even within the same national boundaries, there are profound cultural differences in what qualifies as taboo.

How Many Can You Marry?

Culture plays a role in determining who you can marry, and it also dictates how many. Those of us who were reared in monogamous societies tend to think that our form of marriage is natural, normal and nice. It comes as a surprise when we learn that monogamy is a cultural outlier. 86 per cent of societies are polygynous, which means one man can have multiple wives.[33] A small fraction are polyandrous, which gives one woman multiple husbands. Monogamy is more common than polyandry, but polygyny seems to be the default. That doesn't mean that everyone in the polygynous societies is living a *Big Love* lifestyle. Most men can't support multiple wives. But the option is allowed. In some sense, plural love is allowed in our society as well. If we don't like one spouse, we can divorce and get another. What determines these variations?

The prevalence of polygyny may reflect a familiar theme in this chapter: male dominance. Suppose you live in a society where women can't compete in the open market for jobs. If you are a woman, you will have a strong incentive to find a husband to support you. If you are a man who enjoys female company, you will want to find a wife, and, as an added perk, she can work as a domestic servant at home, bearing children, cooking, cleaning, mending clothes and growing foodstuffs. If you happen to be a very wealthy man, you might even want several wives, because it's nice to have multiple lovers and multiple servants. Having multiple wives may actually save you money, since they could cultivate a small farm on your estate and provide food for the family. If you are a woman, you might not be so keen on the idea at first, but it has some advantages. Life may be better with a wealthy husband than a poor one, and having female companions at home may be better than the social isolation imposed by a monogamous husband who wants to keep you cut off from the rest of the world. In addition, housework is easier with help, and you can team up with your co-wives to help keep your husband in line. Given the prevalence of extreme male dominance, then, it's not surprising that polygyny is permitted in the vast majority of societies on record. As predicted, polygyny correlates with male dominance and it also correlates with economic systems in which female domestic labour is profitable, such as societies in which women can help cultivate the family farm.

Male dominance also explains why polyandry is rare. In fact, it may seem remarkable that polyandry ever exists. Why would men ever want to give up the opportunity to have multiple wives? One answer is economic necessity. Consider Tibet, which traditionally practised a system called fraternal polyandry. Here, brothers would all marry the same wife. To understand why, first recall that Tibet is in the Himalayas, where farmable land is limited. In societies that have private property, family land is passed on to each successive generation. Sometimes it is divided between the sons, and dividing up a parcel of land can result in smaller and smaller plots, which eventually become unusable. In medieval Europe, they addressed this problem by a system of primogeniture, wherein the whole estate went to the firstborn son, and other male offspring could go off and seek their fortune on their own. That's not a workable solution in Tibet, because there simply isn't enough land for these male castaways to occupy. Consequently, there is strong pressure to have all sons share an estate living together in a single household. Now suppose that each of three brothers who are sharing an estate finds a wife. Those wives will bear a number of children. Without adequate birth control that number could be high – perhaps three sons for each wife. If three brothers share the family estate, and each has three sons, then there will be nine heirs between them who will have to share the estate when the three brothers die. Suppose those nine sons marry different women, who each bear three male offspring. Now the estate will be given over to twenty-seven heirs, and so on exponentially. Within a few generations, the family farm will feel like clowns in a phone booth, with insufficient yield to support the whole extended family. Polyandry offers a solution. The number of children a woman can bear does not increase if she happens to have more husbands. She can bear only so many. If three brothers marry one woman, she will end up giving birth to about three sons, and if those three marry one woman, that woman will also bear about three sons. The population will remain constant. With finite land, this is an optimal solution. The family estate never needs to be divided, and it supports the same number of people from generation to generation.

This same pattern was practised by Himalayan groups in Arunachal Pradesh, Buthan, Ladakh, Nepal, Uttarakhand and Zanskar. Polyandry was also practised by the Copper and Netsilik Inuit, who faced similar resource constraints living in Arctic conditions. Polyandry was also traditionally practised in Sri Lanka. There, having multiple husbands in one household was useful because there was a slash and burn economy in which families survived by cultivating their own land, and labour on these family farms was intensive. There were also periods in Sri Lankan history where some men had to spend long periods away from home, doing military service or building irrigation systems. During these times, husbands may have wanted to leave their wives with men who could protect them and help with farming. Having another husband at home, especially a relative, is a solution to this problem.

What about monogamy? Why have we all come to think that having one spouse is the ideal number, when this has been so anomalous historically? It's tempting to credit the rise in monogamy to the decrease in male dominance, but this can't be right, since monogamy comes on the scene in ancient Greece and Rome at a time when power was still in the hands of men. One might infer that the Graeco-Roman world introduced monogamy because of a growth of humanism and democratic ideals. This too would be a mistake. Democratic ideals were narrowly constrained in the ancient world and certainly didn't extend to women. Indeed, monogamy may have a very ugly origin. The classicist Walter Scheidel has argued that its emergence as a normative ideal coincides with the emergence of another institution: chattel slavery.[34] When Greeks started capturing slaves and trading them as property, they also started favouring monogamy. Why? The answer is simple. Polygyny is attractive to men because they can have multiple sex partners and multiple domestic servants. But wives have rights, families who can look out for them and children who are legitimate heirs. With chattel slavery, men suddenly have a class of women they can dominate more completely: slave girls. Sex with enslaved women was widespread in ancient Greece and Rome, and it allowed men to have the benefits of polygyny without the costs, which may be why polygyny was banned.

Early monogamy was probably a setback for women. Under polygyny, women were free to pursue any men, whether married or not. If a wealthy man had a wife, that was no obstacle for an ambitious woman. Under monogamy, wealthy men could have just one legitimate wife, so women who weren't slaves ended up with a more limited marriage market. Plus, they could do nothing about their husbands' philandering because sex with slaves was not considered adultery, and they could not increase political power in the household by banding together with co-wives.

Things changed quite a bit with the rise of Christianity. The Church retained Graeco-Roman monogamy, but made a number of major marriage reforms, including a broadening of adultery norms to include all sex outside the marriage pair. In Greece and Rome, monogamy norms concerned marriage laws, not sex. Wealthy men were de facto polygynists with many sexual partners. The Church tried to stop that. This reform reflects the improved status of women in the early Church, which may be both a cause and an effect of the fact that they played a disproportionate role in spreading the religion. But the intensified commitment to monogamy also coincided with a number of other changes that had little to do with improving women's lot. Over time, the Church banned divorce, greatly restricted remarriage, banned cousin marriage and other forms of incest legal under Roman law, banned adoption and defined premarital sex as a mortal sin. On the face of it, these sweeping reforms seem to have little in common, but the anthropologist Jack Goody has identified a common consequence: they all reduce the probability of consolidating wealth within a family.[35] In Rome, there was quite a lot of marriage within families, and most marriages were arranged to keep careful control over an estate. Individuals could gain power this way and join the Roman elite. Roman families always had heirs to their estates as well. If a man's wife was infertile, he could have a child with a slave and free the child and make him a legitimate heir. Or he could divorce his wife and find another. Or he could adopt an heir. Widowers and widows were encouraged to remarry and start new families if they were still able. Just about every avenue was available to make sure that each man could pass on his estate in the next generation. The Church changed all this.

By prohibiting cousin marriage and other forms of incest, the Church made it harder to consolidate wealth within a family. By extending this prohibition to the seventh degree, they made it likely that many people would accidentally marry distant kin, and offspring of those marriages could be rendered illegitimate. By banning divorce, childless couples were prevented from finding fertile partners, and if a spouse died before having children, remarriage was discouraged and heavily restricted. Childless couples couldn't find an heir by adoption, and if they weren't sexually active before marriage, they couldn't track down bastard sons to take over the estate. In this context, the significance of Christian monogamy does not look like an early effort at women's liberation. Rather, it was part of a comprehensive strategy to increase the chances of heirlessness. If a man couldn't sleep with slaves, co-wives or mistresses, then there could be no way to have a male child with anyone but his legal marriage partner. If that partner was infertile, or if the children died young, there would be no heir to the estate. With no heir, the estate would be taken over by the Church. Goody estimates that infertility and high infant mortality resulted in heirlessness rates of up to 20 per cent in the early centuries of Christianity, and, as a result, the Church quickly went from having no property to being the largest landowner in Europe. Later the Church even banned marriage among priests so that they couldn't take Church-owned property and pass it on to their children. Whether intentional or not, the confluence of marriage reforms in Christianity amounted to one of the biggest land grabs in history.

If this story is right, our own commitment to monogamy today has little to do with moral enlightenment. It arose out of slave rape and was passed on through institutions of ecclesiastical plunder. The Church promoted monogamy and gained power as a result. The spread of Christianity made this unusual system of kinship into the preferred form in large swathes of the civilized world. Wherever the Church went, monogamy followed and most who subscribe to it today simply assume that anything else would be unnatural.

Natural Unions?

These history lessons show that nuptial norms vary greatly across time and space. Culture has a huge impact on our beliefs about how many marriage partners a person can have, and radically different customs can be found across the globe. This leaves us with a final question. Is any one of these arrangements more natural? Does human biology favour one kinship system over any other?

In recent years, there has been an effort to answer this question by looking to apes. The problem is, apes behave very differently from each other. Chimps have a sexual free-for-all, where both males and females can have multiple sexual partners, and there is no long-term bonding. Common chimps exhibit female exogamy, wherein young females leave their birth group and find sexual partners elsewhere, and bonobos exhibit male exogamy. Gorillas have a harem system, where one male has sexual control over several females. Among orangutans, there are two kinds of males: flanged (the ones with the big cheeks) and unflanged. Both find sexual partners while roaming through the forest, but flanged males find willing partners, whom they compete for aggressively, and unflanged males engage in forced copulation – what looks very much like rape. Gibbons, those lesser apes, are monogamous.

Given this variety, there can be no simple inference about what's natural to us from what other apes do. This has led to some clever detective-work on the part of comparative biologists. They have noted that, relative to body size, there is considerable cross-species variation in testicle size among apes.[36] Chimps have enormous testicles, and gorilla testicles are charmingly small. The reason for this is presumed to be that male chimps need to produce more sperm than male gorillas because they are competing with other males to pass genes into the next generation. Perhaps we can infer what our natural mating system would be by comparing the size of human testicles to those of these other species. One problem with this approach is that men have testicles that fall somewhere between chimps' and gorillas', in terms of size. It's very hard to know what that means. Does it mean we are polyamorous, but a bit less libertine than chimps? Does it mean we have slight tendencies towards polygyny, like gorillas, but can be content with monogamy? To make matters more complicated, our testicle size is a bit closer to orangutans', but their mating pattern is different across the two kinds of males, and our species only has one. In addition, we're not that closely related to orangutans. Once we bring them into the mix, we might as well consider other primates.

When monkeys are brought to the picture, there is even more evidence for the correlation between large testicles and polyamory (males and females having multiple partners). The problem is human males still fall in the middle of the testicle spectrum. And if we look at species whose testicles are about the size of ours, relative to body weight, mating patterns run the gamut. Hamadryas baboons are polygynous, spider monkeys are polyamorous, and common marmosets are monogamous. There is no clear algorithm to infer family size from family jewels.

In the end, I think the effort to figure out what comes to us naturally is a fool's errand. It's not just that it's difficult to do the necessary science. Rather, the whole premise is confused. Human beings are not naturally monogamous, or polygamous, or anything else. We are naturally flexible. The distinctive mark of our species is that we can adopt many different forms of social arrangement. Even if there were some default behaviour for human beings, it would have limited significance. Culture can clearly exert a massive influence on how we mate, and variation in matrimonial practices will never be fully illuminated by biology. Those who want to understand our preferences will learn more from history books than from chimpanzee troops in Gombe.

CONCLUSION: UNNATURAL ACTS

Biology ensures that most human beings will seek out sexual partners. If that weren't the case, we'd have become extinct long ago. In that sense, sex is a natural behaviour. But human beings do natural things in very unnatural ways. We build customs, laws and institutions that regulate our most instinctive behaviours. Sex may be more important to human social organization than anything else, so the norms governing intimacy are among the most elaborate and culturally specific. From love letters to lip plates and lap dances, we have reshaped the sexual landscape. Culture revises the laws of attraction, turns relationships into business ventures and skews the range of viable partners, contracting mercilessly in some cases, and expanding beyond biological discretion in others. Biology can help explain why we are more likely to flirt with a person than a potato, but that's just where the story begins. To explain the massive human variation that exists within our evolved constraints, we need to look at the history of social innovations, power struggles and revolutions that have taken us out of the savannah and into citadels and skyscrapers. The story of sex is the story of our species. Here as elsewhere, we are always moving beyond human nature.

Afterword

Throughout this book, I have tried to argue that human beings transcend nature: we are products of culture and experience, not just biology. Therefore, to study human beings, we must move beyond genetics and evolution, and we must recognize that the behaviour by priviledged undergraduates observed in one psychology lab may not reflect the whole of humanity. Culture and history are essential to an understanding of who we are and how we think and act. By way of conclusion, we can ask whether the idea of human nature has any place in the human sciences. To answer that question, it is helpful to distinguish different things one might mean by that popular phrase.

First, human nature might refer to things that are *uniquely* human. There is long-standing interest in what distinguishes us from other creatures. Aristotle advocated a classification system on which each species could be defined by some distinctive feature. Human beings were defined, in his scheme, as rational animals, the idea being that rationality sets us apart and is the essence of our distinctive human nature. Contemporary psychology, however, shows that human beings are often irrational, and other creatures are far more rational than we might have wanted to believe. In more recent times, researchers have tried to identify more specific psychological or behavioural traits that set us apart. It was once thought that human beings are the only animal to use tools, though that has been debunked; chimps use sticks to collect termites, for example. Some authors claim that humans are the only ones to use languages, with complex phrase structures and inflections; research on great apes and dolphins is sometimes presented as a challenge. Recently there has been a debate about whether human beings have unique capacities for social cognition: we learn through imitation and we regard other members of our species as thinking things, rather than seeing them as inanimate objects that happen to move through the world. But this capacity is evident in some primates, and perhaps even in dogs. This suggests that the search for human uniqueness may be based on a mistake. The difference between us and other animals is more quantitative than qualitative.[1] Of all the quantitative differences, the greatest may be our capacity for cultural learning. Thus, the search for what makes us human leads directly to what makes us less constrained by nature than any other species.

The second conception of human nature focuses not on what is uniquely human but on what is universally human. To say that something is a universal, in this sense, doesn't mean that it is found in every single person – there are unusual or pathological individuals. Rather, the idea is that some things are found in all human societies, or nearly all, that have been documented. The anthopologist Donald Brown lists 200 such traits in his book *Human Universals*.[2] Some of these may be uniquely human, such as marriage, religion and art. Others, such as cooperation and memory, are shared by some non-human animals. The scientific search for universals is surely interesting, but also fraught with risks. The chief difficulty is that apparent universals often belie human variation. Consider art. It's true that most human cultures have what we might call art, but this label may fail to pick out any single thing. Some cultures have music and dance, but not paintings and sculptures, for instance, and cultures that have paintings and sculptures may use them for radically different purposes. Is an effigy of an ancestor used for funerary rites really an artwork? Moreover, calling human universals an aspect of human nature is really a misnomer. Many of the items of Brown's list are unlikely to be things we do in virture of our biological make-up. Art again is an illustration. Do we have a painting gene or a sculpture centre in the brain? Unlikely. Art may be an accidental byproduct of other abilities. Indeed, it may be a recent invention. Modern humans have existed for about 200,000 years, but painting and sculpture have only been around for about 35,000 years. This human universal is only universal in recent millennia, and that suggests it is not a human instinct.

The first two conceptions of human nature place emphasis on the word 'human'. The first one draws attention to uniquely human traits, and the second encompasses traits that might be collected in a complete description of what is characteristic of our species. But there is a third conception of human nature, which places emphasis on the word 'nature'. There is a question about what we do naturally, where that means: what aspects of human behaviour owe directly to our biology as opposed to some more general capacity for learning and discovery? In this guise, the study of human nature is the study of our *innate* traits. The idea of innate traits differs from the idea of universal traits. Some human universals are not innate. For example, all human societies have fire, clothing and shelter. But our capacity to use these things may derive from more general capacities rather than specialized innate mechanisms. Bees build hives instinctively, but humans may design and build homes by means of a more general capacity for tool use. Those who study innate traits are also often interested in human variation, rather than human universals. They try to find cases where differences between individuals and groups have a biological basis. Some proposed differences, such as gender differences in spatial cognition, are not presumed to be uniquely human, so this conception of human nature also differs from the first conception. This third notion of human nature has been my primary focus in the book. I have been at pains to say that many researchers have exaggerated the biological contribution to human behaviour. What gets attributed to human nature is often a result of nurture.

That said, nurture doesn't work on its own. Nurture could not affect us if we didn't have the biology that we do. Every cultural trait is really a biocultural trait – every trait that we acquire through learning involves an interaction between biology and the environment. Thus, we cannot simply jettison biology when studying human beings. But it is crucial that we don't study the biological bases of behaviour in lieu of culture. Rather, we should understand our biological endowment as a set of mechanisms that allow us to change with experience. In this picture, there is no sharp contrast between nature and nurture. Nurture depends on nature, and nature exists in the service of nurture. This means we must give up on approaches to social science that try to articulate how humans act or think by nature. Nature alone determines no pattern of behaviour. Rather, the investigation of our natural constitution should be directed at explaining human plasticity. We can call that the study of human nature, but the label is misleading. It carries with it the dubious idea that there is a natural way for human beings to be. This is not the case. By nature, we transcend nature.

Notes

PREFACE

1. Hobbes, T. (1651/1996). *Leviathan.* Cambridge: Cambridge University Press.
2. Rousseau, J.-J. (1754/1984). *A discourse on inequality.* London: Penguin.
3. Hume, D. (1739/1978). *A treatise of human nature.* Oxford: Oxford University Press.
4. Wrangham, R. (1996) *Demonic males: Apes and the origins of human violence.* Boston: Houghton Mifflin.
5. Wright, R. (1995). *The moral animal.* New York: Vintage.
6. Hume, D. (1748/1987). Of national characters. In *Essays, moral, political, and literary.* Indianapolis, IN: Hackett.

CHAPTER 1: THE NATURE–NURTURE DEBATE

1. See, for example, David Buss's (1999) *Evolutionary psychology* (Boston: Allyn and Bacon); Doreen Kimura's (1999) *Sex and cognition* (Cambridge, MA: MIT Press); William Wright's (1999) *Born that way* (London: Routledge); Geoffrey Miller's (2001) *The mating mind* (New York: Anchor); Steven Pinker's (2002) *The blank slate* (New York: Viking); and Marc Hauser's (2006) *Moral minds* (New York: HarperCollins).
2. For example, Henrich, J., Heine, S. J., and Norenzayan, A. (2010). The weirdest people in the world? *Behavioural and Brain Sciences, 33,* 61–135.
3. Mulcaster, R. (1582/1925). *Mulcaster's Elementarie.* London: Clarendon Press.
4. For a sustained critique, see Buller, D. J. (2005). *Adapting minds: Evolutionary psychology and the persistent quest for human nature.* Cambridge, MA: MIT Press.
5. Pinker, S. (1994). *The language instinct.* New York: HarperCollins.

6. Elman, J. L., Bates, E. A., Johnson, M. H., Karmiloff-Smith, A., Parisi, D., and Plunkett, K. (1996). *Rethinking innateness: A connectionist perspective on development.* Cambridge, MA: MIT Press.
7. Gray, J. (1992). *Men are from Mars, women are from Venus.* New York: HarperCollins; Kimura, D. (1999). *Sex and cognition.* Cambridge, MA: MIT Press; Brizendine, L. (2006). *The female brain.* New York: Broadway Books.
8. Fausto-Sterling, A. (1985). *Myths of gender: Biological theories about women and men.* New York: Basic Books; Jordan-Young, R. (2010). *Brainstorm: The flaws in the science of sex differences.* Cambridge, MA: Harvard University Press.

CHAPTER 2: PUTTING THE GENOME BACK IN THE BOTTLE

1. For an accessible review, see Ridley, M. (2003). *Nature via nurture: Genes, experience, and what makes us human.* London: Fourth Estate.
2. Pines, M. (2001). *The genes we share with yeast, flies, worms, and mice: New clues to human health and disease.* Chevy Chase: Howard Hughes Medical Institute.
3. Sacks, O. (1997). *The island of the colorblind.* New York: Knopf.
4. Siira, V. (2007). Interaction of genetic vulnerability to schizophrenia and Communication Deviance of adoptive parents associated with MMPI schizophrenia vulnerability indicators of adoptees. *Nordic Journal of Psychiatry,* 61, 418–26.
5. Spauwen, J., Krabbendam, L., Lieb, R., Wittchen, H. U., Van Os, J. (2006). Evidence that the outcome of developmental expression of psychosis is worse for adolescents growing up in an urban environment. *Psychological Medicine,* 36, 407–15.
6. Goodwin, D. W. (1973). Alcohol problems in adopters raised apart from biological parents. *Archives of General Psychiatry,* 28, 230–43.
7. Brunner, H. G., Nelen, M., Breakefield, X. O., Ropers, H. H., and Oost, B. A. (1993). Abnormal behavior associated with a point mutation in the structural gene for monoamine oxidase A. *Science,* 262, 578–80.
8. Yong, E. (2010). Dangerous DNA: The truth about the 'warrior gene'. *New Scientist,* 2,755, 34.
9. Bouchard, T. J., Lykken, D. T., McGue, M., Segal, N. L., and Tellegen, A. (1990). Sources of human psychological differences: the Minnesota Study of Twins Reared Apart. *Science,* 250, 223–8.

10. Segal, N. L. (2000). *Entwined lives: Twins and what they tell us about human behavior.* New York: Plume.

11. For a useful overview, see Block, N. (1996). How heritability misleads about race. *The Boston Review,* 20 (6), 30–35.

12. Loehlin, J. C., and Nichols, R. C. (1976). *Heredity, environment and personality.* Austin: University of Texas.

13. Scarr, S., and Carter-Saltzman, L. (1979). Twin method: Defense of a critical assumption. *Behavior Genetics,* 9, 527–42.

14. For a systematic critique of twin research, see: Joseph, J. (2003). *The gene illusion: Genetic research in psychiatry and psychology under the microscope.* Ross on Wye: PCCS Books.

15. Bouchard, T. J. (1996). The genetics of personality. In K. Blum and E. P. Noble (eds.), *Handbook of psychiatric genetics.* Boca Rotan, FL: CRC-Press.

16. Haan, N., Millsap, R., and Hartka, E. (1986). As time goes by: Change and stability in personality over fifty years. *Psychology and Aging,* 1, 220–32.

17. Roberts, B. W., and DelVecchio, W. F. (2000). The rank-order consistency of personality from childhood to old age: A quantitative review of longitudinal studies. *Psychological Bulletin,* 126, 3–25.

18. Roberts, B. W., Walton, K., and Viechtbauer, W. (2006). Patterns of mean-level change in personality traits across the life course: A meta-analysis of longitudinal studies. *Psychological Bulletin,* 132, 1–25.

19. Schmitt, D. P., Allik, J., McCrae, R. R., and Benet-Martínez, V. (2007). The geographic distribution of Big Five personality traits. *Journal of Cross-Cultural Psychology,* 38, 173–212.

20. Levine, R., Norenzayan, A., and Philbrick, K. (2001). Cross-cultural differences in helping strangers. *Journal of Cross-Cultural Psychology,* 32, 543–60.

21. Milgram, S. (1974). *Obedience to authority.* New York: Harper and Row.

22. Mantell, D. (1971). The potential for violence in Germany. *Journal of Social Issues,* 27, 101–12.

23. Kilham, W., and Mann, L. (1974). Level of destructive obedience as a junction of transmitter and executant roles in the Milgram obedience paradigm. *Journal of Personality and Social Psychology,* 29, 696–702.

24. Joseph, J. (2001). Is crime in the genes? A critical review of twin and adoption studies of criminality and antisocial behavior. *The Journal of Mind and Behavior,* 22, 179–218.

25. Mednick, S. A., Gabrielli, W. F., and Hutchings, B. (1984). Genetic influences in criminal convictions: Evidence from an adoptive cohort. *Science,* 24, 891–4.

26. *Seventh United Nations survey of crime trends and operations of criminal justice systems, covering the period 1998–2000.* New York: United Nations Office on Drugs and Crime, Centre for International Crime Prevention.

27. McGue, M., and Lykken, D. T. (1992). Genetic influence on divorce. *Psychological Science*, 3, 368–73.

28. Koenig, L. B., McGue, M., Krueger, R. F., and Bouchard, T. J. (2005). Genetic and environmental influences on religiousness: Findings for retrospective and current religiousness ratings. *Journal of Personality*, 73, 471–88.

29. Alford, J. R., Funk, C. L., and Hibbing, J. R. (2005). Are political orientations genetically transmitted? *American Political Science Review*, 99, 153–67.

30. Harris, J. R. (1998). *The nurture assumption: Why children turn out the way they do.* New York: Free Press.

CHAPTER 3: GET SMART

1. Reproduced from Pearson, K. (1924). *The life, letters and labours of Francis Galton*, vol. 2. Cambridge: Cambridge University Press. For an archive of Galton's images and writings, as well as Pearson's biography, visit www.galton.org.

2. Galton, F. (1909). *Essays in eugenics.* London: Macmillan.

3. Pearson, op. cit.

4. Sanger, M. (1921). The eugenic value of birth control propaganda. *Birth Control Review*, 5.

5. Brown, R. A., and Armelagos, G. J. (2001). Apportionment of racial diversity: A review. *Evolutionary Anthropology*, 10 (10), 34–40.

6. Parra, E. J., Kittles, R. A., and Shriver, M. D. (2004). Implications of correlations between skin color and genetic ancestry for biomedical research. *Nature Genetics*, 36, S54–S60.

7. Thangaraj, K., Singh, L., Reddy, A. G., Rao, V. R., Sehgal, S. S., Underhill, P. A., Pierson, M., Frame, I. G., and Hagelberg, E. (2003). Genetic affinities of the Andaman Islanders, a vanishing human population. *Current Biology*, 13, 86–93.

8. Tishkoff, S. A., Reed, F. A., Friedlaender, F. R., Ehret, C., Ranciaro, A., Froment, A., Hirbo, J. B., Awomoyi, A. A., Bodo, J. M., Doumbo, O., Ibrahim, M., Juma, A. T., Kotze, M. J., Lema, G., Moore, J. H., Mortensen, H., Nyambo, T. B., Omar, S. A., Powell, K., Pretorius, G. S., Smith, M. W., Thera, M. A., Wambebe, C., Weber, J. L., and Williams, S. M. (2009). The

genetic structure and history of Africans and African Americans. *Science*, 324, 1,035–44.

9. Shriver, M. D., Parra, E. J., Dios, S., Bonilla, C., Norton, H., Jovel, C., Pfaff, C., Jones, C., Massac, A., Cameron, N., Baron, A., Jackson, T., Argyropoulos, G., Jin, L., Hoggart, C. J., McKeigue, P. M., and Kittles, R. A. (2003). Skin pigmentation, biogeographical ancestry, and admixture mapping. *Human Genetics*, 112, 387–99.

10. Rosenberg, N. A., Pritchard, J. K., Weber, J. L., Cann, H. M., Kidd, K. K., Zhivotovsky, L. A., and Feldman, M. W. (2002). Genetic structure of human populations. *Science*, 298, 2,381–5.

11. Kittles, R. A., and Weiss, K. M. (2003). Race, ancestry, and genes: Implications for defining disease risk. *Annual Review of Genomics and Human Genetics*, 4, 33–67.

12. Watson, J. B. (1930). *Behaviorism*. Chicago: University of Chicago Press.

13. Breland, K., and Breland, M. (1961). The misbehavior of organisms. *American Psychologist*, 16, 681–4.

14. Devlin, B., Daniels, M., and Roeder, K. (1997). The heritability of IQ. *Nature*, 388, 468–71.

15. Bouchard, T. J., and McGue, M. M. (1981). Familial studies of intelligence: A review. *Science*, 212, 1,055–9. The number is similar (.26) in a more recent study: Segal, N. L., McGuire, S., Havlena, J., Gill, P., and Hershberger, S. L. (2007). Intellectual similarity of virtual twin pairs: Developmental trends. *Personality and Individual Differences*, 42, 1,209–19.

16. Turkheimer, E., Haley, A., Waldron, M., D'Onofrio, B., and Gottesman, I. I. (2003). Socioeconomic status modifies heritability of IQ in young children. *Psychological Science*, 14, 623–8.

17. Wahlsten, D. (1997). The malleability of intelligence is not constrained by heritability. In B. Devlin, S. E. Fienberg, D. P. Resnick and K. Roeder (eds.), *Intelligence, genes, and success: Scientists respond to the Bell Curve* (pp. 71–87). New York: Springer.

18. Lewontin, R. (1974). The analysis of variance and the analysis of causes. *American Journal of Human Genetics*, 26, 400–411.

19. Bertrand, M., and Mullainathan, S. (2004). Are Emily and Greg more employable than Lakisha and Jamal? A field experiment on labor market discrimination. *American Economic Review, American Economic Association*, 94, 991–1,013.

20. Steele, C. M., and Aronson, J. (1995). Stereotype Threat and the intellectual test-performance of African-Americans. *Journal of Personality and Social Psychology*, 69, 797–811.

21. Cochran, G., Hardy, J., and Harpending, H. (2006). Natural history of Ashkenazi intelligence. *Journal of Biosocial Science*, 38, 659–93.

22. Winship, C., and Korenman, S. (1997). Does staying in school make you smarter? The effect of education on IQ in *The Bell Curve*. In B. Devlin, S. Fienberg, D. Resnick, and K. Roeder (eds.), *Intelligence, genes, and success: Scientists respond to the Bell Curve* (pp. 215–34). New York: Springer.

23. Green, R. L., Hoffman, L. T., Morse, R., Hayes, M. E., and Morgan, R. F. (1964). *The educational status of children in a district without public schools* (Co-Operative Research Project No. 2321). Washington, DC: Office of Education, US Department of Health, Education, and Welfare.

24. Heyns, B. (1978). *Summer learning and the effects of schooling*. San Diego, CA: Academic Press.

25. Ogbu, J. U. (2002). Cultural amplifiers of intelligence: IQ and minority status in crosscultural perspective. In J. M. Fish (ed.), *Race and intelligence: Separating science from myth*. Mahwah, NJ: Erlbaum.

26. Myerson, J., Rank, M. R., Raines, F. Q., and Schnitzler, M. A. (1998). Race and general cognitive ability: The myth of diminishing returns to education. *Psychological Science*, 9, 139–42.

27. Flynn, J. R. (2009). *What is intelligence: Beyond the Flynn Effect*. Cambridge: Cambridge University Press.

28. Dijksterhuis, A., and van Knippenberg, A. (1998). The relation between perception and behavior or how to win a game of Trivial Pursuit. *Journal of Personality and Social Psychology*, 74, 865–77.

29. Gould, S. J. (1996). *The mismeasure of man*. New York: Norton.

30. Sternberg, R. J. (1988). *The triarchic mind: A new theory of human intelligence*. New York: Viking.

31. Gardner, H. (1999). *Intelligence reframed: Multiple intelligences for the 21st century*. New York: Basic Books.

32. Chorney, M. J., Chorney, K., Seese, N., Owen, M. J., Daniels, J., McGuffin, P., Thompson, L. A., Detterman, D. K., Benbow, C., Lubinski, D., Eley, T., and Plomin, R. (1998). A quantitative trait locus associated with cognitive ability in children. *Psychological Science*, 9, 159–66.

33. Dorman, C. (1991). Microcephaly and intelligence. *Developmental Medicine and Child Neurology*, 33, 267–72.

34. Duncan, J., Seitz, R. J., Kolodny, J., Bor, D., Herzog, H., Ahmed, A., Newell, F. N., and Emslie, H. (2000). A neural basis for general intelligence. *Science*, 289, 457–60.

CHAPTER 4: WHAT BABIES KNOW

1. Baillargeon, R. (1987). Object permanence in 3½ and 4½ month old infants. *Developmental Psychology*, 23, 655–64.

2. Keil, F. C. (1989). *Concepts, kinds, and cognitive development*. Cambridge, MA: MIT Press.

3. Mandler, J. M., and McDonough, L. (1998). Studies in inductive inference in infancy. *Cognitive Psychology*, 37, 60–96.

4. Woodward, A., Phillips, A., and Spelke, E. S. (1993). Infants' expectations about the motions of inanimate vs. animate objects. *Proceedings of the Cognitive Science Society*. Hillsdale, NJ: Erlbaum.

5. Carey, S. (1985). *Conceptual change in childhood*. Cambridge, MA: MIT Press.

6. Inagaki, K., and Hatano, G. (2002). *Young children's naive thinking about the biological world*. New York: Psychology Press.

7. Hirschfeld, L. A. (1995). Do children have a theory of race? *Cognition*, 54, 209–52.

8. Xu, F., and Carey, S. (1996). Infants' metaphysics: the case of numerical identity. *Cognitive Psychology*, 30, 111–53.

9. Keil, op. cit.

10. Gergely, G., Nadasdy, Z., Csibra, G., and Biro, S. (1995). Taking the intentional stance at 12 months of age. *Cognition*, 56, 165–93.

11. Woodward, A. L. (1998). Infants selectively encode the goal object of an actor's reach. *Cognition*, 69, 1–34.

12. Surian, L., Caldi, S and Sperber, D. (2007). Attribution of beliefs by 13-month old infants. *Psychological Science*, 18, 580–86.

13. Xu, F., and Spelke, E. S. (2000). Large number discrimination in 6-month-old infants. *Cognition*, 74, B1–B11.

14. Starkey, P., Spelke, E. S., and Gelman, R. (1983). Detection of intermodal numerical correspondences by human infants. *Science*, 222, 179–81.

15. Wynn, K. (1990). Children's understanding of counting. *Cognition*, 36, 155–93.

16. Clearfield, M., and Westfahl, S. (2006). Familiarization in infants' perception of addition problems. *Journal of Cognition and Development*, 7, 27–43.

17. Scholl, B. J., and Leslie, A. M. (1999). Explaining the infant's object concept: Beyond the perception/cognition dichotomy. In E. Lepore and Z. Pylyshyn (eds.), *What is cognitive science?* (pp. 26–73). Oxford: Blackwell.

18. Mix, K. S., Levine, S. C., and Huttenlocher, J. (1997). Numerical abstraction in infants: Another look. *Developmental Psychology*, 33, 423–8.

CHAPTER 5: SENSIBLE IDEAS

1. Fodor, J. (1981). The present status of the innateness controversy. In *Representations: Philosophical essays on the foundations of cognitive science*. Cambridge, MA: MIT Press.

2. Pecher, D., Zeelenberg, R., and Barsalou, L. W. (2004). Sensorimotor simulations underlie conceptual representations: Modality-specific effects of prior activation. *Psychonomic Bulletin and Review*, 11, 164–7.
3. Borghi, A. M., Glenberg, A. M., and Kaschak, M. P. (2004). Putting words in perspective. *Memory and Cognition*, 32, 863–73.
4. Vermeulen, N., Corneille, O., and Niedenthal, P. M. (2008). Sensory load incurs conceptual processing costs. *Cognition*, 109, 287–94.
5. Kaschak, M. P., Madden, C. J., Therriault, D. J., Yaxley, R. H., Aveyard, M. E., Blanchard, A. A., and Zwaan, R. A. (2005). Perception of motion affects language processing. *Cognition*, 94, B79–B89.
6. Stanfield, R. A., and Zwaan, R. A. (2001). The effect of implied orientation derived from verbal context on picture recognition. *Psychological Science*, 12, 153–6.
7. Yaxley, R. H., and Zwaan, R. A. (2007). Simulating visibility during language comprehension. *Cognition*, 150, 229–36.
8. Chao, L. L., Haxby, J. V., and Martin, A. (1999). Attribute-based neural substrates in posterior temporal cortex for perceiving and knowing about objects. *Nature Neuroscience*, 2, 913–19.
9. Simmons, W. K., Pecher, D., Hamann, S. B., Zeelenberg, R., and Barsalou, L. W. (2003). fMRI evidence for modality-specific processing of conceptual knowledge on six modalities. *Meeting of the Society for Cognitive Neuroscience*, New York, March 2003.

CHAPTER 6: THE GIFT OF THE GAB

1. Brown, R., and Hanlon, C. (1970). Derivational complexity and order of acquisition in child speech. In J. R. Hayes (ed.), *Cognition and the development of language*. New York: Wiley.
2. MacWhinney, B. (2004). A multiple process solution to the logical problem of language acquisition. *Journal of Child Language*, 31, 883–914.
3. Ramscar, M., and Yarlett, D. (2007). Linguistic self-correction in the absence of feedback: A new approach to the logical problem of language acquisition. *Cognitive Science*, 31, 927–60.
4. Crain, S. C., and Nakayama, M. (1987). Structure dependence in grammar formation. *Language*, 63, 522–43.
5. Scholz, B. C., and Pullum, G. K. (2006). Irrational nativist exuberance. In Robert J. Stainton (ed.), *Contemporary debates in cognitive science*. London: Blackwell.
6. Reali, F., and Christiansen, M. H. (2005). Uncovering the richness of the

stimulus: Structural dependence and indirect statistical evidence. *Cognitive Science*, 29, 1,007–28.

7. Singleton, J. L., and Newport, E. L. (2004). When learners surpass their models: The acquisition of American Sign Language from inconsistent input. *Cognitive Psychology*, 49, 370–407.

8. Bickerton, D. (1990). *Language and Species*. Chicago: University of Chicago Press.

9. Kegl, J. (2002). Language emergence in a language-ready brain: Acquisition issues. In G. Morgan and B. Woll (eds.), *Language acquisition in signed languages*. Cambridge: Cambridge University Press.

10. Saygin, A., Dick, F., and Bates, E. (2001). Linguistic and non-linguistic auditory processing in aphasia. *Brain and Language*, 79, 143–5.

11. Hubel, D. H., and Wiesel, T. N. (1970). The period of susceptibility to the physiological effects of unilateral eye closure in kittens. *Journal of Physiology*, 206, 419–36.

12. Newport, E. L. (1990). Maturational constraints on language learning. *Cognitive Science*, 14, 11–28.

13. Elman, J. L. (1993). Learning and development in neural networks: The importance of starting small. *Cognition*, 48, 71–99.

14. Read, D. (2008). Working memory: A cognitive limit to non-human primate recursive thinking prior to hominid evolution. *Evolutionary Psychology*, 6, 603–38.

CHAPTER 7: WORDS AND WORLDS

1. McBrearty, S., and A. Brooks (2000). The revolution that wasn't: A new interpretation of the origin of modern human behavior. *Journal of Human Evolution*, 39, 453–563.

2. Frank, M. C., Fedorenko, E., and Gibson, E. (2008). Language as a cognitive technology: English-speakers match like Pirahã when you don't let them count. *Proceedings of the 30th Annual Meeting of the Cognitive Science Society*.

3. Newton, A. M., and de Villiers, J. G. (2007). Thinking while talking: Adults fail nonverbal false-belief reasoning. *Psychological Science*, 18, 574–9.

4. Hermer-Vazquez, L., Moffet, A., and Munkholm, P. (2001). Language, space, and the development of cognitive flexibility in humans: The case of two spatial memory tasks. *Cognition*, 79, 263–99.

5. Seeley, W. (2010). Imagine crawling home: A case study in cognitive science and aesthetics. *Review of Philosophy and Psychology*, 1, 407–26.

6. Glucksberg, S., and Weisberg, R. W. (1966). Verbal behavior and problem solving: Some effects of labelling in a functional fixedness problem. *Journal of Experimental Psychology*, 71, 659–64.
7. Whorf, B. (1956). Science and linguistics. In J. B. Carrol (ed.), *Language, thought, and reality: Selected writings of Benjamin Lee Whorf.* Cambridge, MA: MIT Press.
8. Lucy, J. (1992). *Grammatical categories and cognition: A case study of the linguistic relativity hypothesis.* Cambridge: Cambridge University Press.
9. Gordon, P. (2004). Numerical cognition without words: Evidence from Amazonia. *Science*, 306, 496–9.
10. Kay, P., and Kempton, W. (1984). What is the Sapir–Whorf hypothesis? *American Anthropologist*, 86, 65–78.
11. Winawer, J., Witthoft, N., Frank, M.,Wu, L., Wade, A., and Boroditsky, L. (2007). The Russian Blues reveal effects of language on color discrimination. *Proceedings of the National Academy of Science*, 104, 7,780–85.
12. Boroditsky, L., Schmidt, L., and Phillips, W. (2003). Sex, syntax and semantics. In D. Gentner and S. Goldin-Meadow (eds.), *Language in mind: Advances in the study of language and thought* (pp. 61–80). Cambridge, MA: MIT Press.
13. Pinker, S. (1994). *The language instinct.* New York: HarperCollins.

CHAPTER 8: THE TAO OF THOUGHT

1. Triandis, H. C. (1995). *Individualism and collectivism.* San Francisco, CA: Westview Press.
2. Witkin, H. A., Moore, C. A., Goodenough, D. R., and Cox, P. W. (1977). Field dependent and field independent cognitive styles and their educational implications. *Review of Educational Research*, 47, 1–64.
3. Witkin, H. A., and Berry, J. W. (1975). Psychological differentiation in cross-cultural perspective. *Journal of Cross-Cultural Psychology*, 1, 5–87.
4. Masuda, T., and Nisbett, R. E. (2001). Attending holistically versus analytically: Comparing the context sensitivity of Japanese and Americans. *Journal of Personality and Social Psychology*, 81, 922–34.
5. Masuda, T., and Nisbett, R. E. (2006). Culture and change blindness. *Cognitive Sciences*, 30, 381–99.
6. Reproduced with permission from Masuda and Nisbett, op. cit. in note 4.
7. Nisbett, R. E., Peng, K., Choi, I., and Norenzayan, A. (2001). Culture and systems of thought: Holistic vs. analytic cognition. *Psychological Review*, 108, 291–310.

8. Peng, K., and Nisbett, R. E. (1999). Culture dialectics and reasoning about contradiction. *American Psychologist*, 54, 741–54.
9. Briley, D. A., Morris, M., and Simonson, I. (2000). Reasons as carriers of culture: Dynamic vs. dispositional models of cultural influence on decision making. *Journal of Consumer Research*, 27, 157–78.
10. Nelson, M. R., and Shavitt, S. (2002). Horizontal and vertical individualism and achievement values: A multimethod examination of Denmark and the United States. *Journal of Cross-Cultural Psychology*, 33, 439–58.
11. Marsh, H. W., and Kleitman, S. (2003). School athletic participation: Mostly gain with little pain. *Journal of Sport and Exercise Psychology*, 25, 205–28.

CHAPTER 9: GENDER AND GEOMETRY

1. Martineau, H. (1823). On female education. *Monthly Repository*, 18, 77–81.
2. Based on Shepard, R. N., and Metzler, J. (1971). Mental rotation of three-dimensional objects. *Science*, 171, 701–3.
3. Steinpreis, R., Anders, K. A., and Ritzke, D. (1999). The impact of gender on the review of the curricula vitae of job applicants and tenure candidates: A national empirical study. *Sex Roles*, 41, 509–28.
4. Spencer, S. J., Steele, C. M., and Quinn, D. M. (1999). Stereotype threat and women's math performance. *Journal of Experimental Social Psychology*, 35, 4–28.
5. Inzlicht, M., and Ben-Zeev, T. (2000). A threatening intellectual environment: Why females are susceptible to experiencing problem-solving deficits in the presence of males. *Psychological Science*, 11, 365–71.
6. Haier, R. J., Jung, R. E., Yeo, R. A., Head, K., and Alkire, M. T. (2005). The neuroanatomy of general intelligence: Sex matters. *NeuroImage*, 25, 320–27.
7. Tannen, D. (1994). *Talking from 9 to 5*. New York: William Morrow.
8. Swann, J. (1988). Talk control: An illustration from the classroom of problems in analyzing male dominance of conversation. In J. Coates and D. Cameron (eds.), *Women in their speech communities*. London: Longman.
9. Tenenbaum, H. R., and Leaper, C. (2003). Parent–child conversations about science: The socialization of gender inequities? *Developmental Psychology*, 39, 34–47. See also: Simpkins, S. D., Davis-Kean, P. E., and Eccles, J. S. (2006). Math and science motivation: A longitudinal examination of the links between choices and belief. *Developmental Psychology*,

42, 70–83; Jacobs, J. E., Davis-Kean, P., Bleeker, M., Eccles, J. S. and Malanchuk, O. (2005). I can, but I don't want to: The impact of parents, interests, and activities on gender differences in math. In A. Gallagher and J. Kaufman (eds.), *Gender differences in mathematics*. Cambridge: Cambridge University Press.

10. Lee, V. E., and Bryk, A. S. (1986). Effects of single sex secondary schools on student achievement and attitudes. *Journal of Educational Psychology*, 78, 381–95.

11. Evans, L., and Davies, K. (2000). No sissy boys here: A content analysis of the representation of masculinity in elementary school reading textbooks. *Sex Roles*, 42, 255–70.

12. Will, J. A., Self, P. A., and Dantan, N. (1976). Maternal behavior and perceived sex of infant. *American Journal of Orthopsychiatry*, 46, 135–9; Vogel, D. A., Lake, M. A., Evans, S., and Karraker, K. H. (1991). Children's and adults' sexstereotyped perception of infants. *Sex Roles*, 24, 605–16.

13. Alexander, G. M., and Hines, M. (2002). Sex differences in response to children's toys in nonhuman primates (*Cercopithecus aethiops sabaeus*). *Evolution and Human Behavior*, 23, 467–79.

14. Ji, L., Peng, K., and Nisbett, R. E. (2000). Culture, control, and perception of relationships in the environment. *Journal of Personality and Social Psychology*, 78, 943–55.

15. Sinha, D. (1980). Sex differences in psychological differentiation among different cultural groups. *International Journal of Behavioral Development*, 3, 455–66.

16. Pontius, A. A. (1989). Color and spatial error in block design in stone-age Auca Indians: ecological underuse of occipital-parietal system in men and of frontal lobes in women. *Brain and Cognition*, 10, 54–75.

17. Roberts, J. E., and Bell, M. A. (2000). Sex differences on a computerized mental rotation task disappear with computer familiarization. *Perceptual and Motor Skills*, 91, 1,027–36.

CHAPTER 10: FEAR AND LOATHING IN MICRONESIA

1. James, W. (1884). What is an emotion? *Mind*, 9, 188–205.

2. Strack, F., Martin, L. L., and Stepper, S. (1988). Inhibiting and facilitating conditions of facial expressions: A nonobtrusive test of the facial feedback hypothesis. *Journal of Personality and Social Psychology*, 54, 768–77.

3. Zajonc, R. B., Murphy, S. T., and Inglehart, M. (1989). Feeling and facial efference: Implications of the vascular theory of emotion. *Psychological Review*, 96, 395–416.

4. Dutton, D. G., and Aron, A. P. (1974). Some evidence for heightened sexual attraction under conditions of high anxiety. *Journal of Personality and Social Psychology*, 30, 510–17.

5. Trivers, R. L. (1971). The evolution of reciprocal altruism. *Quarterly Review of Biology*, 46, 35–57.

6. Frank, R. H. (1988). *Passion within reason: The strategic role of the emotions*. New York: Norton.

7. Naab, P. J., and Russell, J. A. (2007). Judgments of emotion from spontaneous facial expressions of New Guineans. *Emotion*, 7, 736–44.

8. Izard, C. E. (1971). *The face of emotion*. New York: Appleton-Century-Crofts.

9. Shimoda, K., Argyle, M., and Ricci-Bitti, P. (1978). The intercultural recognition of emotional expressions by three national racial groups: English, Italian, and Japanese. *European Journal of Social Psychology*, 8, 169–79.

10. Galati, D., Scherer, K. R., and Ricci-Bitti, P. E. (1997). Voluntary facial expression of emotion comparing congenitally blind with normally sighted encoders. *Journal of Personality and Social Psychology*, 73, 1,363–79.

11. Ekman, P. (1972). Universals and cultural differences in facial expressions of emotion. In J. Cole (ed.), *Nebraska Symposium on Motivation, 1971* (pp. 207–83). Lincoln, NE: University of Nebraska Press.

12. Lutz, C. (1988). *Unnatural emotions: Everyday sentiments on a Micronesian atoll and their challenge to Western theory*. Chicago: University of Chicago Press.

13. Nemeroff, C., and Rozin, P. (1992). Sympathetic magical beliefs and kosher dietary practice: The interaction of rules and feelings. *Ethos*, 20, 96–115.

CHAPTER 11: GLADNESS AND MADNESS

1. Brickman, P., Coates, D., and Janoff-Bulman, R. (1978). Lottery winners and accident victims: is happiness relative? *Journal of Personality and Social Psychology*, 36, 917–27.

2. Lacasse, J. R, and Leo, J. (2005). Serotonin and depression: A disconnect between the advertisements and the scientific literature. *PLoS Med*, 2: e392.

3. Kirsch, I., Deacon, B. J., Huedo-Medina, T. B., Scoboria, A., Moore, T. J., and Johnson, B. T. (2008). Initial severity and antidepressant benefits:

A meta-analysis of data submitted to the Food and Drug Administration. *PLoS Med*, 5: e45.

4. Nesse, R. M. (1998). Emotional disorders in evolutionary perspective. *British Journal of Medical Psychology*, 71, 397–415; Price, J., Sloman, L., Gardner, R., Gilbert, P., and Rohde, P. (1994). The social competition hypothesis of depression. *British Journal of Psychiatry*, 164, 309–15.

5. Wakefield, J. C. (1992). Disorder as harmful dysfunction: A conceptual critique of DSM-III-R's definition of mental disorder. *Psychological Review*, 99, 232–47.

CHAPTER 12: COPING WITH CANNIBALISM

1. Greene, J. D., Sommerville, R. B., Nystrom, L. E., Darley, J. M., and Cohen, J. D. (2001). An fMRI investigation of emotional engagement in moral judgment. *Science*, 293, 2,105–8.

2. Koenigs, M., Young, L., Adolphs, R., Tranel, D., Cushman, F., Hauser, M., and Damasio, A. (2007). Damage to the prefrontal cortex increases utilitarian moral judgments. *Nature*, 446, 908–11.

3. Moll, J., de Oliveira-Souza, R., Bramati, I., and Grafman, J. (2002). Functional networks in emotional moral and nonmoral social judgments. *NeuroImage*, 16, 696–703.

4. Heekeren, H. R., Wartenburger, I., Schmidt, H., Schwintowski, H. P., and Villringer, A. (2003). An fMRI study of simple ethical decision-making. *NeuroReport*, 14, 1,215–19.

5. Harenski, C. L., and Hamann, S. (2006). Neural correlates of regulating negative emotions related to moral violations. *NeuroImage*, 30, 313–24.

6. Berthoz, S., Armony, J. L., Blair, R. J. R., and Dolan, R. J. (2002). An fMRI study of intentional and unintentional (embarrassing) violations of social norms. *Brain*, 125, 1,696–708.

7. Sanfey, A. G., Rilling, J. A., Aronson, J. K., Nystrom, L., and Cohen, J. D. (2003). The neural basis of economic decision making in the Ultimatum Game, *Science*, 300, 1,755–7.

8. King, J. A., Blair, R. J. R., Mitchell, D. G. V., Dolan, R. J., and Burgess, N. (2006). Doing the right thing: A common neural circuit for appropriate violent or compassionate behavior. *NeuroImage*, 30, 1,069–76.

9. Schnall, S., Haidt, J., Clore, G. L., and Jordan, A. (2008). Disgust as embodied moral judgment. *Personality and Social Psychology Bulletin*, 34, 1,096–1,109.

10. Eskine, K. J., Kacinik, N. A., and Prinz, J. J. (2011). A bad taste in the

mouth: Gustatory disgust influences moral judgment. *Psychological Science*, 22, 295–99.

11. Valdesolo, P., and DeSteno, D. (2006). Manipulations of emotional context shape moral judgment. *Psychological Science*, 17, 476–7.

12. Wheatley, T., and Haidt, J. (2005). Hypnotically induced disgust makes moral judgments more severe. *Psychological Science*, 16, 780–84.

13. Haidt, J. (2001). The emotional dog and its rational tail: A social intuitionist approach to moral judgment. *Psychological Review*, 108, 814–34.

14. Blair, R. J. R. (1995). A cognitive developmental approach to morality: Investigating the psychopath. *Cognition*, 57, 1–29.

15. Cleckley, H. M. (1941). *The mask of sanity: An attempt to reinterpret the so-called psychopathic personality*. St Louis, MO: The C. V. Mosby Company.

16. Haidt, J., Koller, S., and Dias, M. (1993). Affect, culture, and morality, or is it wrong to eat your dog? *Journal of Personality and Social Psychology*, 65, 613–28.

17. Rozin, P., Lowery, L., Imada, S., and Haidt, J. (1999). The CAD triad hypothesis: A mapping between three moral emotions (contempt, anger, disgust) and three moral codes (community, autonomy, divinity). *Journal of Personality and Social Psychology*, 76, 574–86.

18. Shweder, R. A., Much, N. C., Mahapatra, M., and Park, L. (1997). The 'Big Three' of morality (autonomy, community, divinity), and the 'Big Three' explanations of suffering. In P. Rozin and A. Brandt (eds.), *Morality and health* (pp. 119–72). New York: Routledge.

19. Wrangham, R. (2004). Killer species. *Daedalus*, 133, 25–35.

20. Henrich, J., Boyd, R., Bowles, S., Camerer, C., Fehr, E., and Gintis, H. (eds.) (2004). *Foundations of human sociality: Economic experiments and ethnographic evidence from fifteen small-scale societies*. Oxford: Oxford University Press.

21. Bahry, D. L., and Wilson, R. K. (2006). Confusion or fairness in the field? Rejection in the Ultimatum Game under the strategy method. *Journal of Economic Behavior and Organization*, 60, 37–54.

22. Kay, A. C., Wheeler, S. C., Bargh, J. A., and Ross, L. (2004). Material priming: The influence of mundane physical objects on situational construal and competitive behavioral choice. *Organizational Behavior and Human Decision Processes*, 95, 83–96.

23. Cosmides, L., and Tooby, J. (1992). Cognitive adaptations for social exchange. In J. H. Barkow, L. Cosmides and J. Tooby (eds.), *The adapted mind: Evolutionary psychology and the generation of culture* (pp. 163–228). New York: Oxford University Press.

24. Stone, V., Cosmides, L., Tooby, J., Kroll, N., and Knight, R. (2002). Selective impairment of reasoning about social exchange in a patient with bilateral limbic system damage. *Proceedings of the National Academy of Sciences*, 99, 11,531–6.
25. Blair, op. cit.
26. Cleckley, op. cit.
27. De Waal, F. B. E. (1996). *Good natured: The origins of right and wrong in humans and other animals*. Cambridge, MA: Harvard University Press.
28. Warneken, F., and Tomasello, M. (2006). Altruistic helping in human infants and young chimpanzees. *Science*, 311, 1301–3.
29. Silk, J. B., Brosnan, S. F., Vonk, J., Henrich, J., Povinelli, D. J., Richardson, A. F., Lambeth, S. P., Mascaro, J., and Schapiro, S. J. (2005). Chimpanzees are indifferent to the welfare of other group members. *Nature*, 435, 1,357–9.
30. Boyer, P. (2001). *Religion explained: The human instincts that fashion gods, spirits and ancestors*. New York: Basic Books.
31. Harris, M. (1986). *Good to eat: Riddles of food and culture*. New York: Simon and Schuster.
32. Harner, M. (1977). The ecological basis for Aztec sacrifice. *American Ethnologist*, 4, 117–35.
33. Nisbett, R. E., and Cohen, D. (1996). *Culture of honor: The psychology of violence in the South*. Boulder, CO: Westview Press.
34. Edgerton, R. B. (1971). *The individual in cultural adaptation: A study of four East African peoples*. Berkeley, CA: University of California Press.

CHAPTER 13: IN BED WITH DARWIN

1. Horvath, T. (1981). Physical attractiveness: The influence of selected torso parameters. *Archives of Sexual Behavior*, 10, 21–4; Penton-Voak, I. S., Perrett, D. I., Castles, D. L., Kobayashi, T., Burt, D. M., and Murray, L. K. (1999). Menstrual cycle alters face preference. *Nature*, 399, 741–2.
2. Frost, P. (1994). Preference for darker faces in photographs at different phases of the menstrual cycle: Preliminary assessment of evidence for a hormonal relationship. *Perception and Motor Skills*, 79, 507–14.
3. Jones, D. (1995). Sexual selection, physical attractiveness, and facial neoteny: Cross-cultural evidence and implications. *Current Anthropology*, 36, 723–48.
4. Tovée, M. J., Reinhardt, S., Emery, J. L., and Cornelissen, P. L. (1998). Optimum body-mass index and maximum sexual attractiveness. *Lancet*, 352, 548.

5. Langlois, J. H., and Roggman, L. A. (1990). Attractive faces are only average. *Psychological Science*, 1, 115–21.

6. Winkielman, P., Halberstadt, J., Fazendeiro, T., and Catty, S. (2006). Prototypes are attractive because they are easy on the mind. *Psychological Science*, 17, 799–806.

7. Yu, D., and Shepard, G. H. (1998). Is beauty in the eyes of the beholder? *Nature*, 396, 321–2.

8. Tovée, M. J., Swami, V., Furnham, A., and Mangalparsad, R. (2006). Changing perceptions of attractiveness as observers are exposed to a different culture. *Evolution and Human Behavior*, 27, 443–56.

9. Freese, J., and Meland, S. (2002). Seven tenths incorrect: Heterogeneity and change in the waist-to-hip ratios of *Playboy* centerfold models and Miss America pageant winners. *Journal of Sex Research*, 39, 133–8.

10. Hamer, D. H., Hu, S., Magnuson, V. L., Hu, N., and Pattatucci, A. M. (1993). A linkage between DNA markers on the X chromosome and male sexual orientation. *Science*, 261, 321–7.

11. Herdt, G. H. (1993). *Ritualized homosexuality in Melanesia*. Berkeley, CA: University of California Press.

12. Kelly, R. C. (1977). *Etoro social structure: A study in structural contradiction*. Ann Arbor: University of Michigan Press.

13. Boswell, J. (1980). *Christianity, social tolerance and homosexuality*. Chicago: University of Chicago Press.

14. Vassilev, G. (1994). Traces of the Bogomil movement in English. *Etudes Balkaniques*, 3, 85–94.

15. Buss, D. M., and Schmitt, D. P. (1993) Sexual strategies theory: A contextual evolutionary analysis of human mating. *Psychological Review*, 100, 204–32.

16. Alexander, M. G., and Fisher, T. D. (2003). Truth and consequences: Using the bogus pipeline to examine sex differences in self-reported sexuality. *The Journal of Sex Research*, 40, 27–35.

17. Brown, G. R., Laland, K. N., and Mulder, M. B. (2009). Bateman's principles and human sex roles. *Trends in Ecology and Evolution*, 24, 297–304.

18. Pedersen, W. C., Miller, L. C., Putcha-Bhagavatula, A., and Yang, Y. (2002). Evolved sex differences in the number of partners desired? The long and the short of it. *Psychological Science*, 13, 157–61.

19. Pillsworth, E. G. (2008). Mate preferences among the Shuar of Ecuador: Trait rankings and peer evaluations. *Evolution and Human Behavior*, 29 (4), 256–67.

20. Marlowe, F. W. (2004). Mate preferences among Hadza hunter-gatherers. *Human Nature*, 2004, 365–76.

21. Eagly, A. H., and Wood, W. (1999). The origins of sex differences in human behavior: Evolved dispositions versus social roles. *American Psychologist*, 54, 408–23.

22. Kenrick, D. T., and Keefe, R. C. (1992). Age preferences in mates reflect sex differences in human reproductive strategies. *Behavioral and Brain Sciences*, 15, 75–133.

23. Buss, D. M., Larsen, R. J., Westen, D., and Semmelroth, J. (1992). Sex differences in jealousy: Evolution, physiology, and psychology. *Psychological Science*, 3, 251–5.

24. DeSteno, D., Bartlett, M., Braverman, J., and Salovey, P. (2002). Sex differences in jealousy: Evolutionary mechanism or artifact of measurement? *Journal of Personality and Social Psychology*, 83, 1,103–16.

25. Harris, C. R. (2000). Psychophysiological responses to imagined infidelity: The specific innate modular view of jealousy reconsidered. *Journal of Personality and Social Psychology*, 78, 1,082–91.

26. Harris, M. (1993). The Evolution of human gender hierarchies: A trial formulation. In B. D. Miller (ed.), *Sex and gender hierarchies*. Cambridge: Cambridge University Press.

27. See Harris, M. (1989). *Our kind*. New York: HarperCollins.

28. Thornhill, N. W. (1991). An evolutionary analysis of rules regulating human inbreeding and marriage. *Behavioral and Brain Sciences*, 14, 247–9.

29. Shaw, B. (1992). Explaining incest: Brother-sister marriage in Graeco-Roman Egypt. *Man*, 27, 267–99.

30. Wolf, A. P. (1970). Childhood association and sexual attraction: A further test of the Westermarck hypothesis. *American Anthropologist*, 72, 503–15.

31. Lieberman, D., Tooby, J., and Cosmides, L. (2003). Does morality have a biological basis? An empirical test of the factors governing moral sentiments regarding incest. *Proceedings of the Royal Society, London B*, 270, 819–26.

32. Slotkin, J. S. (1947). On a possible lack of incest regulations in old Iran. *American Anthropologist*, 49, 612–17.

33. Murdock, G. P. (1981). *Atlas of world cultures*. Pittsburgh, PA: University of Pittsburgh Press.

34. Scheidel, W. (2009). Sex and empire: a Darwinian perspective. In I. Morris and W. Scheidel (eds.), *The dynamics of ancient empires: State power from Assyria to Byzantium* (pp. 255–324). New York: Oxford University Press.

35. Goody, J. (1983). *The development of the family and marriage in Europe*. Cambridge: Cambridge University Press.

36. Dixson, A., and Anderson, M. (2001). Sexual selection and the comparative anatomy of reproduction in monkeys, apes, and human beings. *Annual Review of Sex Research*, 12, 121–44.

AFTERWORD

1. Douglas, K. (2008). Six 'uniquely' human traits now found in animals. *New Scientist*, 17, 11.
2. Brown, D. E. (1991). *Human universals*. New York: McGraw-Hill.

Index

References to figures and pictures are given in italics

Kellogg, John Harvey 277–8, 281
Kempton, Willett 186
Kenrick, Douglas 341
King, John 298
Kirsch, Irving 273
Kleitman, Sabina 211
Koenigs, Michael 297
Koko (gorilla) 166
Korea 206
koro (anxiety) 286
Kovalevskaya, Sofia 216
!Kung people 348
kuru 323

Lacasse, Jeffrey 272–3
language(s)
American Sign (ASL) 155
Bermino 187
Chinese 183
cognitive theory of 175–9
counting and problem-solving 175
critical period of acquisition
162–5, 168
deficits 158–61
and Empiricism 137, 175, 179–80
expressive theory of 173–6
and genes 160
Hawaiian Creole 154, 156
Hebrew 187
and infants/children 137–53
influences on perception 187–90
innateness of 149, 153, 156–8
interactive theory of 177, 180
invention of 171–3
Japanese 187
and limitations of apes 166–8
Nicaraguan Sign 155–7
Nootka 181
pidgin 154, 156
Pirahã 186
Russian 187
shaping thought 180–82
sign 154, 157, 166

Specific Language Impairment
(SLI) 159–61
theory of 137–43, 144–6
Tzeltal 184–6
Yucatec Mayan 185–6
language of thought 119–24, 126,
128, 131, 173
latah (hyperstartle disorder) 286
Le Bon, Gustave 215
Leibniz, Gottfried 117
Lenneberg, Eric 161–2, 165
Leo, Jonathan 272–3
Lerner, Jennifer 299
lesbianism 338
Leslie, Alan 106
Lever Case, the 295–8, 296
Levine, Robert 42
Levinson, Stephen 184
Lewontin, Richard 64
Li-Jun Ji 234
liberty 326
Lieberman, Deborah 354
liget 265
Limbaugh, Rush 351
lingua mentis 117
linguistic relativity 170, 184
LISP (language-programming code)
118
Locke, John 84, 109, 113, 116–17,
119, 174–6, 179–80
Loehlin, John 37
logic 207, 209
Lombroso, Cesare 43–4
long-distance binding 164–5
Lorenz, Konrad 161
love 249, 253–6, 262–3, 265, 345;
see also sex
Lucy, John 185
Luo people 233
Lutz, Catherine 263

McBrearty, Sally 173
McDonough, Laraine 94–5